RESEARCH METHODOLOGY IN CHEMICAL SCIENCES

Experimental and Theoretical Approach

RESEARCH METHODOLOGY IN CHEMICAL SCIENCES

Experimental and Theoretical Approach

Edited by

Tanmoy Chakraborty, PhD,
Lalita Ledwani, PhD

Apple Academic Press Inc.	Apple Academic Press Inc.
3333 Mistwell Crescent	9 Spinnaker Way
Oakville, ON L6L 0A2	Waretown, NJ 08758
Canada	USA

© 2016 by Apple Academic Press, Inc.

First issued in paperback 2021

Exclusive worldwide distribution by CRC Press, a member of Taylor & Francis Group

No claim to original U.S. Government works

ISBN-13: 978-1-77463-549-0 (pbk)
ISBN-13: 978-1-77188-127-2 (hbk)

Library and Archives Canada Cataloguing in Publication

Research methodology in chemical sciences : experimental and theoretical approach / edited by Tanmoy Chakraborty, PhD, Lalita Ledwani, PhD.

Includes bibliographical references and index.
Issued in print and electronic formats.
ISBN 978-1-77188-127-2 (hardcover).—ISBN 978-1-4987-2860-7 (pdf)

1. Chemistry—Research. 2. Chemistry—Methodology. I. Chakraborty, Tanmoy, author, editor II. Ledwani, Lalita, author, editor

QD40.R48 2016	540.72	C2016-900387-6	C2016-900388-4

Library of Congress Cataloging-in-Publication Data

Names: Chakraborty, Tanmoy. | Ledwani, Lalita.
Title: Research methodology in chemical sciences : experimental and theoretical approach / [edited by] Tanmoy Chakraborty, PhD, Lalita Ledwani, PhD.
Description: Toronto : Apple Academic Press, 2016. | Includes bibliographical references and index.
Identifiers: LCCN 2016002072 (print) | LCCN 2016005025 (ebook) | ISBN 9781771881272 (hardcover : alk. paper) | ISBN 9781498728607 ()
Subjects: LCSH: Chemistry—Research.
Classification: LCC QD40 .R434 2016 (print) | LCC QD40 (ebook) | DDC 540.72/4—dc23
LC record available at http://lccn.loc.gov/2016002072

Apple Academic Press also publishes its books in a variety of electronic formats. Some content that appears in print may not be available in electronic format. For information about Apple Academic Press products, visit our website at **www.appleacademicpress.com** and the CRC Press website at **www.crcpress.com**

About the Editors

Tanmoy Chakraborty, PhD

Tanmoy Chakraborty, PhD, is now working as Associate Professor in the Department of Chemistry at Manipal University Jaipur, India. He has been working in the challenging field of computational and theoretical chemistry for the last six years. He completed his PhD from the University of Kalyani, West-Bengal, India, in the field of application of QSAR/QSPR methodology in the bioactive molecules. He has published many international research papers in peer-reviewed international journals with high impact factors. Dr. Chakraborty is serving as an international editorial board member of the *International Journal of Chemoinformatics and Chemical Engineering* and he is also reviewer of the *World Journal of Condensed Matter Physics* (WJCMP). Dr. Tanmoy Chakraborty is the recipient of prestigious Paromeswar Mallik Smawarak Padak, from Hooghly Mohsin College, Chinsurah (University of Burdwan) in 2002.

Lalita Ledwani, PhD

Lalita Ledwani, PhD, is currently Associate Professor in the Department of Chemistry at Manipal University Jaipur, India. She has been both an Assistant Professor and Senior Lecturer at Pandit Deendayal Petroleum University (PDPU) in Gandhinagar, Gujarat, India, as well as a Lecturer at a private degree college in Jaipur, affiliated with Rajasthan Technical University, Rajasthan. A member of several professional organizations, including the American Chemical Society, she has published over a dozen papers in international refereed journals and has presented at many conferences and invited talks. She is the author of the forthcoming book *Petroleum Industrial Chemistry*.

She currently guides several students toward their PhD and postgraduate degrees. She received her PhD from Dr. Bhim Rao Ambedkar University, formerly Agra University, in Uttar Pradesh India.

Contents

List of Contributors

Sankaralingam Arunachalam
School of Chemistry, Bharathidasan University, Tiruchirappalli, Tamil Nadu, India,
E-mail: arunasurf@yahoo.com

Ved Prakash Arya
DESM, Regional Institute of Education, Ajmer, India, E-mail: aryavedp@gmail.com

Samita Basu
Chemical Science Division, Saha Institute of Nuclear Physics, 1/AF Bidhannagar, Kolkata, India,
E-mail: samita.basu@saha.ac.in

Nitu Bhatnagar
Department of Chemistry, Manipal University Jaipur, India, E-mail: nitu.bhatnagar@jaipur.manipal.edu

Shabori Bhattacharya
Manipal University, Jaipur, Centre for Converging Technology, University of Rajasthan, India

Sharad Bohra
Department of Chemistry, Poornima University, Jaipur, Rajasthan, India

Tanmoy Chakrborty
Department of Chemistry, Manipal University Jaipur, India, E-mail: tanmoychem@gmail.com

Amrish Chandra
Amity Institute of Pharmacy, Amity University, Noida, India, E-mail: amrish_chandra@yahoo.com

Sanjan Choudhary
Department of Chemistry, Manipal University Jaipur, India

Anshu Dandia
Center of Advanced Studies, Department of Chemistry, University of Rajasthan, Jaipur, India,
E-mail: dranshudandia@yahoo.co.in

Hemen Dave
Department of Chemistry, Manipal University Jaipur, Jaipur, India

Nisha Devi
Department of Chemistry, Dr. B. R. Ambedkar National Institute of Technology, Jalandhar, India

SeemaDhail
Department of Chemistry, Manipal University Jaipur, India

P. J. John
Department of Zoology, University of Rajasthan, India

Ajay Kumar
Department of Mechatronics Engineering, Manipal University Jaipur, Dehmi-Kalan, Jaipur, India

P. S. Anil Kumar
Department of Physics, Indian Institute of Science, Bangalore, India

Sukhbeer Kumari
Center of Advanced Studies, Department of Chemistry, University of Rajasthan, Jaipur, India

Lalita Ledwani
Department of Chemistry, Manipal University, Jaipur, India, E-mail: lalitaledwani@gmail.com

Shuchi Maheshwari
Center of Advanced Studies, Department of Chemistry, University of Rajasthan, Jaipur, India

Preeti Mehta
Department of Chemistry, Sangam University, Bhilwara, Rajasthan, India,
E-mail: preetimehta461@gmail.com

Rajeev Mehta
Department of Chemistry, Sangam University, Bhilwara, Rajasthan, India

Rekha N. Nair
Department of Chemistry, Poornima University, Jaipur, Rajasthan, India, E-mail: rekha_124@yahoo.co.in

Selvan Nehru
Department of Physical Chemistry, School of Chemical Sciences, University of Madras, Guindy Campus, Chennai, Tamil Nadu, India

S. K. Nema
FCIPT, Institute for Plasma Research, Gandhinagar, India

A. Pandey
Department of Mechanical Engineering, School of Engineering and Technology, Manipal University, Jaipur, Rajasthan, India, E-mail: anand.pandey@jaipur.manipal.edu

Ram Prakash
Plasma Devices Technology Team, Microwave Tubes Division, CSIR-Central Electronics Engineering Research Institute (CEERI), Pilani 333031, Rajasthan, India, E-mail: ramprakash@ceeri.ernet.in, rplavania@yahoo.com

V. Prasad
Department of Physics, Indian Institute of Science, Bangalore, India

Prabhat Ranjan
Department of Electronics and Communication Engineering, Manipal University, Jaipur, Dehmi-Kalan, Jaipur, India

Ravindra K. Rawal
Department of Pharmaceutical Chemistry, Indo-Soviet Friendship College of Pharmacy, Moga, Punjab, India

Amrit Sarmah
Department of Chemistry, Birla Institute of Technology and Science (BITS), Pilani, Rajasthan, India, E-mail: amritjorhat2009@gmail.com

Deepali Saxena
Amity Institute of Pharmacy, Amity University, Noida, India

Banabithi Koley Seth
Chemical Science Division, Saha Institute of Nuclear Physics, 1/AF Bidhannagar, Kolkata, India

Rahul Shrivastava
Department of Chemistry, Manipal University, Jaipur, India, E-mail: rahul.shrivastava@jaipur.manipal.edu

Virender Singh
Department of Chemistry, Dr. B. R. Ambedkar National Institute of Technology, Jalandhar, India,
E-mail: singhv@nitj.ac.in/singhvirender010@gmail.com

Pragya Soni
Center of Advanced Studies, Department of Chemistry, University of Rajasthan, Jaipur, India

Summan Swami
Department of Chemistry, Manipal University, Jaipur, India

Selvakumar Veeralakshmi
School of Chemistry, Bharathidasan University, Tiruchirappalli, Tamil Nadu, India

Srujana Venigalla
Department of Chemistry, Manipal University Jaipur, Dehmi-Kalan, Jaipur, India

Preface

Recent Methodology in Chemical Sciences provides an eclectic survey of contemporary problems in experimental, theoretical, and applied chemistry. This book covers recent trends of research in different domains of chemical sciences.

In Chapter 1, the effect of the size of a Schiff base complex on the interradical distance and the subsequent spin-flipping tendency are examined. The structures of Schiff base copper complexes can be tuned to design site-specific electron transfer along with the magnetic field effect and DNA hopping phenomena by modulating DNA base dynamics and selective synthesis, which are very much functional in DNA technology.

In Chapter 2, the synthesis of a new water-soluble single- and double-chain surfactant–cobalt (III) complex is reported. The structural features are characterized by different modern spectroscopic techniques. Observed parameters of this report nicely explain different interaction processes of single- and double-chain systems. The conformational and environmental change of instant compound is also reported.

Chapter 3 is a review of chemoselective techniques of spiro heterocycles invoking 1,3-dipolar cycloaddition reaction. Dandia et al. have mentioned in this chapter the advantages of selective synthesis of various spiro heterocyclic compounds and emphasized their recent works in this particular domain.

A one-pot multicomponent synthetic procedure of novel heterocyclic framework is presented in Chapter 4. The yield of this cycloaddition reaction proves its efficacy. To explore stereochemical features of this synthetic process, a theoretical study in terms of density functional theory (DFT) has been performed.

In Chapter 5, the importance and wide range of plasma chemistry have been depicted. Several spectroscopic techniques have been discussed to explore the various basic plasma parameters. The accuracy of spectral techniques always depends on the availability of establishment of equilibrium as well as available atomic data.

In Chapter 6, the novel strategy of synthesis of β-carboline derivatives has been reported. This methodology is useful for large-scale preparation of instant compounds. The authors have mentioned further exploration of carboline derivatives on the basis of the reported method.

In Chapter 7, an overview of different application strategies of treatment on cotton textile is reported. Cotton is a soft, fluffy natural vegetable fiber with great economic importance as a raw material for textile. Since plasma exposure of polymers enhances its surface properties without altering bulk properties, the pretreatment and finishing of textile fabrics by plasma received enormous attention as a solution for environmental problems of textiles.

In Chapter 8, utilization of 1-butyl-3-methylimidazolium fluoride[bmim] F as an activator of the organosilanes with simple handling, storage, and workup in contrast to traditional fluorine source such as tetrabutylammonium fluoride (requisite for Hiyama coupling) is reported.

Acyclic and macrocyclic Schiff-based chelating ligands for uranium ion (UO^{2+}) complexation are reported in Chapter 9. Schiff-based chelators can form stable non-toxic complex with uranyl ion. Supramolecular chemistry of Schiff base ligands and their reduced homologs is rapidly growing due to a wide range of complexation.

A study of effluent from dyeing and the influence of operational parameters on eliminating Azo dyes from textile effluent by advanced oxidation technology are reported in Chapter 10. The effect of various parameters on the photocatalytic degradation of commercially available textiles' azo dye in aqueous heterogeneous suspension has been studied.

The effect of PGRs in in vitro callus culture for production of secondary metabolites is reported in Chapter 11. The study was conducted to explore the hidden potential of natural products synthesized in the medicinal plant *Tinospora cordifolia*.

In Chapter 12, a vivid description of the DFT has been given. DFT is the computationally cost-effective solution for higher-level computation on relatively large systems. Applications of DFT associated with approximate functionals significantly improve the performance of theoretical computation over a wide realm chemical science. In this review article, the author has nicely explained the theory, improvement of methodologies, and applications of DFT in the real field.

A review of asbestos carcinogenicity and its bioremediation is reported in Chapter 13. Detailed insight into the carcinogenic effects of asbestos is envisaged by studies on animal models along with some of the probable detoxification or bioremediation strategies have been reported in the review.

In Chapter 14, the causes and mechanism of aluminum corrosion and its subsequent prevention have been mentioned. In this article, the use of eco-friendly inhibitors on corrosion protection of aluminum has been emphasized. These inhibitors are organic compounds that are absorbed on metallic sites to prevent corrosion. All the inhibitors discussed in this article are nontoxic in nature.

A synthetic technique of vertically aligned carbon nanotubes has been reported in Chapter 15. This method invokes simple pyrolysis without any predeposition of catalyst particles. The authors have claimed their reported synthetic process as simple and economic.

The wide range of applications of electrochemical process in machining is a well-known fact. This process has gained importance due to its promising commercial utilization in manufacturing sector. The commercial terminology of this process is referred as electrochemical machining. In Chapter 16, a brief review has been described on this particular process. This article has depicted the details of principle of process, process capability, and modern-day applications.

Applications of mathematical aspects in different chemical equations have been discussed in Chapter 17. Mathematical chemistry is an important domain in the chemical sciences, and it carries a long history behind it. In this report, the authors have tried to explore different mathematical techniques in explaining simple chemical equations.

Quantitative structure–activity relationship (QSAR) is an emerging field of research in the domain of drug-designing processes. This popular attempt has a wide range of industrial applications. In Chapter 18, a review of different QSAR techniques has been presented. The authors have tried to jot down several theoretical QSAR methodologies and their applications in the real field.

In Chapter 19, the authors have described the toxic effects of lead in a lucid manner. Adverse effects of lead metal and its mechanistic pathway on living systems have been mentioned in this article. The impact of flavonoids on human health is also discussed in this article. Nowadays, this organic compound is very popular due to its many fold applications. Structural features of flavonoids have been also described here.

In Chapter 20, a theoretical analysis on bimetallic nanoalloy clusters has been studied invoking DFT methodology. In this study, the authors have employed several conceptual DFT-based descriptors to correlate experimental properties of Ag–Au alloy with theoretical counterparts. A nice qualitative correlation is reported in this survey.

In Chapter 21, a review article is been presented on pesticide residues in vegetables and fruits in India. The study reveals the flaws of several analytical techniques used in identification of pesticide residues. The authors also have highlighted the features of ultra-performance liquid chromatography–time-of flight mass spectrometry in the domain of pesticide residue analysis on the basis of its high sensitivity and selectivity.

CHAPTER 1

Magnetic Field Effect on Photoinduced Interactions: Its Implications in Distance-Dependent Photoinduced Electron Transfer Between CT-DNA and Metal Complex

Banabithi Koley Seth and Samita Basu*

Chemical Science Division, Saha Institute of Nuclear Physics, 1/AF Bidhannagar, Kolkata, India
*Email: samita.basu@saha.ac.in

CONTENTS

ABSTRACT

The steady-state and time-resolved absorption and fluorescence help to identify the steady-state products and transient intermediates, respectively, generated through photoinduced electron transfer (PET), which may be one of the plausible phenomena in drug–protein/DNA interactions. However, the importance of application of low magnetic field of the order of 0.01–0.02 T lies in its ability to identify initial spin state, one of the deciding factors for ultimate product formation, as well as to assess the intermediate distance in geminating spin-correlated radical ion pairs/radical pairs produced as transients, an useful technique to study "distance-dependent" interactions in biomacromolecules. We have synthesized and studied five new copper(II) Schiff base complexes with differently substituted heterocyclic ligands, $[CuL^1]\cdot 2ClO_4$, $[CuL^2]\cdot 2ClO_4$, $[CuL^3]\cdot 2ClO_4$, $[CuL^4]\cdot 2ClO_4$, and $[CuL^5]\cdot 2ClO_4$, among which the first two metal complexes with N_2O_2 donor set of atoms and the other three metal complexes with N_4 donor set of atoms with different aliphatic substitutions, to understand their effect on interaction with calf thymus DNA (CT-DNA). Laser flash photolysis coupled with an external magnetic field has helped to assess the efficiency of PET from CT-DNA to the complexes. The possibility of PET in triplet state between CT-DNA and the metal complexes having N_2O_2 donor set of atoms, CuL^1 and CuL^2, is insignificant due to the presence of oxygen as ligand atom. However, the other three complexes with N_4 donor set atoms undergo PET with CT-DNA. The extent of PET is much more prominent with pyrrole containing complexes, CuL^4 and CuL^5, compared to pyridine-substituted complex, CuL^3. The increase in the yield of radical ions in the presence of magnetic field depicts the initial spin correlation of the geminate radical ion pair as triplet. The difference between experimental and calculated $B_{1/2}$ values that determines the extent of hyperfine interactions present in the system is much higher for unsubstituted pyrrole copper complex, CuL^4, compared to the substituted one, CuL^5, since the former due to its smaller structure can approach DNA with greater proximity which leads to much more "through-space" hole hopping for intrastrand and interstrand DNA bases. However, the superexchange interaction, which reduces the hole-hopping rate on increasing the size of the nucleobases' bridge, becomes much more prominent leading to a decrease in experimental $B_{1/2}$ value for methyl-substituted pyrrole–DNA system.

1.1 INTRODUCTION

Conventional spectroscopic techniques such as UV–visible (UV–vis) absorption, fluorescence, and circular dichroism used in the study of drug–protein/DNA interactions can yield useful information about ground-state and excited-state phenomena. However, photoinduced electron transfer (PET) may be a possible phenomenon in the drug–protein/DNA interaction, which may go unnoticed if only conventional spectroscopic observations are taken into account. Laser flash photolysis coupled with an external magnetic field (MF) can be utilized to confirm the occurrence of PET and authenticate the initial spin states of the radicals/radical ions formed. Actually most of the photochemical and photobiological reactions involve radical ion pairs (RIPs) and radical pairs (RPs) as transient intermediates generated through PET and hydrogen abstraction or bond cleavage, respectively. Some of the UV–vis spectroscopic techniques such as steady-state and time-resolved absorption and fluorescence serve as efficient tools to identify the transient intermediates and pathways of such reactions. The geminate RIPs/RPs may recombine or separate out to form free radical ions/radicals during reaction. Utilization of these escaped products will be effective if the initial pair maintains spin correlation between two free electrons as triplet, otherwise singlet spin-correlated pairs will undergo recombination leading to initial reactants. Therefore, to avoid recombination of RIPs/RPs, it is necessary to identify their initial spin states. Since individual radical ion/radical contains free electron, application of either internal or external MF can flip or rephrase the electron spin, which leads to intersystem crossing (ISC) between singlet and triplet of the geminate spin-correlated RIPs/RPs. However, utmost ISC will be obtained when radical ions/radicals of geminate RIPs/RPs are separated out by a certain distance where exchange interaction becomes negligible. An internal MF, that is, hyperfine interaction (HFI), present in the system in the order of 0.01–0.02 T is large enough to induce ISC. Application of an external MF in competition with HFI can reduce ISC by introducing Zeeman splitting in triplet sublevels leading to an increase in recombination product or free ion formation depending on the initial spin state of RIPs/RPs as singlet or triplet, respectively. Thus, MF acts as an efficient tool to identify "initial spin state" of the RIPs. Moreover, it can signify the importance of "optimum separation distance" that provides maximum spin flipping and formation of free ions or recombination products. Most of the workers in the field of "spin chemistry" are used to apply low or high MF to identify radical ions or radicals. The distance-dependent magnetic field effect (MFE) has been studied using linked system where radical ions/radicals are separated by varying chain length. Our objective is to study not only the use of MF as a tool to identify the initial spin states of RIPs/RPs but also the effect of structure of molecules that plays crucial role in controlling the optimum separation distance especially for intermolecular RIPs/RPS on MFE.[1–28]

Previously, in our laboratory, we carried out several works on interactions of therapeutically important drugs with biomacromolecules in the presence of external MF.[16–28] The biological systems that had been highlighted in these works were mainly some important model proteins and DNA along with its nucleobases, nucleosides, and nucleotides. Dealing with intra-and intermolecular electron transfer in such elementary biological units helps to unravel the modes of interactions of DNA and proteins with small drug-like molecules, which is of high pharmacological importance. While studying the interaction of anticancer drugs menadione (2-methyl-1,4-naphthoquinone) and 4-nitroquinoline-1-oxide with lysozyme protein, we observed PET from tryptophan residue of the model protein to individual drug in the excited state without any

MFE. However, in the study of interaction between the model protein human serum albumin (HSA) with acridine derivatives, acridine yellow (AY) and proflavin (PF$^+$), appreciable MFE was observed along with electron transfer. Owing to its distance dependence, MFE gave an idea about the proximity of the radicals/radical ions (PF$^·$, AY$^{·-}$, TrpH$^{·+}$, Trp$^·$) during interaction in the system and also helped to elucidate the reaction mechanism. A prominent MFE was observed for this system in homogeneous buffer medium owing to the pseudoconfinement of the radicals/radical ions provided by the complex structure of the HSA protein and also predicted the separation distance between the donor and acceptor in-between 10 Å and 17 Å, which was further supported by docking analyses.[28] On the other hand, in the interactions between two quinone drugs (2-methyl-1,4-naphthoquinone or more commonly known as menadione and its higher homologue 9,10-anthraquinone), which serve a good purpose as anticancer agents being efficient electron acceptors, with DNA and RNA bases, that is, adenine, thymine, guanine, cytosine, and uracil and their corresponding nucleosides, adenosine, thymidine, guanosine, cytidine, and uridine, in both homogeneous acetonitrile/water mixture and heterogeneous micellar medium, electron transfer has been found to be competitive with hydrogen atom transfer.

A prominent MFE was observed for the triplet-born radicals during the interaction of a transition metal complex, [Cu(phen)$_2$]$^{2+}$, with DNA even in homogeneous aqueous medium, which is a rare phenomenon. This process of partial intercalation of the complex within DNA might be responsible for the observation of MFE in the homogeneous medium. MFE was also observed in organized assemblies, for example, reverse micelles instead of water as reaction medium; however, it is not very much prominent due to large distance of separation between the component radicals of the geminate RIPs. In extension with ternary metal complexes comprising aromatic amino acids, for example, tyrosine and tryptophan and as a second ligand that contains an aromatic ring such as 2,2 -bipyridyl or 1,10-phenanthroline, [Cu(phen)(Htyr)]ClO$_4$ and [Cu(phen)(Htrp)]ClO$_4$ (Htyr: L-tyrosinato and Htrp: L-tryptophanato) predict the occurrence of electron transfer reactions with calf thymus (CT) DNA. It was observed that in both the complexes, intramolecular electron transfer occurs from amino acids to phen moiety on photoexcitation. However, in the presence of CT-DNA, intermolecular electron transfer occurs between DNA and complexes. The occurrence of partial intercalation of the complexes within DNA helps in maintaining the proper interradical distance between the RIPs generated through PET, so that spin correlation exists between them and MFE could be observed. Therefore, not only organic compounds but also inorganic copper complexes take part in PET with DNA and shows prominent MFE from where the drug–protein/DNA separation distance may be predicted. The versatile coordination behavior of metal complexes, especially transition metal complexes, with variable ligands and metal ions makes them excellent probes exhibiting high selectivity in PET reactions along with MFE.[29–34]

These works motivated us to investigate the role of copper Schiff base complexes along with their structural dependence in PET coupled with external MFE with

CT-DNA in detail. To carry out this investigation, five different copper Schiff base complexes have been used; two metal complexes with N_2O_2 donor set of atoms and the other three with N_4 donor set of atoms with different aliphatic substitutions. The laser flash photolysis coupled with external MF has been utilized to identify the efficiency of charge/electron transfer between CT-DNA and reacting copper complexes having different substituted Schiff base ligands as well as to authenticate the spin state where it initially occurs. Schiff base ligands have been used because of their easy and inexpensive syntheses, versatile metal coordination behaviors with different sets of donor atoms, and biological applications.[35,36] Moreover, $B_{1/2}$ value, the field at which half the saturation of the field effect reaches, has also been calculated to predict the extent of HFI present in the system.[37]

1.2 EXPERIMENTAL

1.2.1 Materials

All the chemicals and solvents used for syntheses of the complexes are of analytical grade. The chemicals 1,2-diaminopropane, 1,3-daminopropane, 2-pyridinecarboxaldehyde, 2-pyrrolecarboxaldehyde, 2-acetylpyridine, and 2-acetylpyrrol have been purchased from Aldrich Chemical Co., USA. The highly polymerized CT-DNA has been purchased from Sisco Research Laboratory, India, and Tris buffer, sodium chloride, and hydrochloric acid (AR) have been purchased from Merck, Germany. All the reagents have been used without further purification. Triple distilled water has been used for the preparation of all aqueous solutions. Solvents required for syntheses and spectroscopic studies have been purchased from SRL, India, and Spectrochem, India, respectively. Copper perchlorate has been prepared as before.[38] CT-DNA solutions have been prepared in Tris–HCl/NaCl buffer maintaining biological pH 7.4. All the complexes have been dissolved in minimum volume of dimethyl sulfoxide and then diluted with Tris–HCl/NaCl buffer solution.

1.2.2 Syntheses

1.2.2.1 Synthesis of Ligand L1, L2, L3, L4, and L5

The ligands L^1 and L^2 have been resynthesized[38] by refluxing a 50 mL methanolic solution of 1,2-diaminopropane (5 mmol) (for L^1)/1,3-diaminopropane (for L^2) with salicylaldehyde (10 mmol) for ~1.5 h at 35°C. Ligands L^3 and L^4 have also been prepared[39] by refluxing a 50 mL methanolic solution of 1,2-diaminopropane (5 mmol) and 2-pyrridine carboxaldehyde, and 2-pyrrolecarboxaldehyde, respectively, whereas for L^5, 2-acetylpyrrole (10 mmol) has been used and refluxed for 6 h at 35°C. Thus, obtained Schiff base ligands have been used directly for complexes syntheses.

1.2.2.2 Syntheses of Complexes CuL^1, CuL^2, CuL^3, CuL^4, and CuL^5

The ligand solutions (each 1 mmol, 10 mL) have been added dropwise to the methanolic solutions of $Cu(ClO_4)_2 \cdot 6H_2O$ (each 1 mmol, 0.499 g) and kept at undisturbed condition for crystal growth. All the solutions obtained from L^{1-5} solutions yield crystalline complexes CuL^1, CuL^2, CuL^3, CuL^4, and CuL^5 after 1, 1, 7, 5, and 14 days, respectively. The elemental analyses, IR, UV–vis, and Mass data of the ligand and the complexes have been matched with our earlier data.[38,39] For CuL^1, the corresponding spectroscopic data have been reported below. The chemical structures of the complexes have been shown in Figure 1.1.

Ligand L^1: Anal. Calc. for ligand L^1 ($C_{17}H_{18}N_2O_2$): C, 72.34%; H, 6.38%; N, 9.93%, O, 11.35%; UV–vis: λ_{max} (nm) (ε_{max} ($dm^3\ mol^{-1}\ cm^{-1}$)) (methanol), 220 (23,560), 330 (8500).

CuL^1: Yield: 0.364 (73% with respect to metal perchlorate). Anal. Calc. for CuL^1 ($C_{17}H_{16}N_2O_2Cu$): C, 59.38%; H, 4.66%; N, 8.16%; O, 9.32%; Cu, 18.49%; Found: C, 58.93%; H, 4.03%; N, 8.02%. Main FT-IR bands (KBr, cm^{-1}): $\nu(Cu-N)$ 425 cm^{-1}, $\nu(C=N)$ 1613 cm^{-1}. UV–vis: λ_{max} (nm) (ε_{max}($dm^3\ mol^{-1}\ cm^{-1}$)) (methanol), 227 (10,060), 273 (5195), 356 (3031), 564 (281).

1.2.3 Physical Measurements

Elemental analyses (carbon, hydrogen, nitrogen microanalyses) of the complexes have been carried out by Perkin-Elmer 2400 series II CHN analyzer. The Fourier transform infrared spectra have been taken using a Perkin Elmer Spectrum 100 FT-IR Spectrometer in the range 400–4000 cm^{-1} with a solid KBr disc. The mass spectra of the complexes have been recorded with a Qtof Micro YA mass spectrophotometer. The absorption spectra have been recorded on a Jasco V-650 absorption spectrophotometer over a wavelength range 200–800 nm with 1 cm quartz cuvette, whereas the transient absorption spectra measurements of CT-DNA complex systems have been performed by nanosecond laser flash photolysis (Applied Photophysics) using a Nd:YAG laser (Lab series, Model Lab 150, Spectra Physics).

1.2.4 Laser Flash Photolysis

In nanosecond flash photolysis setup having an Nd:YAG laser, the sample has been excited by 266 nm laser light (10 mJ) with full width at half maximum (FWHM) $\approx$ 8 ns. Absorption of light from a pulsed Xe lamp (150 W) at right angle to the laser beam has been used to detect the newly generated transient species in the system. The photomultiplier (R928) output has been fed into an Agilent Infiniium oscilloscope (DSO8064A, 600 MHz, 4 Gs/s), and the data have been transferred to a computer through IYONIX software. The MFEs on the transient absorption spectra have been explored by passing direct current through a pair of electromagnetic coils placed inside the sample chamber and the strength of MF has been varied from 0.0 to 0.08 T. All the samples have been deaerated properly by argon gas before experiments to avoid quenching. No degradation of the samples has been observed during the experiment. The software Origin 8.0 has been used for curve fitting.

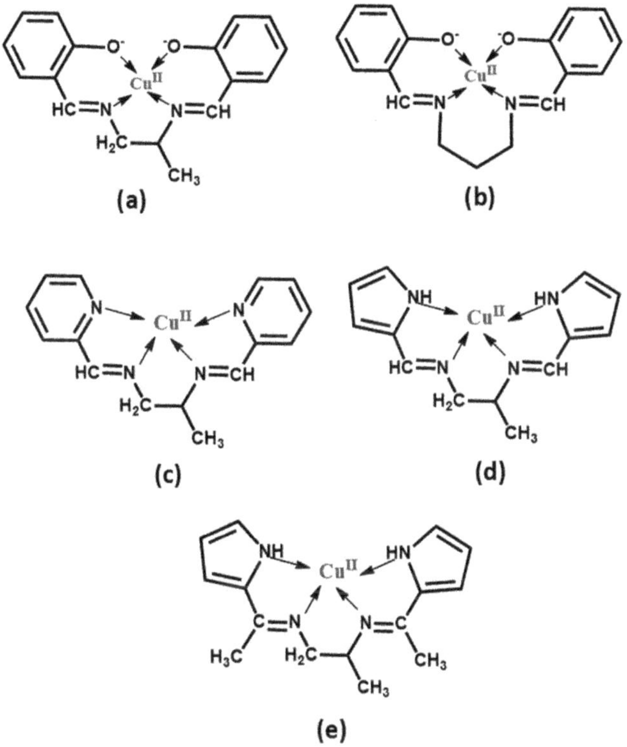

FIGURE 1.1   Chemical Structures of Metal Complexes (a) CuL^1, (b) CuL^2, (c) CuL^3, (d) CuL^4, and (e) CuL^5.

1.3   RESULTS AND DISCUSSION

The interactions between five different copper Schiff base complexes and CT-DNA have been investigated in triplet state. All the complexes are shown in Figure 1.1. The metal complexes having N_2O_2 donor set of atoms, CuL^1 (Fig. 1.1a) and CuL^2 (Fig. 1.1b), do not show any significant transient absorption spectra. Therefore, the possibilities of PET in triplet state between CT-DNA and these complexes become zero/ insignificant. Free oxygen is a very good quencher of triplet state.[40] Hence, the presence of oxygen may quench the possibility of PET in these cases. Therefore, to identify the role of metal complexes in PET, metal complexes with N_4 donor set of atoms have been used instead of those with N_2O_2 donor set of atoms. Initially, two different complexes, pyridine- and pyrrole-substituted complexes ([Fig. 1.1c] CuL^3 and [Fig. 1.1d] CuL^4), with N_4 donor set of atoms have been utilized. The complexes have been excited separately by 266 nm laser light (10 mJ) with FWHM $\approx$ 8 ns, and several newly generated transient species in the system have been detected. The pyridine-substituted complex shows very weak characteristic absorption peak, which has been quenched rapidly in the presence of CT-DNA. However, the pyrrole-substituted complex shows prominent characteristic absorption peaks at 330, 460, and 540 nm, shown in Figure 1.2. On gradual addition of CT-DNA, the spectra of this CuL^4 complex show some changes that indicate the existence of some interactions between the complexes and CT-DNA in triplet state. Earlier it has been found that DNA is a good carrier for long range electron transfer and the DNA bases, adenine, guanine, thymine, and cytosine, possess potential electron donating capability.[16–20] Among the four bases, guanine is the most efficient electron donor. Dey et al.[21–24] have shown that PET between copper complexes and CT-DNA commences through charge transfer from the guanine moiety of CT-DNA to the copper phenanthroline complex. Similarly, the possibility of ET from guanine and other bases of CT-DNA to pyrrole containing Schiff base complexes has been formed out on gradual addition of CT-DNA to complex system, which shows gradual quenching of characteristic absorption peak as well as gradual rising of a new peak around 370, 420, and 480–500 nm. The peaks at 370, 420, and 480–500 nm arise mainly for the formation of DNA radical cations of guanine, adenine, thymine, and cytosine, respectively. Therefore, the results suggest the occurrence of PET from DNA to complexes. Further, the methyl substituted pyrrole complex [(e) CuL^5] has also been used to investigate in detail about the potentiality of pyrrole Schiff base copper complexes toward PET. The experimental results depict that this complex imposes more pronounced effect on PET compared to other pyrrole complex CuL^4 (Fig. 1.2).

Further, external MF has been employed to envisage the initial spin state, either singlet or triplet, as well as the initial separation distance between the components of spin-correlated RPs/RIPs produced as transient intermediates of CT-DNA–metal complex reacting system. In our system, the increases in the yields of the transient ions in the presence of external MF for the pyrrole complexes–DNA systems, shown in Figure 1.3, confirm the occurrence of PET in triplet state. From the variation of absorbance in the presence of external MF, the values of $B_{1/2}$, the MF at which half the saturation of its effect reaches, have also been calculated for the pyrrole–DNA system following the theoretical expression for quantitative correlation of $B_{1/2}$ with the HFI energy of the individual RP established by Weller et al,[37]

$$B_{1/2} = \frac{2(B_1^2 + B_2^2)}{B_1 + B_2},$$

where B_i represents the effective nuclear MF at the unpaired electron in each radical. Some of our previous works on MFE on PET show prominent discrepancies between experimental and calculated $B_{1/2}$ values. The higher experimental $B_{1/2}$ value compared to the calculated value may be due to hopping or lifetime broadening through frequent re-encounter within the RIP.

While working with N-ethyl carbazole (ECZ)-1,4-dicyanobenzene (DCB) and pentamethyl carbazole (PMC)–DCB systems,[10–14] we found that the experimental and calculated $B_{1/2}$ values show good resemblance to each other. Moreover, with increase in the concentration of the nonfluorescent acceptor, DCB, which enhances electron hopping within geminate RIPs, there is no significant change in experimental $B_{1/2}$ value since the maximum contribution to HFI originates from the fluorophore itself, that is, ECZ or PMC and not from the DCB molecules. However, in other systems such as pyrene-N,N-dimethylaniline (Py-DMA), 9-cyanophenanthrene-trans-anethole (CNP-AN), the experimental $B_{1/2}$ value increases with the concentration of nonfluorescent donor, DMA or AN possessing significant HFI, because of shortening of lifetime of a particular RIP owing to electron hopping from one donor to other, leading to a broadening in the S–T energy levels. To overcome this, the energy broadening higher field is required to get the saturation, and hence $B_{1/2}$ increases.

The MFE on the PET between phenazine (PZ) and the amines, DMA, N,N-diethylaniline (DEA), 4,4'-bis(dimethylamino)diphenylmethane (DMDPM), and triethylamine (TEA), was studied in micelles, reverse micelles, and small unilamellar vesicles (SUVs). The differential behavior of the amines can be explained in terms of their confinement in different zones of the organized assemblies depending on their bulk, hydrophobic, and electrostatic effects. The structure of the assembly is found to greatly affect the PET dynamics and hence the MF behavior of all the acceptor–donor systems. The MF behavior in micelles is consistent with the hyperfine mechanism, but higher experimental $B_{1/2}$ values compared to calculated values were obtained with PZ–DMA and PZ–DMDPM systems, which can be ascribed to hopping and lifetime broadening since both donor and acceptor remain in hydrophobic region. However, for PZ–TEA system, the calculated and experimental $B_{1/2}$ values are almost same because TEA remains in hydrophilic and PZ is in hydrophobic regions. Therefore, separation of radical ions in different zones of the heterogeneous media reduces the effect of electron hopping within geminate RIPs on the experimental values of $B_{1/2}$.

Unlike simple PZ molecule, one of its derivatives, dibenzo[a,c]phenazine (DBPZ), forms a charge transfer complex in the triplet state (^{3}ECT) with different amines, for example, DMA, DMDPM, and TEA.[24,25] The RIPs are much more abundant in the cases of DMA and DMDPM rather than in TEA. Interestingly, a prominent MFE is observed in nonviscous medium; this was explained by considering the extended planar structure of DBPZ and interradical hydrogen bonding mediated by the intervening water molecules in both the cases of ^{3}ECT and RIPs in homogeneous acetonitrile–water $(MeCN/H_2O)$ mixture. The MF behavior is consistent with the hyperfine mechanism; however, low $B_{1/2}$ value for DBPZ–TEA system is ascribed to fast electron exchange due to close proximity of the corresponding radical ions. On the other hand, the bulky size of the DMDPM molecule hinders the approach of other DMDPM molecules toward DBPZ and the corresponding intermolecular electron hopping among the DMDPM molecules, which makes experimental $B_{1/2}$ almost similar to calculated $B_{1/2}$.

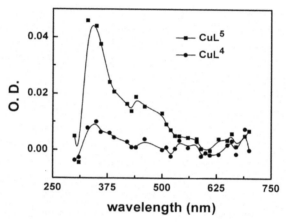

FIGURE 1.2 Transient Absorption Spectra of CuL^4 (25 μM) and CuL^5 (25 μM) in Tris–HCl/NaCl Buffer at 1 μs after the Laser Flash at 266 nm.

Therefore, the discrepancy between the magnitude of $B_{1/2}$ values of presently considering two pyrrole–DNA systems, shown in Figure 1.4, where the smaller complex (CuL^4) shows larger $B_{1/2}$ value compared to the larger one (CuL^5), indicates the possibility of the "through-space" hole hopping for intrastrand and interstrand DNA bases. However, the superexchange interaction is much more prominent for intrastrand base pairs, which reduces the hole-hopping rate on increasing the size of the nucleobases bridge.[41] The drop-off of $B_{1/2}$ value of methyl-substituted pyrrole–DNA system compared to unsubstituted system is owing to negative effect of superexchange, which reduces the effective HFI present in the system. According to Schulten,[42] the sterically fixed intramolecular system can even show the reduction of effective HFI by 50% due to the presence of super exchange compared to the intermolecular system, which was later experimentally verified by Petrov et al.[43] The experimental $B_{1/2}$ value is higher than the calculated value because of the presence of more number of DNA bases in CT-DNA polymer than that in oligonucleotide.

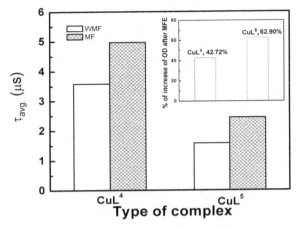

FIGURE 1.3 Changes in OD and τ_{avg} Values of CuL^4 and CuL^5 in the Absence and Presence of 0.08 T MF at a Delay at 1 µS after the Laser Pulse at 266 nm in Tris Buffer.

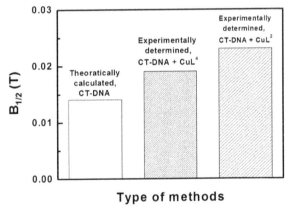

FIGURE 1.4 Theoretically and Experimentally Determined $B_{1/2}$ Values of CuL^4 and CuL^5 Complexes.

1.4 CONCLUSION

The N_2O_2 donor set of atoms containing copper Schiff base complexes do not contribute in PET with CT-DNA. However, N_4 donor set of atoms containing copper Schiff base complexes impose pronounced effect on PET. And in the case of two N_4 containing complexes, pyrrole complexes exhibit prominent PET with CT-DNA, whereas that of pyridine complexes is almost insignificant. The MFE in PET reactions can serve as an efficient tool in the identification of the initial spin state of the geminate

RIPs formed due to electron transfer. The essential features for observation of MFEs are the diffusion, spin flipping, and recombination or free ion formation depending on the singlet or triplet spin states, respectively, of the spin correlated geminate RIPs. If the participating radical ions are very close to each other, the exchange interaction will hinder spin conversion, whereas a large distance of separation between them will destroy the spin correlation and their geminate characteristics. Both the phenomena will reduce MFE. Therefore, MFE indirectly serves as a tool to estimate the separation distance between geminate radical ions, and maximum field effect is obtained at an optimum interradical distance where maximum spin flipping and consecutive phenomena could take place. The shape and size of CuL^5 complex favor itself to maintain the optimum interradical distance exhibiting maximum spin flippling and consecutive phenomena compared to other pyrrole complex (CuL^4), which shows maximum hole hopping with CT-DNA due to smaller size. So, the structures of Schiff base copper complexes can be tuned further to design site-specific ET along with MFE and DNA hopping phenomena by modulating DNA base dynamics and selective synthesis, which are very much functional in DNA technology.

ACKNOWLEDGMENTS

This research work is supported by funding from the Biomolecular Assembly, Recognition and Dynamics (BARD) project, SINP of Department of Atomic Energy (DAE), Government of India. B. Koley Seth would like to acknowledge University Grant Commission (UGC), India, for her research fellowship. Authors would like to thank Mrs. Chitra Raha and Mr. Ajay Das for their kind assistance and technical support.

KEYWORDS

- CT-DNA
- Schiff base metal complex
- laser flash photolysis
- photoinduced electron transfer
- magnetic field effect
- DNA hopping
- superexchange interaction

REFERENCES

1. Steiner, U. E.; Ulrich, T. Magnetic field effects in chemical kinetics and related phenomena. *Chem. Rev.* 1989, 89, 51–147.
2. Bhattacharya, K.; Chowdhury, M. Environmental and magnetic field effects on exciplex and twisted charge transfer emission. *Chem. Rev.* 1993, 93, 507–535.
3. Nagakura, S.; Hayashi, H.; Azumi, T. *Dynamic Spin Chemistry Magnetic Controls and Spin Dynamics of Chemical Reactions*; Kodansha Ltd: Tokyo, 1998.
4. Gould, I. R.; Turro, N. J.; Zimmt, M. B. Magnetic field and isotope effects on the products of organic reactions. *Adv. Phys. Org. Chem.* 1984, 20, 1–53.
5. Tanimoto, Y.; Fujiwara, Y. *Effects of high magnetic fields on photochemical reactions*. In *Handbook of Photochemistry and Photobiology, Vol. 1: Inorganic Chemistry*; Nalwa, H. S., Ed.; American Scientific Publishers: Stevenson Ranch, CA, 2003; p 413.
6. Grissom, C. B. Magnetic field effects in biology: a survey of possible mechanisms with emphasis on radical-pair recombination. *Chem. Rev.* 1995, 95, 3–24.
7. Boxer, S. G.; Chidsey, C. E. D.; Roelofs, M. G. Magnetic field effects on reaction yields in the solid state: an example from photosynthetic reaction centers. *Ann. Rev. Phys. Chem.* 1983, 34, 389–417.
8. Nath, D.; Chowdhury, M. Effect of environment on the magnetic field modulation of exciplex luminescence. *Chem. Phys. Lett.* 1984, 109, 13–17.
9. Buchachenko, A. L. MIE versus CIE: comparative analysis of magnetic and classical isotope effects. *Chem. Rev.* 1995, 95, 2507–2528.
10. Aich, S.; Basu, S. Laser flash photolysis studies and magnetic field effect on a new heteroexcimer between N-ethyl carbazole and 1,4- dicyanobenzene in homogeneous and heterogeneous media. *J. Chem. Soc. Faraday Trans.* 1995, 91, 1593–1600.
11. Aich, S.; Basu, S. Enhancement of intersystem crossing by substitution: assignment of a long-lived triplet state by magnetic field effects in a non-viscous medium. *Chem. Phys. Lett.* 1997, 281, 247–253.
12. Aich, S.; Basu, S. Magnetic field effect: a tool for identification of spin state in a photoinduced electron-transfer reaction. *J. Phys. Chem. A.* 1998, 102, 722–729.
13. Dutta Choudhury, S.; Basu, S. Magnetic field effect on N-ethylcarbazole-dimethylterephthalate and N-ethylcarbazole-1,4-dicyanobenzene: a comparative study focusing on steric effect. *Chem. Phys. Lett.* 2005, 408, 274–278.
14. Dutta Choudhury, S.; Basu, S. Exploring the extent of magnetic field effect on intermolecular photoinduced electron transfer in different organized assemblies. *J. Phys. Chem. A.* 2005, 109, 8113–8120.
15. Sengupta, T.; Dutta Choudhury, S.; Basu, S. Medium-dependent electron and H atom transfer between 2'-deoxyadenosine and menadione: a magnetic field effect study. *J. Am. Chem. Soc.* 2004, 126, 10589–10593.
16. Bose, A.; Dey, D.; Basu, S. Interactions of guanine and guanosine hydrates with quinones: a laser flash photolysis and magnetic field effect study. *J. Phys. Chem. A.* 2008, 112, 4914–4920.
17. Bose, A.; Basu, S. Laser flash photolysis and magnetic field effect studies on the interaction of uracil and its derivatives with menadione and 9,10-anthraquinone. *J. Phys. Chem. A.* 2008, 112, 12045–12053.
18. Bose, A.; Sarkar, A. K.; Basu, S. Interaction of 9,10-anthraquinone with adenine and 2'-deoxyadenosine. *Biophysical Chemistry.* 2008, 136, 59–65.
19. Bose, A.; Dey, D.; Basu, S. Laser flash photolysis and magnetic-field-effect studies on interaction of thymine and thymidine with menadione: role of sugar in controlling reaction pattern. *Sci. Technol. Adv. Mater.* 2008, 9, 024205–024210.
20. Bose, A.; Basu, S. Medium-dependent interactions of quinones with cytosine and cytidine: a laser flash photolysis study with magnetic field effect. *Biophysical Chemistry.* 2009, 140, 62–68.
21. Chakraborty, B.; Basu, S. Study of interaction of proflavin with triethylamine in homogeneous and micellar media: photoinduced electron transfer probed by magnetic field effect. *Chem. Phys. Lett.* 2009, 477, 382–387.

22. Chakraborty, B.; Basu, S. Interaction of proflavin with aromatic amines in homogeneous and micellar media: photoinduced electron transfer probed by magnetic field effect. *Chem. Phys. Lett.* 2010, 487, 51–57.

23. Sarangi, M. K.; Dey, D.; Basu, S. Associated electron and proton transfer between Acridine and Triethylamine in AOT reverse micelles probed by laser flash photolysis with magnetic field. *Chem. Phys. Lett.* 2011, 506, 205–210.

24. Dey, D.; Bose, A.; Chakraborty, M.; Basu, S. Magnetic field effect on photoinduced electron transfer between dibenzo[a,c]phenazine and different amines in acetonitrile-water mixture. *J. Phys. Chem. A.* 2007, 111, 878–884.

25. Dey, D.; Bose, A.; Bhattacharyya, D.; Ghosh, S.; Maity, S. S.; Basu, S. Dibenzo[a,c]phenazine: a polarity-insensitive hydrogen bonding probe. *J. Phys. Chem. A.* 2007, 111, 10500–10506.

26. Dey, D.; Bose, A.; Pramanik, N.; Basu, S. Magnetic field effect on photoinduced electron transfer between [Cu(phen)2]2+ and DNA. *J. Phys. Chem. A.* 2008, 112, 3943–3946.

27. Dey, D.; Pramanik, N.; Basu, S. Exploring the mechanism of electron transfer between DNA and a ternary copper complex. *J. Phys. Chem. B.* 2009, 113, 8689–8694.

28. Chakraborty, B.; Singha Roy, A.; Dasgupta, S.; Basu, S. Magnetic field effect corroborated with docking study to explore photoinduced electron transfer in drug-protein interaction. *J. Phys. Chem. A.* 2010, 114, 13313–13325.

29. Hackl, E. V.; Galkin, V. L.; Blagoi, Y. P. DNA interaction with biologically active divalent metal ions: binding constants calculation. *Int. J. Biol. Macromol.* 2004, 34, 245–250.

30. Jiao, K.; Wang, Q. X.; Sun, W.; Jian, F. F. Synthesis, characterization and DNA-binding properties of a new cobalt(II) complex:Co(bbt)₂Cl₂. *J. Inorg. Biochem.* 2005, 99, 1369–1375.

31. Gaur, R.; Khan, R. A.; Tabassum, S.; Shah, P.; Siddiqi, M. I.; Mishra, L. Interaction of a ruthenium(II)–chalcone complex with double stranded DNA: spectroscopic, molecular docking and nuclease properties. *J. Photochem. Photobiol. A: Chem.* 2011, 220, 145–152.

32. Li, Y.; Yang, Z. Y.; Liao, Z. C.; Han, Z. C.; Liu, Z. C. Synthesis, crystal structure, DNA binding properties and antioxidant activities of transition metal complexes with 3-carbaldehyde-chromone semicarbazone. *Inorg. Chem. Commun.* 2010, 13, 1213–1216.

33. Budagumpi, S.; Kulkarni, N. V.; Kurdekar, G. S.; Sathisha, M. P.; Revankar, V. K. Synthesis and spectroscopy of CoII, NiII, CuII and ZnII complexes derived from 3,5-disubstituted-1H-pyrazole derivative: A special emphasis on DNA binding and cleavage studies. *European J Med. Chem.* 2010, 45, 455–462.

34. Joyner, J. C.; Reichfield, J.; Cowan, J. A. Factors influencing the DNA nuclease activity of iron, cobalt, nickel, and copper chelates. *J. Am. Chem. Soc.* 2011, 133, 15613–15626.

35. Mahalakshmi, N.; Rajavel, R. Synthesis, synthesis, spectroscopic characterization, DNA cleavage and antimicrobial activity of binuclear copper(II), nickel(II) and oxovanadium(Iv) Schiff base complexes. *Asian J Biochem. Pharm. Res.* 2011, 1, 525–543.

36. Xu, Z.; Xi, P.; Chen, F.; Liu, X.; Zeng, Z. Synthesis, characterization, and DNA-binding properties of copper(II), cobalt(II), and nickel(II) complexes with salicylaldehyde 2-phenylquinoline-4-carboylhydrazone. *Transition Met. Chem.* 2008, 33, 267–273.

37. Weller, A.; Staerk, H.; Treichel, R. Magnetic-field effects on geminate radical-pair recombination. *Faraday Discuss. Chem. Soc.* 1984, 78, 271–278.

38. Roy, A.; Koley Seth, B.; Pal, U.; Basu, S. Nickel(II)-Schiff base complex recognizing domain II of bovine and human serum albumin: spectroscopic and docking studies. *Spectrochemica Acta Part A.* 2012, 92, 164–174.

39. Koley Seth, B.; Ray, A.; Saha, A.; Saha, P.; Basu, S. Potency of Photoinduced Electron Transfer and antioxidant efficacy of pyrrole and pyridine based Cu(II)-Schiff complexes while binding with CT-DNA. *J Photochem. Photobiol. B.* 2014, 132, 72–84.

40. Grewer, C.; Brauer, H. D. Mechanism of the triplet-state quenching by molecular oxygen in solution. *J. Phys. Chem.* 1994, 98, 4230–4235.

41. Voityuk, A. A.; Jortner, J.; Bixon, B.; Rosch, N. Electronic coupling between Watson–Crick pairs for hole transfer and transport in desoxyribonucleic acid. *J. Chem. Phys.* 2001, 114, 5614–5620.
42. Schulten, K. The effect of intramolecular paramagnetic–diamagnetic exchange on the magnetic field effect of radical pair recombination. *J. Chem. Phys.* 1985, 82, 1312–1317.
43. Petrov, N. K.; Alfimov, M. V.; Budyka, M. F.; Gavrishova, T. N.; Staerk, H. Intramolecular electron hopping in double carbazole molecules studied by the fluorescence-detected magnetic field effect. *J. Phys. Chem. A.* 1999, 103, 9601–9604.

CHAPTER 2

Role of Hydrophobicity of Some Single- and Double-Chain Surfactant– Cobalt(III) Complexes on the Interaction with Bovine Serum Albumin

Selvakumar Veeralakshmi[1], Selvan Nehru[1,2], Sankaralingam Arunachalam[1*]

[1]School of Chemistry, Bharathidasan University, Tiruchirappalli, Tamil Nadu, India
[2]Department of Physical Chemistry, School of Chemical Sciences, University of Madras, Guindy Campus, Chennai, Tamil Nadu, India
*Email: arunasurf@yahoo.com

CONTENTS

ABSTRACT

A new water-soluble single- and double-chain surfactant–cobalt(III) complexes, $[Co(dien)(TA)Cl_2]ClO_4$ (**1**) and $[Co(dien)(TA)_2Cl](ClO_4)_2$ (**2**), where dien =is diethylenetriamine and TA is =tetradecylamine, have been synthesized. The structure of the complexes was characterized by UV–visible (UV–vis), Fourier transform infrared, NMR, and electrospray ionization mass spectrometry. Hydrophobicity of these surfactant–cobalt(III) complexes was investigated by partition-coefficient method. The critical micelle concentration (CMC) values of these surfactant metal complexes in aqueous solution were obtained from conductivity measurements at five different temperatures. The biophysical interaction of these amphiphilic molecules with bovine serum albumin (BSA) has been examined by fluorescence, synchronous, three-dimensional (3D) fluorescence, UV–vis, and circular dichroism (CD) techniques at pH 7.4. The results of hydrophobicity and CMC values indicate that double-chain surfactant–cobalt(III) complex has more hydrophobicity compared to single-chain surfactant–cobalt(III) complex. The fluorescence titration at three different temperatures has shown that the interaction between surfactant–cobalt(III) complexes and BSA was mainly a static quenching process. Interestingly, on increasing temperature, binding constant and number of binding sites get decreased for single-chain system whereas increased for double-chain system, due to the changes in the mode of protein–complex interaction. The observed thermodynamic parameters clearly showed that surfactant–cobalt(III) complexes with single-chain system prefer electrostatic binding, whereas those with double-chain system prefer hydrophobic interaction. Moreover, the results from UV–vis absorption, synchronous fluorescence, 3D fluorescence, and CD indicate that conformational and some microenvironmental changes occurred in BSA.

2.1 INTRODUCTION

In the past decades, a large number of studies have been dedicated to understand the interactions of biomacromolecules with various ligands, and those can provide useful information of the structural features that determine the therapeutic effectiveness of drugs. Among various biomacromolecules, serum albumins have been intensively studied due to their physiological functions. The most important function of albumin is to serve as a depot and transport protein for a variety of compounds like fatty acids, amino acids, hormones, bilirubin, metal ions, drugs, and pharmaceuticals. Bovine serum albumin (BSA) has been one of the most extensively studied drug carrier protein, particularly because of its structural homologous with human serum albumin (HSA). Binding of small molecules to serum albumin may significantly affect the absorption, distribution, metabolism, and toxicity of drugs. Consequently, it is of great interest to investigate the interactions between bioactive compounds and serum albumin. Such studies are helpful to explain the metabolism and transportation process of bioactive compounds.

Surfactant metal complexes are a new class of coordination complexes in which metal-containing coordination sphere acts as hydrophilic head group, whereas long alkyl chain-containing ligand acts as hydrophobic group. Similar to conventional surfactants, these types of surface-active molecules are able to lower the surface tension of water, and also aggregate into micelles.[1] Uniquely, surfactant metal complexes offer properties such as variable metal center, oxidation state, reactivity, color, multicharged head group ligands, photochemistry, which have attracted the researchers for employing in various applications such as emulsions, catalysis, optoelectronics,[2] templates for mesoporous materials,[3] and metallodrugs.[4]

The employment of designing effective metallodrugs with reduced side effects against human diseases is an active area of research, and the structural modification of metallodrugs can alter their affinity with biomacromolecules, such as nucleic acids, proteins, which are important to consider during the drug designing. Serum albumins are the most abundant proteins in blood plasma and are responsible for the binding and transportation of various endogenous and exogenous ligands such as fatty acids, hormones, and harmful substances.[5] Drug–protein interactions are closely related to drug efficiency in the treatment of diseases because the absorption, transportation, distribution, and metabolism of drugs strongly depend on their binding properties.[6] Generally, the strong binding with protein decreases the concentrations of free drug in plasma, whereas the weak binding leads to shorter lifetime or poor distribution of drugs. Moreover, the investigation of binding of the drugs to serum albumins is of great toxicological and medical importance, and it may afford key information to rational drug design. However, the impact of protein binding of metallodrugs on antimicrobial activities is still not clear. Among these aspects of drug designing, one of the factors, hydrophobicity of metal complexes, plays a major role in the penetration of cell membrane to precede cell death.[7]

In our laboratory, we have been focusing on the design, development, and interactions of surfactant metal complexes with proteins and nucleic acids. Interaction of proteins with surfactants mainly depends on surfactant features like size, charge, chain length, hydrophobicity, and concentration. Several reports have been investigated on the interaction of proteins with conventional surfactants, but those with surfactant metal complexes are limited. Thus, the present study focuses on how the single- and double-chain surfactant–cobalt(III) complexes affect their hydrophobicity, critical micelle concentration (CMC) and its thermodynamic parameters, and the interaction with BSA.

2.2 EXPERIMENTAL SECTION

2.2.1 Materials

BSA (lyophilized powder, essentially fatty acid free, and globulin free ≥99%), HSA (lyophilized powder, fatty acid free, and globulin free ≥99%), and tetradecylamine (TA) were purchased from Sigma Aldrich and used as supplied. The cobaltous chloride and diethylenetriamine were obtained from Rankem, India. All other chemicals were of analytical reagent grade, and doubly distilled water was used throughout the study.

2.2.2 General Methods

Elemental analysis (C, H, and N) was carried out at Perkin-Elmer Series II 2400 CHNS/O Elemental Analyzer. Electrospray ionization mass spectrometry (ESI-MS) analysis was performed in the positive ion mode on a liquid chromatography–ion trap mass spectrometer (LCQ Fleet, Thermo Fisher Instruments Limited, USA). Complexes 1 and 2 were dissolved in water, and the mass scan range was from 100 to 1000 amu. ^1H and ^{13}C NMR measurements were performed on BRUKER 400 MHZ NMR spectrometer using d_6-dimethyl sulfoxide (DMSO) as solvent. Infrared spectra were recorded using Perkin-Elmer FT-IR spectrophotometer with samples prepared as KBr pellets. Absorption measurements were performed on Shimadzu UV-1800 UV–Vis spectrophotometer using cuvettes of 1 cm path length. Circular dichroism (CD) spectra were recorded on a JASCO-J810 spectropolarimeter with a cylindrical cuvette of 0.1 cm path length. Fluorescence experiments were carried out on a thermostatic bath coupled JASCO FP650 spectrofluorometer using a 1 cm quartz cuvette. Conductivity measurements were made with an Elico Conductivity bridge-type CM 82 and dip-type cell with a cell constant of 1.0. The percentage of cobalt content present in the surfactant–cobalt(III) complexes was determined spectrophotometrically by converting the complexes into $[CoCl_4]^{2-}$ whose molar absorbance coefficient is 561 M^{-1} cm^{-1} at 691 nm.

2.2.3 Synthesis of Surfactant–Cobalt(III) Complexes

$[Co(dien)Cl_3]$ was synthesized according to the reported procedure.[8] To a saturated aqueous solution of $[Co(dien)Cl_3]$ (3.2215 g, 0.2825 mmol), ethanolic solution of respective mole ratio of ligand, TA (2.757 mL, 0.1854 mmol for 1; 5.514 mL, 0.3708 mmol for 2), was added drop by drop over a period of 30 min. During this addition, the dark brown color of the solution gradually became light brown color and the resulting mixture was kept at room temperature for 48 h. Afterward, a saturated solution of sodium perchlorate in very dilute perchloric acid was added to the reaction mixture. The obtained precipitate was filtered off and washed with cold ethanol followed by acetone and dried over fused calcium chloride and stored in a vacuum desiccator.

[Co(dien)(TA)Cl$_2$]ClO$_4$ (**1**). Violet color solid, yield: 2.632 g (78%); Anal. cald. for C$_{18}$H$_{44}$Cl$_3$CoN$_4$O$_4$ (Found): C, 39.61 (39.50); H, 8.12 (7.96); N, 10.26 (10.43); Co, 10.80 (10.62). ESI-MS (H$_2$O, m/z): 445.52 [Co(dien)(TA)Cl$_2$]$^+$. ^{1}H NMR (d_6-DMSO, 400 MHz): δ (ppm) 7.8 (7H), 2.78 (4H), 1.53 (4H), 1.23 (26H), 0.84 (3H). ^{13}C NMR (d_6-DMSO, 400 MHz): δ (ppm) 16.63, 24.76, 31.61, 31.72, 33.96, 41.78. IR (KBr, cm^{-1}): 631, 1079, 1218, 1368, 1455, 1729, 2863, 2913, 3245, 3642. UV–visible (UV–vis) in water (λ_{max}, nm) (ε/M^{-1} cm^{-1}): 516 (90), 295 (1320), 211 (15,740).

[Co(dien)(TA)$_2$Cl](ClO$_4$)$_2$ (**2**). Brown color solid, yield: 2.184 g (83%); Anal. cald. for C$_{32}$H$_{75}$Cl$_3$CoN$_5$O$_8$ (Found): C, 46.69 (46.61); H, 9.18 (9.03); N, 8.51 (8.68); Co, 7.16 (7.03). ESI-MS (H$_2$O, m/z): 311.38 [Co(dien)(TA)$_2$Cl]$^{2+}$. ^{1}H NMR (d_6-DMSO, 400 MHz): δ (ppm) 7.53 (6H), 4.83 (3H), 2.27 (4H), 1.28 (4H), 0.9 (52H), 0.61 (6H). ^{13}C NMR (d_6-DMSO, 400 MHz): δ (ppm) 13.89, 22.02, 25.73, 28.44, 28.96, 31.22, 39.32. IR (KBr, cm^{-1}): 614, 1059, 1219, 1361, 1478, 1744, 2839, 2918, 3253, 3634. UV–vis in water (λ_{max}, nm) (ε/M^{-1} cm^{-1}): 681 (40), 511 (80), 211 (8440).

2.2.4 Partition Coefficients Determination

Hydrophobicity of surfactant–cobalt(III) complexes is one of the parameters, which influence its biological activity. The partition coefficients, usually expressed as log P values, were measured by the "Shake flask" method between octanol/water phase partitions as reported earlier.[9] Complexes **1** and **2** were dissolved in a mixture of water and *n*-octanol and shaken for 1 h. The mixture was allowed to settle over a period of 30 min, and the two phases that resulted were collected separately without cross-contamination of one solvent layer into another. The concentration of surfactant–cobalt(III) complexes in each phase was determined by UV–vis absorption spectroscopy at room temperature. The results are given as the mean values obtained from three independent experiments.

2.2.5 Conductivity Measurements

The CMC values of the surfactant–cobalt(III) complexes were determined conductometrically by using a digital conductivity meter (Elico CM 82). After calibrating cell constant with standard KCl solutions of known specific conductivities, conductivity measurements were made in a thermostated water bath, which was maintained at constant temperature ±0.1°C. Specific conductivity values for the aqueous solution of surfactant–cobalt(III) complexes having concentration in the range of 10^{-6}–10^{-2} M^{-1}

were measured at 303, 308, 313, 318, and 323 K. Each reading was noted after thorough mixing and temperature equilibration until no significant change occurred. The CMC values of complexes **1** and **2** were obtained by plotting specific conductance versus concentration of surfactant–cobalt (III) complex.[10]

2.2.6 Protein Binding Studies

Protein binding studies were carried out using Tris–HCl buffer (pH = 7.4), and the concentrations of BSA were determined spectrophotometrically from the respective molar extinction coefficient of 43,800 at 278 nm. The initial setup was made for fluorescence measurements as follows: excitation and emission slits were set at 5 and 3 nm, respectively, and scanning speed was set at 500 nm/min. The protein binding study was performed by fluorescence quenching experiments keeping the concentrations of BSA (10 mM) and varying concentrations of surfactant–cobalt(III) complexes (0–90 μM). The fluorescence emission spectra were recorded in the wavelength range 290–450 nm by exciting at 280 nm. UV–vis experiments were performed by keeping the concentrations of BSA (10 mM) and varying concentrations of surfactant–cobalt(III) complexes (0–90 μM), and the absorbance due to complex itself is nullified by adding in both sample and reference cells. The fluorescence quenching experiments were carried out in a manner that the concentrations of protein and surfactant–cobalt(III) complexes were fixed as those used in the UV–vis studies. To eliminate the inner filter effect, absorbance measurements were performed at the excitation and emission wavelength for each concentration of metal complex (including the protein without metal complex) and then multiply the observed fluorescence value using the following equation[11]:

$$F_{cor} = F_{obs} \times 10^{(A_1 + A_2)/2}$$

where F_{cor} and F_{obs} are the fluorescence intensities corrected and observed, respectively, and A_1 and A_2 are the sum of the absorbance of protein and ligand at the excitation and emission wavelengths, respectively.

The synchronous fluorescence spectra were recorded with $\Delta\lambda$ = 15 nm and $\Delta\lambda$ = 60 nm for tyrosine and tryptophan residues, respectively. The three-dimensional (3D) fluorescence spectra were measured under the following conditions: the emission wavelength was recorded between 250 and 500 nm, the initial excitation wavelength was set to 250 nm with increments of 5 nm, the number of scanning curves was 14, and the emission and excitation slit widths were fixed at 5 nm and 5 nm, respectively.

2.3 RESULTS AND DISCUSSION

2.3.1 Characterization of Single- and Double-Chain Surfactant–Cobalt(III) Complexes

$[Co(dien)(TA)Cl_2]^+$ $[Co(dien)(TA)_2Cl]^{2+}$

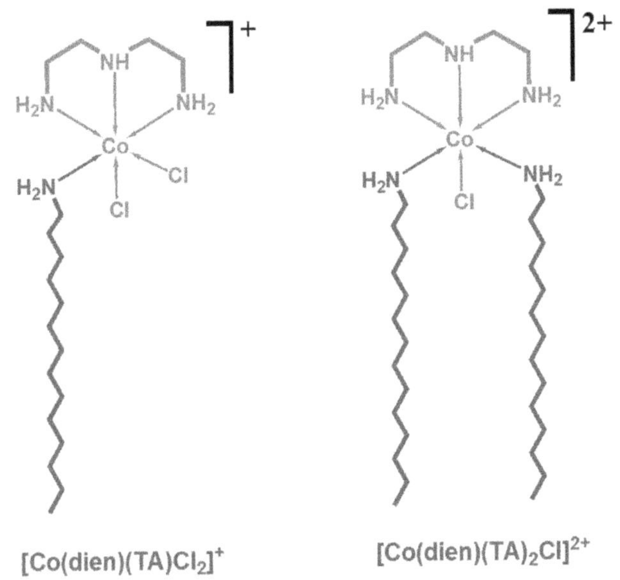

FIGURE 2.1   Structure of the Single- and Double-Chain Surfactant–Cobalt(III) Complexes.

The single- and double-chain surfactant–cobalt(III) complexes were synthesized from $[Co(dien)Cl_3]$ by ligand substitution method in which one or two labile chloride ligands were replaced by one or two amine groups of the alkylamine ligands (Fig. 2.1). The UV–vis absorption spectra of surfactant–cobalt(III) complexes clearly show an intense band around 213–219 nm due to $N(\sigma) \rightarrow Co(III)$ charge transfer and a band around 511–522 nm due to d–d transitions.[12] The IR spectra can afford the characteristic vibrational frequencies for the formation of surfactant–cobalt(III) complexes.[13] The precursor complex $[Co(dien)Cl_3]$ shows that N–H and C–H symmetric and asymmetric stretching vibrational bands around 3615, 2852, and 2921 cm^{-1} were red shifted to 3439, 2849, and 2917 cm^{-1} after coordination with alkylamine in the surfactant–cobalt(III) complexes, respectively. These shifts can be explained by the fact that nitrogen atom of alkylamine ligand donates a pair of electrons to the cobalt center forming a coordinate bond. The band observed around 1113 cm^{-1} can be assigned to perchlorate ionic species; this means that the counterion was not involved in the coordination to cobalt. Furthermore, the bands around 627 and 1088 cm^{-1} can be attributed to the (Co–N) and (C–N) stretching vibrations of surfactant–cobalt(III) complexes.[13] The ^{1}H NMR spectra also resulted in the corroboration of structure of surfactant–cobalt(III) complexes. The methylene protons of the alkylamine chains and dien ligands appeared in the region of 0.99–3.08 ppm for complexes **1** and **2**. It is also noted that a typical triplet signal at 0.84 ppm corresponds to the terminal methyl group of the long aliphatic alkylamine chain. The N–H protons appeared as broad peaks in the region of 4.60–8.00 ppm. The ^{13}C NMR spectra of complexes **1** and **2** gave signals in the region of 22.07–31.21 ppm due to the merging of methylene carbon signals of alkylamine chain and dien ligands. Furthermore, a signal around 13.87 ppm was observed for terminal methyl carbon. ESI-MS spectra of complexes **1** and **2** showed molecular ion peaks at 445.52 and 311.38.

2.3.2   Determination of Partition Coefficient

Hydrophobicity of surfactant–cobalt(III) complexes is an important parameter to analyze penetration behavior across the cell membrane and is compared in terms of partition coefficient (log *P*).[14] Here, the surfactant–cobalt(III) complexes are likely to differ in their hydrophobicity due to the variation in the single- and double-chain systems. Based on the concentration of surfactant–cobalt(III) complexes distributed in the biphasic system (*n*-octanol/water), partition coefficients were calculated by the following equation:

$$\log P = \log \left[\frac{[\text{complex}]_{\text{octanol}}}{[\text{complex}]_{\text{water}}} \right] \tag{2.1}$$

The calculated values of log *P* for complexes **1** and **2** are −1.12 and −0.90; these results indicate that double-chain surfactant–cobalt(III) complexes have higher hydrophobicity than that of single-chain surfactant–cobalt(III) complexes.

2.3.3   Determination of CMC

The surfactant–cobalt(III) complexes tend to aggregate themselves in aqueous medium on increasing their concentration and start to form micelles at a particular concentration called CMC, during which their physical properties, like specific conductivity, are also altered due to change in the mobility of molecules in the system. A typical change in the specific conductivity of surfactant–cobalt(III) complexes of the present study with increase of concentration at five different temperatures (303, 308, 313, 318, and 323 K) was observed, and the values were plotted in Figure 2.2.

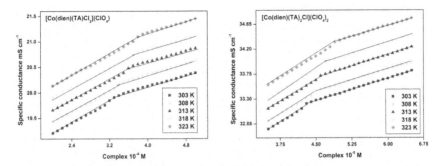

FIGURE 2.2 Plots for Specific Conductivity Versus Concentration of Surfactant–Cobalt(III) Complexes (**1** and **2**) in Aqueous Solution.

The obtained plots for both the surfactant–cobalt(III) complexes showed a sharp change from the premicellar to postmicellar regions due to the reduction in the mobility of molecules by aggregation. It was observed that CMC value increased with increase of temperature due to the disturbance of the water surrounding the hydrophobic group, and this retards micellization leading to a higher CMC value.[15]

TABLE 2.1 Critical Micelles Concentration (CMC) Values of Single-and Double-Chain Surfactant–Cobalt(III) Complexes

Complexes	T (K)	$CMC \times 10^4$ M
$[Co(dien)(TA)Cl_2]ClO_4$	303	3.3053
	308	3.4098
	313	3.6136
	318	3.7121
	323	3.9091
$[Co(dien)(TA)_2Cl](ClO_4)_2$	303	0.4299
	308	0.4516
	313	0.4693
	318	0.4779
	323	0.4976

The obtained CMC values for the surfactant–cobalt(III) complexes are shown in Table 2.1, and it is suggested that these surfactant–cobalt(III) complexes have more capacity to associate themselves in forming micelle aggregates than ordinary synthetic organic surfactants. The CMC values for complexes **1** and **2** clearly show that the double-chain surfactant–cobalt(III) complexes are found to have lower CMC value than the respective single-chain complexes. This is due to the increase in the hydrophobicity of the tail part, which tends to favor aggregation for micellization at lower concentration.

2.3.4 Interaction of Surfactant–Cobalt(III) Complexes with Serum Albumins

The interaction between serum albumins and surfactant–cobalt(III) complexes can give valuable information about structural factor governing protein–drug binding behavior. To develop the efficient surfactant-based metallodrugs, it is important to analyze the process behind the protein–drug complex formation i) whether the drug interact with ground-state protein (static process) or with the excited-state protein (dynamic process),[16] ii) strength and stability of protein–drug complex,[17] iii) binding number of protein–drug complex formation, iv) the nature of binding forces (electrostatic, hydrophobic, hydrogen bonding, and van der Waals interaction) acting upon the protein–drug complex formation, and v) conformational and microenvironmental changes in the protein.

2.3.5 Analysis of Quenching and Binding Parameters

To investigate the quenching and binding nature of surfactant–cobalt(III) complexes with serum albumins, the emission spectra of BSA were monitored in the wavelength range 290–450 nm by exciting the proteins at 280 nm, resulting in a strong fluorescence emission peak at 350 nm for BSA due to their tryptophan residues. The changes in the emission spectra of protein with increase of surfactant–cobalt(III) complex concentration at three different temperatures (278, 293, and 308 K) were recorded and the representative fluorescence emission spectra are shown in Figure 2.3.

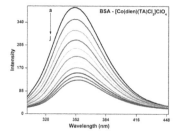
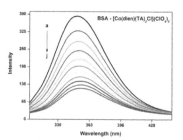

FIGURE 2.3 Fluorescence Emission Spectra of Bovine Serum Albumin (BSA) in the Absence and Presence of Surfactant–Cobalt(III) Complexes (**1** and **2**). [BSA] = 10 μM and [Surfactant–Cobalt(III) Complexes] = 90 μM.

Generally, it is noticed that fluorescence emission intensities of protein is quenched regularly by addition of surfactant–cobalt(III) complexes, indicating the formation of efficient complex between protein and surfactant metal complexes.[18]

In order to understand the quenching and binding behavior, data were analyzed using the following equations (2.2) and (2.3) at 278, 293, and 308 K,[19]

$$F_0 / F = 1 + K_{sv}[Q] = 1 + k_q \tau_0 [Q] \tag{2.2}$$

$$\log[(F_0 - F)/F] = \log K_b + n \log[Q] \tag{2.3}$$

where F_0 and F correspond to the fluorescence intensities of the protein in the absence and presence of the quencher, respectively, $[Q]$ is the total concentration of the quencher, τ_0 is the average lifetime of protein in the absence of quencher, K_{sv} is Stern–Volmer quenching constant, k_q is quenching rate constant, K_b is the binding constant showing the extent of interaction between protein and surfactant–cobalt(III) complex, and n is the binding number per albumin molecule. The values of K_{sv} and k_q can be obtained from the slope of the plot between (F_0/F) and $[Q]$ (Fig. 2.4). Similarly, the values of K_b and n were evaluated from the intercepts and slopes of double logarithm regression curve by plotting $\log(F_0 - F/F)$ versus $\log[Q]$ (Fig. 2.5). The obtained results were analyzed through equations (2.2) and (2.3), and the values for K_{sv}, k_q, K_b, and n are summarized in Table 2.2.

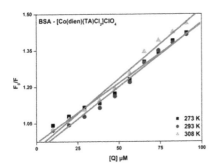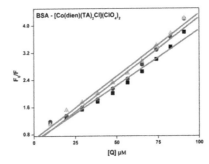

FIGURE 2.4   Stern–Volmer Plots for Quenching of Bovine Serum Albumin by Surfactant–Cobalt(III) Complexes **1** and **2** at Three Different Temperatures (273, 293, and 308 K).

As all the Stern–Volmer plots are linear, it is concluded that the existence of a single type of quenching mechanism, either static or dynamic, can be distinguished from the temperature dependence of K_{sv} and k_q values.

TABLE 2.2 The Stern–Volmer Quenching Constant (K_{sv}), Quenching Rate Constant (k_q), Binding Constant (K_b), Binding Number (n), and the Thermodynamic Parameters ($\Delta H°$, $\Delta G°$, and $\Delta S°$) for the Interaction of Bovine Serum Albumin (BSA) with Surfactant–Cobalt(III) Complexes at Different Temperatures

Complexes	T (K)	$K_{sv} \times 10^{-4}$ (M^{-1})	$K_q \times 10^{-12}$ $(M^{-1}s^{-1})$	$K_b \times 10^{-4}$ (M^{-1})	n	$\Delta H°$ (kJ mol^{-1})	$\Delta G°$ (kJ mol^{-1})	$\Delta S°$ (J mol^{-1} K^{-1})
BSA– complex 1	278	0.4799	0.4799	11.130	1.3471	−18.956	−26.857	+28.086
	293	0.4869	0.4869	6.375	1.2858		−26.949	
	308	0.4986	0.4986	4.471	1.2389		−27.982	
BSA– complex 2	278	3.2881	3.2881	137.69	1.4676	+40.117	−32.671	+262.696
	293	3.6956	3.6956	474.22	1.5229		−37.446	
	308	3.7035	3.7035	948.64	1.5907		−41.138	

It is known that higher temperature is likely to result in the decrease of quenching constant values for static process due to weakening of ground-state complex stability, whereas in the increase of quenching constant values for dynamic process due to the faster diffusion of the excited molecules.[20] Thus, the increasing mannerism of the quenching constant values (K_{sv} and k_q) with respect to temperature (Table 2.2) indicates the stimulation of dynamic quenching process upon BSA by the surfactant–cobalt(III) complexes.

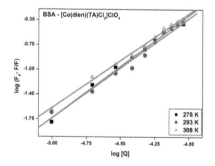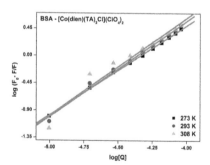

FIGURE 2.5 Double Logarithmic Plots for the Quenching of Bovine Serum Albumin by Surfactant–Cobalt(III) Complexes (**1–4**) at Three Different Temperatures (273, 293, and 308 K).

However, the obtained quenching rate constant values ($k_q \approx 10^{11}$–10^{12} M^{-1} s^{-1}) are 10–100 times higher than the maximum value possible for diffusion-controlled dynamic quenching (i.e., 2.0×10^{10} M^{-1} s^{-1}). This observation can be explained based on the existence of a special process in which surfactant–cobalt(III) complexes probably quench the BSA via initiation of static mechanism rather than dynamic process.[16,21] Similarly, the values of binding number per albumin molecule (n) is around 1, indicating strong and independent binding site granted by BSA to the surfactant–cobalt(III) complexes. It is also observed that the extent of binding (K_b) between protein and surfactant–cobalt(III) complexes with respect to temperature decreased for single-chain system, whereas this has increased for double-chain systems. This may be because the increase of temperature diminishes the electrostatic attraction between protein and single-chain surfactant–cobalt(III) complexes, thereby decreasing the stability of protein–surfactant cobalt(III) complexes. However, the increase of temperature tights the hydrophobic attraction between protein and double-chain surfactant–cobalt(III) complexes, thereby increasing the stability of protein–surfactant metal complexes.

2.3.6 Thermodynamic Parameters and Nature of the Binding Force

In order to analyze the nature of the binding forces (hydrophobic, electrostatic, hydrogen bonding, and van der Waals interactions) existing between protein and surfactant–cobalt(III) complexes, the sign and magnitude of the thermodynamic parameters such as ΔG° (free energy change), ΔH° (enthalpy change), and ΔS° (entropy change) for the interaction process were calculated by using the equations (2.4) and (2.5).

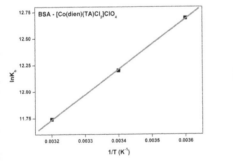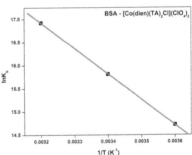

FIGURE 2.6 van't Hoff Plots for the Interaction of Surfactant–Cobalt(III) Complexes **1** and **2** with Bovine Serum Albumin.

Ross and Subramanian have summarized the following equations (2.4) and (2.5), which can be used to analyze the type of binding involved in the thermodynamic parameters of protein interaction[22]

$$\ln K = -\Delta H^\circ / RT + \Delta S^\circ / R \tag{2.4}$$

$$\Delta G^\circ = -RT \ln K_b \tag{2.5}$$

where K_b is analogous to the associative binding constant at the corresponding temperature and R is the gas constant. By plotting the binding constant versus temperature, the enthalpy change ($\Delta H°$) and entropy change ($\Delta S°$) can be calculated from the slope and intercept of the van't Hoff relationship, respectively. The free energy ($\Delta G°$) change was estimated based on the binding constants at three different temperatures (Fig. 2.6).

Ross and Subramanian[22a] have studied various models to explain the existence of principal binding forces in the protein association process using thermodynamic parameters and the results showed that (i) both positive $\Delta H°$ and $\Delta S°$ resulted by hydrophobic forces, (ii) both negative $\Delta H°$ and $\Delta S°$ resulted by van der Waals interaction and hydrogen bond formation, and (iii) negative $\Delta H°$ and positive $\Delta S°$ resulted by electrostatic interaction. As seen from Table 2.2, the negative free energy change values for the interaction between surfactant–cobalt(III) complexes and BSA indicate that the binding process is spontaneous. The positive $\Delta H°$ and $\Delta S°$ values indicate that hydrophobic interactions played a dominant role in the interactions between double-chain surfactant–cobalt(III) complexes and BSA.[23] Whereas the negative $\Delta H°$ and positive $\Delta S°$ values for the interaction between single-chain surfactant–cobalt(III) complexes and BSA indicate that the electrostatic interaction plays a major role.

2.3.7 UV–vis Absorption Studies

UV–vis absorption spectroscopic measurement is a simple and effective method to explore the structural changes and to establish the complex formation.[24] The absorption spectra of BSA in the absence and presence of surfactant–cobalt(III) complexes are shown in Figure 2.7.

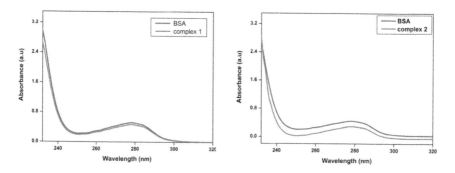

FIGURE 2.7   UV–Visible Absorption Spectra of Bovine Serum Albumin (BSA) in the Absence and Presence of Surfactant–Cobalt(III) Complexes (**1** and **2**). [BSA] = 10 μM and [Surfactant–Cobalt(III) Complexes] = 90 μM.

Generally, it is noticed that the addition of surfactant–cobalt(III) complexes to the protein solution results in hypochromism around 210 nm responsible for α-helix contents of protein and hyperchromism around 280 nm responsible for aromatic residues of proteins. The observed hypochromism shows the occurrence of structural influence in the α-helix of protein, whereas hyperchromism shows the occurrence of alteration in the microenvironment around the aromatic acid residues probably through the extending into the aqueous environment. These changes in the absorbance of protein by the interaction with surfactant–cobalt(III) complexes can be the evidence for existence of static quenching process. This result indicates the ground-state complex formation.

2.3.8 Synchronous Fluorescence Studies

Synchronous fluorescence spectroscopy is used to monitor the microenvironmental and conformational changes around the vicinity of tryptophan (Trp) and tyrosine (Tyr) chromophores in the protein, following their extent of quenching and shift in the emission maximum by the addition of surfactant–cobalt(III) complexes. The characteristic emission for the tyrosine and tryptophan residues of the protein can be obtained by maintaining the wavelength interval ($\Delta\lambda$) as 15 and 60 nm, respectively.[20] For the investigated concentration range, the changes in the synchronous fluorescence spectra of BSA/HSA upon increasing the concentration of surfactant–cob alt(III) complexes at $\Delta\lambda$ = 15 and 60 nm are shown in Figure 2.8.

As seen from Figure 2.8, there is no significant shift in the emission maximum of Tyr residues upon addition of complexes **1** and **2**. In contrast, an obvious red shift in the emission of Trp residues was noticed for the single-chain surfactant–cobalt(III) complexes, which indicated that there is enhancement of polarity by the reduction of hydrophobicity. Whereas a slight blue shift in the emission of tryptophan residues was observed for double-chain surfactant–cobalt(III) complexes, which indicates that the polarity around the tryptophan residues was decreased by the increase of hydrophobicity due to conformational changes in BSA. From this result, we can conclude that double-chain surfactant–cobalt(III) complexes should have higher hydrophobic than single-chain surfactant–cobalt(III) complexes.

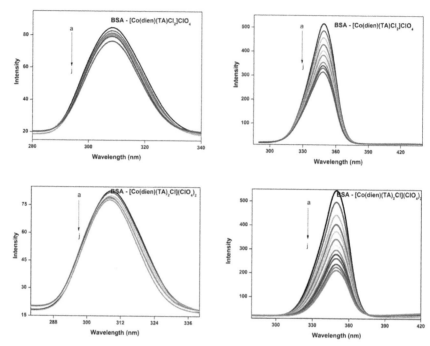

FIGURE 2.8 Synchronous Fluorescence Spectra of Bovine Serum Albumin (BSA) Left at $\Delta\lambda = 15$ nm and Right at $\Delta\lambda = 60$ nm. BSA $= 10\,\mu$M and [Surfactant–Cobalt(III) Complexes] $= 90\,\mu$M.

2.3.9 CD Spectroscopic Studies

CD spectroscopy is a sensitive technique to investigate the changes in the secondary structure of protein upon interaction with metallodrugs. The far-UV CD spectra of BSA exhibit two negative bands at 208 and 222 nm, which are characteristic of the typical α-helical structure of protein and is contributed by the n → π* transfer of the peptide bonds of α-helix.[25] So in order to obtain an insight into the changes in the secondary structure of BSA upon interaction with surfactant–cobalt(III) complexes, the far-UV CD spectra of BSA were recorded in the absence and presence of complexes **1** and **2.**

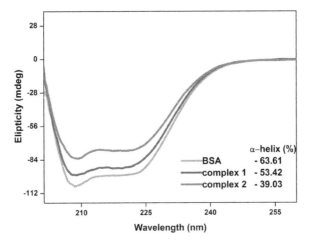

FIGURE 2.9   Circular Dichrosim Spectra of Bovine Serum Albumin (BSA) in the Absence and Presence of Complexes **1** and **2**. [BSA] = 2 μM and [Surfactant–Cobalt(III) Complexes] = 4 μM.

In CD spectral measurements, the concentration of BSA was maintained at 4 μM. The spectra were recorded in the range of 200–270 nm with a scan rate of 200 nm/ min and a response time of 4 s. Three scans were accumulated for each spectrum. The observed ellipticity of CD results (in millidegrees) was expressed in terms of mean residue ellipticity (MRE) in deg cm^2 dmol^{-1} according to the following equation:[26]

$$MRE = \frac{\text{observed CD(m deg)}}{10 n l C_p} \tag{2.6}$$

$$\alpha\text{-helix(\%)} = \frac{-MRE_{208} - 4000}{33,000 - 4000} \times 100 \tag{2.7}$$

where n is the number of amino acid residues in the protein, l is the cell path length, and C_p is the molar concentration of the protein. The α-helix contents of free and combined BSA were calculated from MRE values at 208 nm from equation (2.7). MRE_{208} is the observed MRE value at 208 nm, 4000 is the MRE of the β-form and random coil conformation cross at 208 nm, and 33,000 is the MRE value of a pure α-helix at 208 nm.

As can be seen from Figure 2.9, the negative ellipticity values of BSA decreased by the addition of surfactant–cobalt(III) complex, indicating the unfolding of peptide strands, thereby lowering of the α-helical content in the protein. The extent of decreasing the α-helical content in BSA by the complexes **1** and **2** shows that the binding of surfactant–cobalt(III) complexes with BSA induces conformational changes in BSA which may affect the physiological functions of proteins.

Based on the results present in Figure 2.9, it is found that α-helix content of protein has been reduced to a large extent in the presence of double-chain complexes than the respective single-chain complexes. This is due to the larger hydrophobicity of double-chain than the single-chain surfactant–cobalt(III) complexes.

2.3.10   Three-Dimensional Fluorescence Studies

The 3D fluorescence spectroscopy is a convenient technique to investigate the occurrence of conformational and microenvironmental changes in the protein as a function of excitation and emission wavelengths simultaneously. The intensity and maximum emission wavelength corresponding to the fluorescence peaks of protein residues have a close relationship with the polarity of the environment.[27] The 3D fluorescence spectra of BSA were investigated in the absence and the presence of surfactant–cobalt(III) complexes (Fig. 2.10). As seen, the two peak regions, peak 1 corresponding to Rayleigh scattering peak (lex = lem) and peak 2 corresponding to fluorescence peak (lex < lem, Dl = 60 nm), were observed. Peak 2 shows the spectral behavior of tryptophan and tyrosine residues, whereas fluorescence of phenylalanine is negligible.[28] The fluorescence intensity of peak 2 was dramatically decreased with increasing the concentration of surfactant–cobalt(III) complexes, indicating that conformational changes in BSA were induced as a result of exposure of aromatic residues buried in the hydrophobic microenvironment. These results are in accordance with CD and synchronous spectral studies to support the conformational and microenvironmental changes in BSA occurred upon binding with surfactant–cobalt(III) complexes.

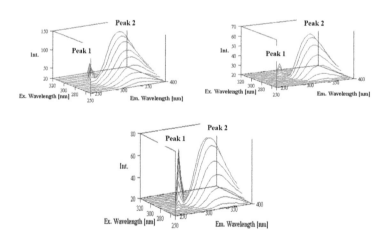

FIGURE 2.10   Three-Dimensional Fluorescence Spectra for Bovine Serum Albumin (BSA) (a), BSA + Complex 1 (b), BSA + Complex 2 (c). [BSA] = 90 mM and [Surfactant–Cobalt(III) Complexes] = 10 mM.

2.4 CONCLUSION

The effect of hydrophobic part in the surfactant–cobalt(III) complexes on CMC behavior and protein interaction was investigated by changing the chain length of alkylamine ligands. These surfactant–cobalt(III) complexes have very low CMC compared to simple classical surfactant, that is, they have more capacity to associate themselves, forming aggregates. On increasing the chain length of the surfactant–cobalt(III) complexes, their CMC values decreased due to increase in the ability of association. All the complexes quenched the fluorescence of protein through static quenching process. Their quenching and binding constant mainly depended upon the extension of alkylamine ligand chain length. Interestingly, binding constant and number of binding sites decreased for single-chain system whereas increased for double-chain system with increase of temperature, due to the changes in the mode of protein–complex interaction. This was further analyzed from thermodynamic parameters and supported by synchronous studies, which observed that single-chain system involves electrostatic interaction, while double-chain system involves hydrophobic interaction. Moreover, UV–vis absorption, 3D fluorescence, and CD studies revealed that appreciable conformational and some microenvironmental changes occurred in the BSA. Significantly, this work deals with the tail tune effect, which is important in future for designing the surfactant-based metallodrugs, having desired binding mode with drug carrier serum albumin. This kind of tuning the hydrophobicity of surfactant metal complexes with suitable tail ligand could be a better pathway for optimizing the condition for effective anticancer drugs, and this attention is going on now.

ACKNOWLEDGMENTS

The authors are grateful to the UGC-SAP and COSIST and DST-FIST programs of the Department of Chemistry, Bharathidasan University. SV acknowledges the UGC-RFSMS for financial support (Junior Research Fellowship). SA thanks the sanction of research schemes from funding agencies, CSIR [grant no. 01(2461)/11/EMR-II] and UGC [grant no. 41-223/ 2012(SR)]. The authors thank Dr. P. Suresh, School of Chemistry, Madurai Kamaraj University, for providing the ESI-MS measurements. It is also acknowledged to Dr. A. Ramu, School of Chemistry, Madurai Kamaraj University, for providing the circular dichroism measurements.

KEYWORDS

- Single- and double-chain surfactant–cobalt(III) complexes
- critical micelle concentration
- hydrophobicity
- BSA
- electrostatic and hydrophobic interaction

REFERENCES

1. (a) Griffiths, P.; Fallis, I.; Chuenpratoom, T.; Watanesk, R. Metallosurfactants: interfaces and micelles. *Adv. Colloid Interface Sci.* 2006, 122(1), 107–117; (b) Griffiths, P. C.; Fallis, I. A.; Tatchell, T.; Bushby, L.; Beeby, A. Aqueous solutions of transition metal containing micelles. *Adv. Colloid Interface Sci.* 2008, 144(1), 13–23.

2. Kimura, E.; Hashimoto, H.; Koike, T. Hydrolysis of lipophilic esters catalyzed by a zinc (II) complex of a long alkyl-pendant macrocyclic tetraamine in micellar solution. *J. Am. Chem. Soc.* 1996, 118(45), 10963–10970.

3. Jervis, H.; Bruce, D.; Raimondi, M.; Seddon, J. Templating mesoporous silicates on surfactant ruthenium complexes: a direct approach to heterogeneous catalysts. *Chem. Commun.* 1999, 20, 2031–2032.

4. Walker, G. W.; Geue, R. J.; Sargeson, A. M.; Behm, C. A. Surface-active cobalt cage complexes: synthesis, surface chemistry, biological activity, and redox properties. *Dalton Trans.* 2003, 15, 2992–3001.

5. (a) Huang, B. X.; Kim, H.-Y.; Dass, C. Probing three-dimensional structure of bovine serum albumin by chemical cross-linking and mass spectrometry. *J. Am. Soc. Mass Spectrom.* 2004, 15(8), 1237–1247; (b) Tian, J.; Liu, J.; Hu, Z.; Chen, X. Interaction of wogonin with bovine serum albumin. *Bioorg. Med. Chem.* 2005, 13(12), 4124–4129.

6. Frostell-Karlsson, Å; Remaeus, A.; Roos, H.; Andersson, K.; Borg, P.; Hämäläinen, M.; Karlsson, R. Biosensor analysis of the interaction between immobilized human serum albumin and drug compounds for prediction of human serum albumin binding levels. *J. Med. Chem.* 2000, 43(10), 1986–1992.

7. Zhao, X.; Sheng, F.; Zheng, J.; Liu, R. Composition and stability of anthocyanins from purple solanum tuberosum and their protective influence on Cr (VI) targeted to bovine serum albumin. *J. Agric. Food Chem.* 2011, 59(14), 7902–7909.

8. Caldwell, S. H.; House, D. Hydrolysis products from trichloro-diethylenetriaminechromium (III) and cobalt (III) complexes. *J. Inorg. Nucl. Chem.* 1969, 31(3), 811–823.

9. (a) Gupta, R. K.; Pandey, R.; Sharma, G.; Prasad, R.; Koch, B.; Srikrishna, S.; Li, P.-Z.; Xu, Q.; Pandey, D. S. DNA Binding and anti-cancer activity of redox-active heteroleptic piano-stool Ru (II), Rh (III), and Ir (III) complexes containing 4-(2-methoxypyridyl) phenyldipyrromethene. *Inorg. Chem.* 2013, 52(7), 3687–3698; (b) Gupta, R. K.; Sharma, G.; Pandey, R.; Kumar, A.; Koch, B.; Li, P.-Z.; Xu, Q.; Pandey, D. S. DNA/protein binding, molecular docking, and in vitro anticancer activity of some thioether-dipyrrinato complexes. *Inorg. Chem.* 2013, 52(24), 13984–13996.

10. Mukerjee, P. The thermodynamics of micelle formation in association colloids. *J. Phys. Chem.* 1962, 66(7), 1375–1376.

11. Van De Weert, M. Fluorescence quenching to study protein-ligand binding: common errors. *J. Fluoresc.* 2010, 20(2), 625–629.

12. Nehru, S.; Arunachalam, S.; Arun, R.; Premkumar, K. Polymer–cobalt(III) complexes: structural analysis of metal chelates on DNA interaction and comparative cytotoxic activity. *J. Biomol. Struct. Dyn.* 2013, 31, 1–13.

13. Kumar, R. S.; Arunachalam, S.; Periasamy, V.; Preethy, C.; Riyasdeen, A.; Akbarsha, M. Surfactant–cobalt (III) complexes: synthesis, critical micelle concentration (CMC) determination, DNA binding, antimicrobial and cytotoxicity studies. *J. Inorg. Biochem.* 2009, 103(1), 117–127.

14. (a) Chang, T.; Lord, M.; Bergmann, B.; Macmillan, A.; Stenzel, M. H. Size effects of self-assembled block copolymers spherical micelles and vesicle on cellular uptake in human colon carcinoma cells. *J. Mater. Chem. B.* 2014, 20, 2883–2891; (b) Jagadeesan, S.; Balasubramanian, V.; Baumann, P.; Neuburger, M.; Häussinger, D.; Palivan, C. G. Water-soluble Co (III) complexes of substituted phenanthrolines with cell selective anticancer activity. *Inorg. Chem.* 2013, 52, 12535–12544.

15. Kumaraguru, N.; Santhakumar, K. Studies on synthesis, determination of CMC values, kinetics and the mechanism of iron (II) reduction of surfactant–Co (III)–ethylenediamine complexes in aqueous acid medium. *Phys. Chem. Liq.* 2010, 48(6), 747–763.

16. Yang, C.-Y.; Liu, Y.; Zheng, D.; Zhu, J.-C.; Dai, J. Luminescence of aniline blue in hydrophobic cavity of BSA. *J. Photochem. Photobiol. A Chem.* 2007, 188(1), 51–55.

17. Zhang, C.; Liu, S.; Zhu, Q.; Zhou, Y. A knowledge-based energy function for protein-ligand, protein-protein, and protein-DNA complexes. *J. Med. Chem.* 2005, 48(7), 2325–2335.

18. Xiao, J.; Suzuki, M.; Jiang, X.; Chen, X.; Yamamoto, K.; Ren, F.; Xu, M. Influence of B-ring hydroxylation on interactions of flavonols with bovine serum albumin. *J. Agric. Food Chem.* 2008, 56(7), 2350–2356.

19. (a) Lakowicz, J. R. *Principles of Fluorescence Spectroscopy*; Springer: Berlin, 2009; (b) Zhang, Y.-Z.; Zhou, B.; Liu, Y.-X.; Zhou, C.-X.; Ding, X.-L.; Liu, Y. Fluorescence study on the interaction of bovine serum albumin with p-aminoazobenzene. *J. Fluoresc.* 2008, 18(1), 109–118; (c) Tarushi, A.; Kljun, J.; Turel, I.; Pantazaki, A. A.; Psomas, G.; Kessissoglou, D. P. Zinc (II) complexes with the quinolone antibacterial drug flumequine: structure, DNA-and albumin-binding. *N. J. Chem.* 2013, 37(2), 342–355.

20. Samari, F.; Hemmateenejad, B.; Shamsipur, M.; Rashidi, M.; Samouei, H. Affinity of two novel five-coordinated anticancer Pt (II) complexes to human and bovine serum albumins: a spectroscopic approach. *Inorg. Chem.* 2012, 51(6), 3454–3464.

21. Shcharbin, D.; Pedziwiatr, E.; Chonco, L.; Bermejo-Martín, J. F.; Ortega, P.; de la Mata, F. J.; Eritja, R.; Gómez, R.; Klajnert, B.; Bryszewska, M. Analysis of interaction between dendriplexes and bovine serum albumin. *Biomacromolecules.* 2007, 8(7), 2059–2062.

22. (a) Ross, P. D.; Subramanian, S. Thermodynamics of protein association reactions: forces contributing to stability. *Biochemistry.* 1981, 20(11), 3096–3102; (b) Divsalar, A.; Bagheri, M. J.; Saboury, A. A.; Mansoori-Torshizi, H.; Amani, M. Investigation on the interaction of newly designed anticancer Pd (II) complexes with different aliphatic tails and human serum albumin. *J. Phys. Chem. B.* 2009, 113(42), 14035–14042.

23. Yu, M.; Ding, Z.; Jiang, F.; Ding, X.; Sun, J.; Chen, S.; Lv, G. Analysis of binding interaction between pegylated puerarin and bovine serum albumin by spectroscopic methods and dynamic light scattering. *Spectrochim. Acta A Mol. Biomol. Spectrosc.* 2011, 83(1), 453–460.

24. Cheng, X.-X.; Lui, Y.; Zhou, B.; Xiao, X.-H.; Liu, Y. Probing the binding sites and the effect of berbamine on the structure of bovine serum albumin. *Spectrochim. Acta A Mol. Biomol. Spectrosc.* 2009, 72(5), 922–928.

25. (a) Paul, B. K.; Samanta, A.; Guchhait, N. Exploring hydrophobic subdomain IIA of the protein bovine serum albumin in the native, intermediate, unfolded, and refolded states by a small fluorescence molecular reporter. *J. Phys. Chem. B.* 2010, 114(18), 6183–6196; (b) Yang, Q.; Liang, J.; Han, H. Probing the interaction of magnetic iron oxide nanoparticles with bovine serum albumin by spectroscopic techniques. *J. Phys. Chem. B.* 2009, 113(30), 10454–10458; (c) Chi, Z.; Liu, R.; Teng, Y.; Fang, X.; Gao, C. Binding of oxytetracycline to bovine serum albumin: spectroscopic and molecular modeling investigations. *J. Agric. Food Chem.* 2010, 58(18), 10262–10269.

26. Zhang, J.; Wang, X.-J.; Yan, Y.-J.; Xiang, W.-S. Comparative studies on the interaction of genistein, 8-chlorogenistein, and 3′, 8-dichlorogenistein with bovine serum albumin. *J. Agric. Food Chem.* 2011, 59(13), 7506–7513.

27. Vignesh, G.; Nehru, S.; Manojkumar, Y.; Arunachalam, S. Spectroscopic investigation on the interaction of some surfactant-cobalt (III) complexes with serum albumins. *J. Lumin.* 2014, 145, 269–277.

28. Hu, X.; Cui, S. Fluorescence studies of interaction between flavonol p-coumaroylglucoside tiliroside and bovine serum albumin. *Spectrochim Acta A Mol Biomol Spectrosc.* 2010, 77(2), 548–553.

A Review on the Selective Synthesis of Spiro Heterocycles Through 1,3-Dipolar Cycloaddition Reactions of Azomethine Ylides

Anshu Dandia*, Sukhbeer Kumari, Shuchi Maheshwari, and Pragya Soni

Center of Advanced Studies, Department of Chemistry, University of Rajasthan, Jaipur, India
*Email: dranshudandia@yahoo.co.in

CONTENTS

ABSTRACT

The 1,3-dipolar cycloaddition is an elegant and atom-efficient process for the synthesis of spiro heterocycles, involving the formation of C–C and C–N bonds in a single step. The construction of a spiro heterocyclic framework has always been a challenging endeavor for synthetic organic chemists as it frequently requires synthetic design based on 1,3-dipolar cycloaddition reactions. This review gives a short summary of the advances for the selective synthesis of various spiro heterocyclic compounds through multicomponent reactions and 1,3-dipolar cycloaddition reactions with the main emphasis on the work done in our laboratory.

3.1 INTRODUCTION

Cycloaddition reactions are one of the most important classes of reactions in synthetic chemistry.[1] Within this class, the 1,3-dipolar cycloaddition reaction has found extensive use as a high-yielding regio- and stereocontrolled method for the synthesis of different heterocyclic compounds.[2] Indeed, the 1,3-dipolar cycloaddition reaction has been described as "the single most important method for the construction of heterocyclic five-membered rings in organic chemistry."[3] Integrating the ylide (dipole) and the alkene or alkyne (dipolarophile) within the same molecule provides direct access to bicyclic (or polycyclic) products of considerable complexity. The proximity of the reactants and the conformational constraints often lead to ready cycloaddition with very high or complete selectivity.

In general, the [π4s + π2s] thermal cycloaddition of 1,3-dipoles with alkene and alkyne dipolarophiles generates five-membered heterocycles and are called 1,3-dipolar cycloadditions because of the dipolar nature of the principal resonance structures and the 1,3-additions which they undergo (Fig. 3.1).[4]

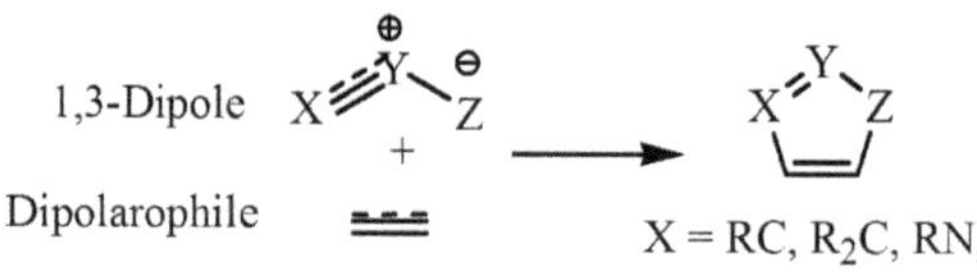

$$1,3\text{-Dipole} \quad X \overset{\oplus}{=\!\!=} Y \overset{\ominus}{\underset{Z}{\diagdown}} \quad \longrightarrow \quad X \overset{Y}{\underset{\diagdown}{\diagup}} Z$$

Dipolarophile ═══

X = RC, R$_2$C, RN, R$_2$N, O
Y = N, NR, O
Z = RC, R$_2$C, RN, R$_2$N, O

FIGURE 3.1   General 1,3-Dipolar Cycloaddition and Possible *X*, *Y*, and *Z* Combinations from First Row Atoms.

1,3-Dipoles have resonance structures that allow them to react as nucleophiles as well as electrophiles. 1,3-Dipolar molecule is a species represented by zwitter ionic octet and sextet structures. It can be represented by two octet structures, in which the positive charge is located on the central atom and the negative charge is distributed over the two terminal atoms, and two sextet structures, wherein two of the four π-electrons are localized at the central atom (Fig. 3.2). Basically, 1,3-dipoles can be divided into two different types: the allyl anion type and the propargyl/allenyl anion type.[5] The three atoms of the 1,3-dipole molecule can be a wide variety of combination of C, O, and N.

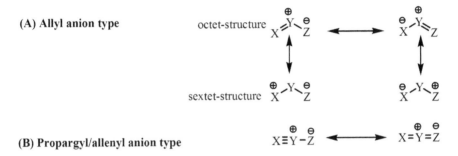

(A) Allyl anion type   octet-structure

sextet-structure

(B) Propargyl/allenyl anion type

FIGURE 3.2   The Basic Resonance Structure of 1,3-Dipoles.

The azomethine ylide represents one of the most reactive and versatile classes of 1,3-dipoles and is readily trapped by a range of dipolarophiles forming substituted pyrrolidines and pyrrolizidines that are prevalent in a variety of biologically active compounds.[6]

Azomethine ylides have four π-electrons spread over the three-atom C–N–C unit, and they can be represented by a zwitter ionic (or diradical) form. Four zwitter ionic resonance forms can be drawn, as shown in Figure 3.3. The most common representation has a positive charge located on the nitrogen atom and a negative charge distributed over the two carbon atoms.[7] The extent of negative charge on each carbon atom is determined by the nature and number of substituents at these carbons. Alternatively, two resonance forms with the positive and negative charges on the carbon atoms can be drawn to represent the 1,3-dipole.

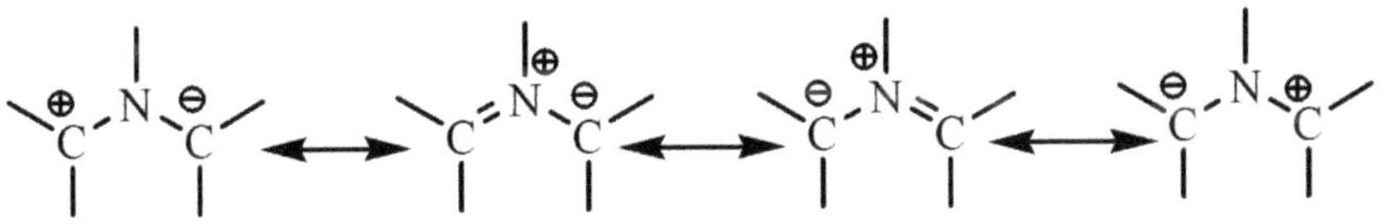

FIGURE 3.3   Zwitter Ionic Resonance Forms of Azomethine Ylide.

Extensive investigations during the last two decades has shown that nonstabilized azomethine ylides can be obtained by i) deprotonation of iminium salts or amine N-oxides,[8] ii) decarboxylation of pyridinium-1-acetic acid salts or amino acids,[9] iii) thermal and photochemical ring opening of aziridines or heterocyclic compounds such as 4-oxazolines,[10] and iv) desilylations of N-(trimethylsilylmethyl)iminium cation.[11] In these methods, the most useful and the most frequently applied are the decarboxylation of amino acids (Fig. 3.4).

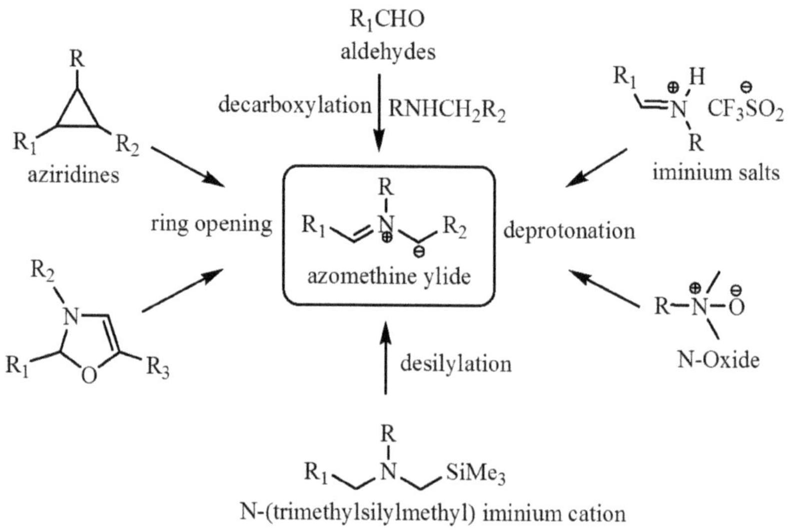

FIGURE 3.4 Generation of Azomethine Ylides.

The synthetic importance of azomethine ylides stems from their use for the preparation of five-membered nitrogen heterocycles, which are present everywhere in nature and often found as subunits of bioactive natural products. It is especially noteworthy that nitrogen-containing heterocycles form the basic skeleton of numerous alkaloids and therapeutic agents.[12] Pyrrolidine, pyrrolizines, and oxindole alkaloids constitute a class of compounds with significant biological activities, such as antimicrobial, antitumor, antiviral, and antibiotic.[13] In addition, they also act as inhibitors of human NK-I receptor activity.[14] Some spiro pyrrolidines are potential antileukemic and anticonvulsant agents and possess local anesthetic activities.[15]

The asymmetric structure of the molecule due to the chiral spiro carbon atom is one of the important criteria of the biological activities. Compounds with spiro skeletons not only constitute subunits in numerous alkaloids and natural products but are also templates for drug discovery and have been used as scaffolds for combinatorial libraries (Fig. 3.5).[16–19]

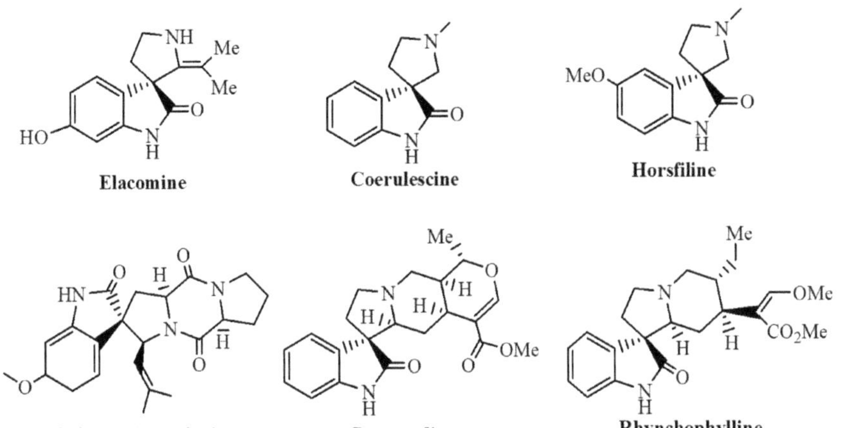
NH
Me
Me
O
HO
N
H
Elacomine
N
O
N
H
Coerulescine
N
MeO
O
N
H
Horsfiline
HN
O
H
O
N
N
H
O
O
Spirotryprostatin A
Me
O
N
H
H
H
OMe
O
O
N
H
Pteropodine
Me
N
OMe
CO2Me
H
H
O
N
H
Rhynchophylline

FIGURE 3.5 Naturally Occurring Pharmaceutically Important Spiropyrrolidines.

The efficiency of a chemical synthesis can nowadays be measured not only by parameters like selectivity and overall yield but also by the raw material, time, human resources, and energy requirements, as well as the toxicity and hazard of the chemicals and the protocols involved.[20] To meet these goals, chemists have been attempting specially to develop multicomponent reactions (MCRs). Since these reactions offer significant advantages over conventional linear-type syntheses, as more often are recognized cost-effective and comparatively fast routes though generating less chemical waste.[21]

This article aims to review the design and selective synthesis of different types of spiro heterocyclic compounds through 1,3-dipolar cycloaddition reaction with special emphasis on the work done by our research group in the last decade. The present review is divided into various sections based on the construction of spiro carbon atom during the product formation.

3.2 SYNTHESIS OF MONOSPIRO HETEROCYCLES

An efficient and diversity-oriented regioselective three-component 1,3-dipolar cycloaddition reaction of acenaphthenequinone, sarcosine, and Knoevenagel adducts for the synthesis of spiro[acenaphthylene-1,2′-pyrrolidine] derivatives (**1**) was described by our research group using aqueous methanol in highly regio- and stereoselective manner (Scheme 3.1).[22] Moreover, three stereogenic centers with one spiro carbon are controlled very well for this intermolecular three-component combinatorial process. There is no evidence for the synthesis of other regioisomer **2**.

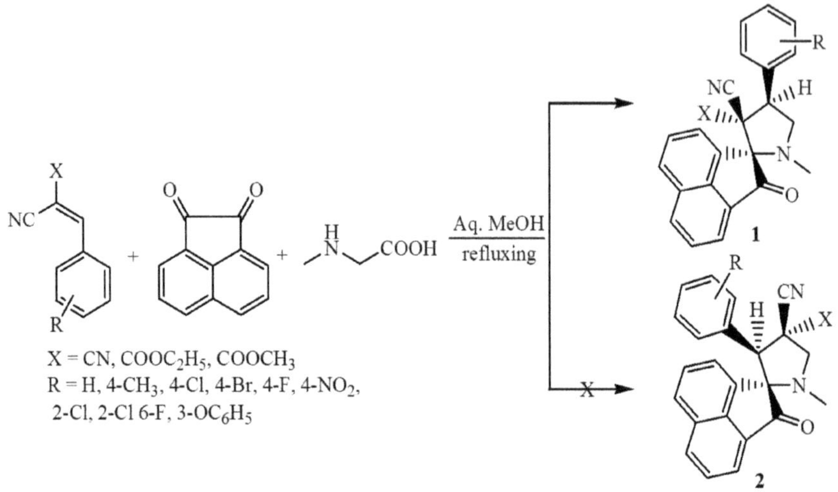

SCHEME 3.1 Synthesis of Spiro[acenaphthylene-1,2′-Pyrrolidine] Derivatives.

The present cycloaddition reaction was also studied in other solvents such as aqueous methanol, ethanol, acetonitrile, dioxane, dichloromethane, and tetrahydrofuran (THF). Lower yield of product was obtained in several hours as compared to aqueous methanol (Table 3.1).

TABLE 3.1 Optimization for the Synthesis of Spiro[acenaphthylene-1,2′-pyrrolidine] Derivatives (1)

Entry	Solvent	Temperature (°C)	Time (min)	Yield (%)
1	Ethanol	78	58	80
2	Aq. methanol (2:8)	65	45	84
3	Dichloromethane	40	270	23
4	Acetonitrile	82	180	68
5	Tetrahydrofuran	66	270	32
6	1,4-Dioxane	101	270	61

When sarcosine was replaced by proline, a series of spiro[acenaphthylene-1,2′-pyrrolizidine] derivatives (**3**) were obtained under the same reaction conditions (Scheme 3.2).

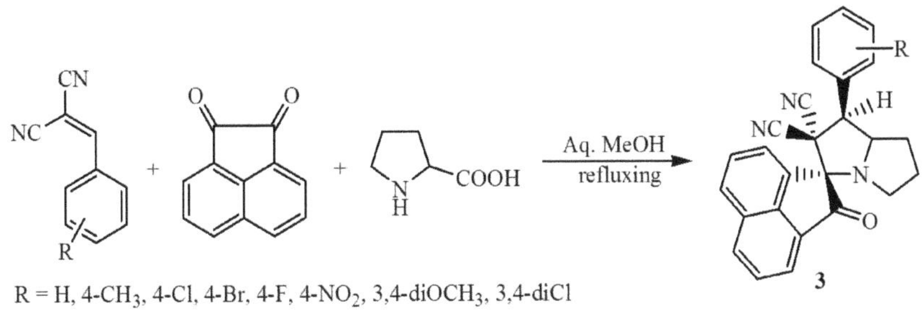

R = H, 4-CH$_3$, 4-Cl, 4-Br, 4-F, 4-NO$_2$, 3,4-diOCH$_3$, 3,4-diCl

SCHEME 3.2   Synthesis of Spiro[acenaphthylene-1,2′-Pyrrolizidine] Derivatives.

To expand the scope of this three-component 1,3-dipolar cycloaddition reaction, the reaction of Knoevenagel adducts and isatin with sarcosine was attempted. To our delight, under the above-optimized conditions, the reactions proceeded smoothly and a variety of the desired spiro[indoline-3,2′-pyrrolidine] derivatives (**4**) were obtained in good yields (Scheme 3.3).

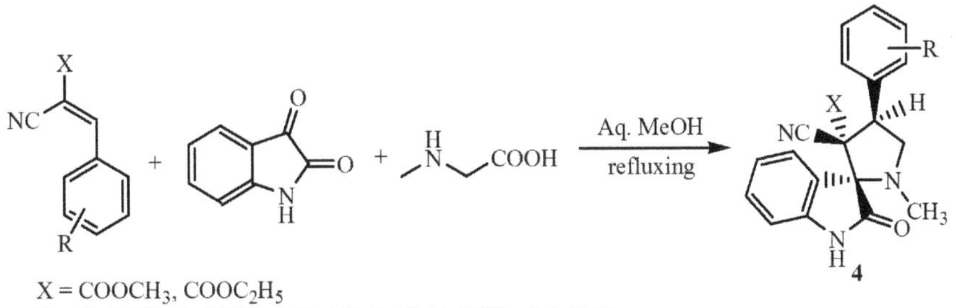

X = COOCH$_3$, COOC$_2$H$_5$
R = H, 4-CH$_3$, 4-OCH$_3$, 4-Cl, 4-F, 3,4,5-tri-OCH$_3$, 3,4-di Cl

SCHEME 3.3   Synthesis of Spiro[indoline-3,2′-Pyrrolidine] Derivatives.

We described the use of ethyl lactate as a valuable bio-based green solvent for the selective synthesis of medicinally privileged spiro[benzo[*f*]pyrrolo[2,1-*a*]isoindole-5,3′-indoline]-2′,6,11-trione derivatives (**5**) via the 1,3-dipolar cycloaddition reaction of azomethine ylide generated in situ by the decarboxylative condensation of substituted isatin and proline with napthoquinone as dipolarophile (Scheme 3.4).[23] The present cycloaddition reaction for the synthesis of antimicrobial spirooxindole derivatives was also studied in ethanol under reflux condition by Bhaskar et al.[24] These conditions have limitations in terms of the no recovery of solvent, longer reaction time, and lower yields. Hence, there is a compelling need to develop an effective synthetic procedure under more ecofriendly conditions. Further, this one-pot multicomponent methodology carried out at room temperature had several advantages such as i) rapid assembly of heterocyclic molecules by a three-component process with minimum generation of waste; ii) the process has high atom economy and is environmentally benign, since only molecules of water and CO_2 are lost; and iii) multiple bonds and stereocenters are formed in a single reaction.

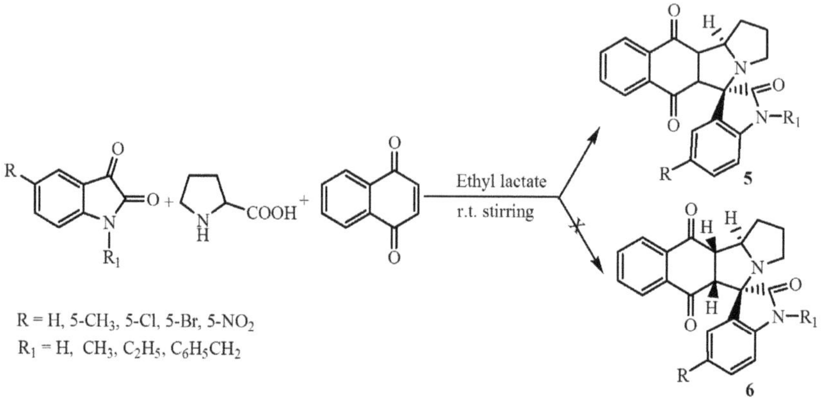

SCHEME 3.4 Synthesis of Spiro[benzo[f]pyrrolo[2,1-a]isoindole-5,3'-Indoline]-2',6,11-Trione Derivatives.

The optimization of the cycloaddition reaction using different solvents such as ethanol, methanol, THF, toluene, acetonitrile, and dimethylformamide, as well as ethyl lactate was screened (Table 3.2). After optimization, the ethyl lactate was found to be the best solvent to obtain the desired product in good yields. But the reaction was found to give comparatively lower yield of the product in other petroleum-based solvents.

TABLE 3.2 Optimization of Reaction Conditions for the Synthesis of Spiro[benzo[f] pyrrolo[2,1-a]isoindole-5,3'-indoline]-2',6,11-triones (5)

Entry	Solvent	Time (h)	Yield (%)
1	Toluene	24	22
2	Acetonitrile	20	16
3	Tetrahydrofuran	14	25
4	Dimethylformamide	18	30
5	Methanol	24	No reaction
6	Ethanol	24	No reaction
7	Ethyl lactate	1	82

Although the detailed mechanism of the above reaction is not fully clarified, the formation of cycloadduct could be explained as follows: decarboxylative condensation of the isatin with proline gives the azomethine ylide **I** that subsequently undergoes 1,3-dipolar cycloaddition reaction with the dipolarophile to afford the cycloadduct **II**, which is tautomerized to the hydronapthoquinone **III**, and rapid oxidation under atmospheric air conditions, resulting in the formation of cycloadduct (**5**) only without any traces of stereoisomer **6** (Scheme 3.5).

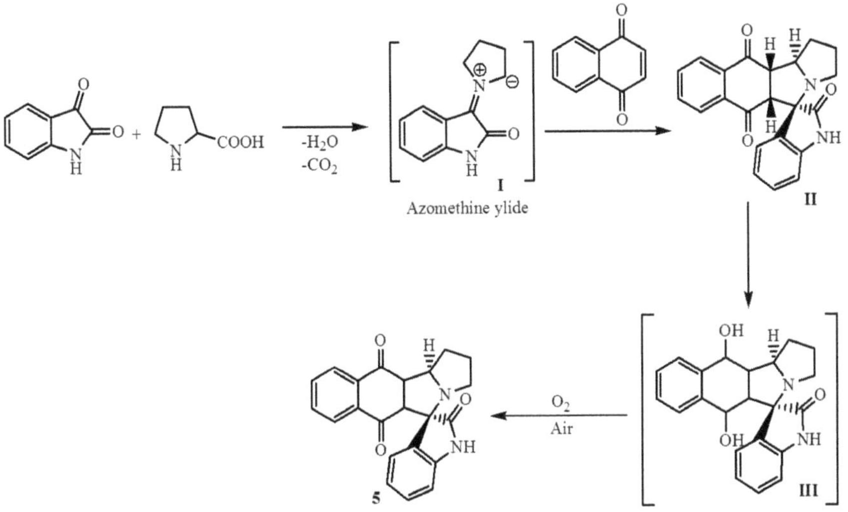

SCHEME 3.5 Mechanism for the Synthesis of Spiro[benzo[f]pyrrolo[2,1-a]isoindole-5,3'-Indoline]-2',6,11-Trione Derivatives.

3.3 SYNTHESIS OF DISPIRO HETEROCYCLES

One-pot synthetic method of pharmaceutically important novel ethyl 5″-chloro-4-cyano-1-methyl-2',2'-dioxodispiro[3'H-indole-3',2-pyrrolidine-3,3″[3″H]-indoline]-4-carboxylate derivatives (**7**) with creation of up to three stereogenic centers concurrently has also been accomplished by us via a facile [3 + 2]-cycloaddition reaction of azomethine ylides, derived from isatin and sarcosine in [bmim]BF$_4$ ionic liquid (Scheme 3.6).[25] This type of cycloaddition reaction was also studied in methanol under reflux condition and montmorillonite K-10 clay in methanol as reported by Lakshmi et al.[26] and Shanmugam et al.,[27] respectively. Both of these methods suffered from various drawbacks such as complicated workup, lower yields, longer reaction period, use of hazardous solvents in reaction procedure. However, further studies are still necessary for the essence of facile, environmental, and economical methodology.

The process described by our research group affords significant synthetic advantages in terms of shorter reaction time, product diversity, and selectivity and simplicity of the reaction procedure with good-to-excellent yields.

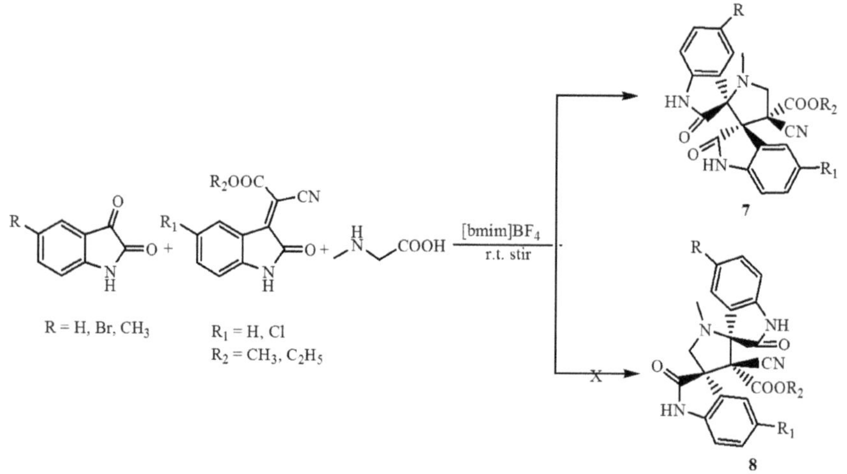

SCHEME 3.6 Synthesis of Ethyl 5$^{2''}$-Chloro-4-Cyano-1-Methyl-2',2$^{2''}$-Dioxodispiro[3'H-Indole-3',2-Pyrrolidine-3,3$^{2''}$[3$^{2''}H$]-Indoline]-4-Carboxylate Derivatives.

The feasibility of the strategy and optimization of the reaction conditions in various solvents such as methanol, ethanol, acetonitrile, THF, and toluene as well as different ionic liquids, such as [bmim]BF$_4$ and [bmim]PF$_6$, were explored. After optimization, [bmim]BF$_4$ was found to be the best solvent in terms of higher yield and shorter reaction times for this reaction (Table 3.3).

TABLE 3.3 Optimization Condition for the Synthesis of Dispiropyrrolidine Bisoxindole Derivatives (7)

Entry	Solvent	Temperature (°C)	Time (min)	Yield (%)
1	Methanol	Reflux	120	82
2	Ethanol	Reflux	140	80
3	Acetonitrile	Reflux	320	69
4	Tetrahydrofuran	Reflux	480	42
5	Toluene	Reflux	520	38
6	[bmim]BF$_4$	80	80	94
7	[bmim]PF$_6$	80	90	90

Although the detailed mechanism of the above reaction is not fully clarified, the formation of regioisomer could be explained as follows: decarboxylative condensation of the isatin with sarcosine gives the azomethine ylide (dipole), which subsequently undergoes 1,3-dipolar cycloaddition reaction with the dipolarophile to afford novel cycloadduct as a single regioisomer. The regioselectivity in the product formation can be explained by considering the secondary orbital interaction (SOI) of the orbital of the carbonyl group of dipolarophile with those of the ylide (Scheme 3.7). Accordingly, the observed regioisomer via path **A** is more favorable due to the presence of SOI, which is not possible in path **B**. Thus, it ruled out the possibility of formation of stereoisomer **8**.

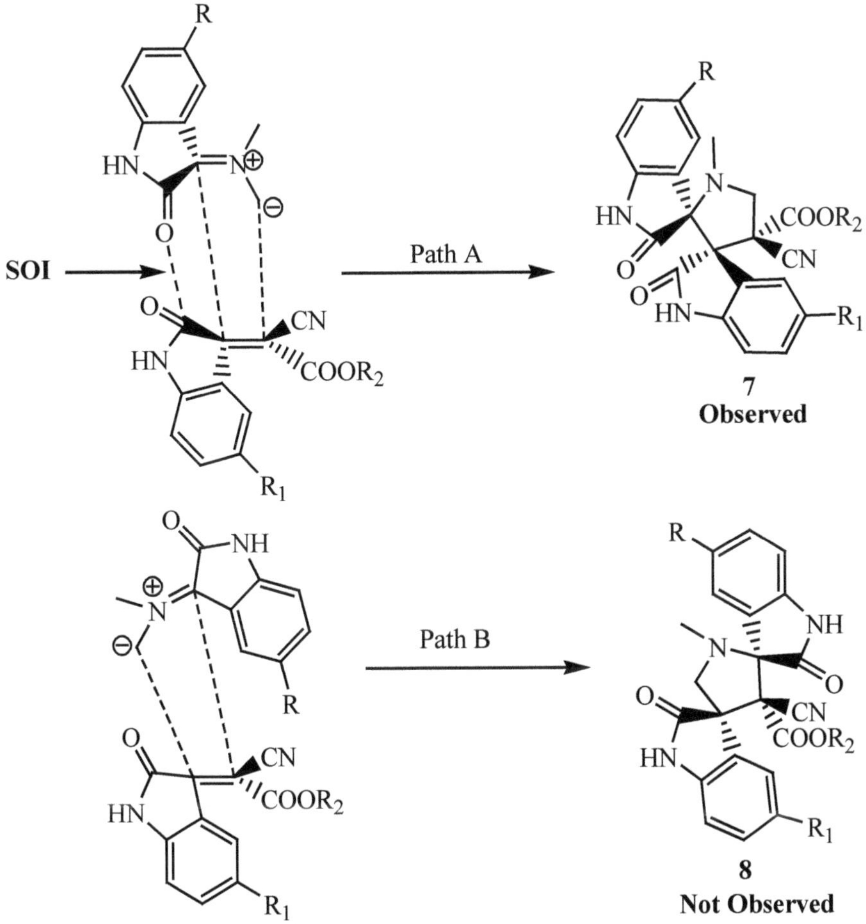

SCHEME 3.7 Mechanism for the Regioselective Synthesis of Ethyl 5″-Chloro-4-Cyano-1-Methyl-2′,2″-Dioxodispiro[3′H-Indole-3′,2-Pyrrolidine-3,3″[3″H]-Indoline]-4-Carboxylate Derivatives.

In the azomethine ylide of sarcosine, **E** isomer of alkyl 2-cyano-2-(2-oxoindo-lin-3-ylidene)acetate is taking part as a dipolarophile producing dispiropyrrolidine bisoxindoles exclusively, while in case of azomethine ylide of proline, **Z** isomer of alkyl 2-cyano-2-(2-oxoindolin-3-ylidene)acetate is participating to generate the ethyl 5″-chloro-4-cyano-3,4,4a,5,6,7-hexahydro-2′,2″-dioxodispiro[3′H-indole-3′,2-pyrrolizine-3,3″[3″H]-indoline]-4-carboxylates (**9**) as single product without using any catalyst in ionic liquid (Scheme 3.8). Good functional group tolerance and broad scope of working substrate are other prominent features of the present methodology with high degree of obtained chemo-, regio-, and stereoselectivity. It is observed that the generated three stereogenic centers can be controlled very well in this intermolecu-lar multicomponent one-pot process.

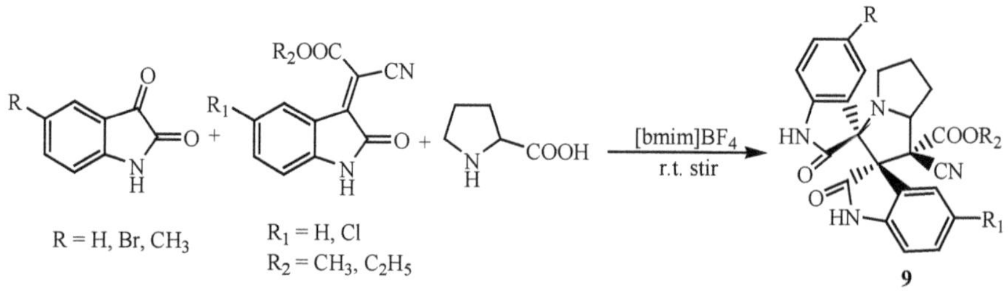

SCHEME 3.8 Synthesis of Ethyl 5″-Chloro-4-Cyano-3,4,4a,5,6,7-Hexahydro-2′,2″-Dioxodispiro[3′H-Indole-3′,2-Pyrrolizine-3,3″[3″H]-Indoline]-4-Cyano-3,4,4a,5,6,7-Hexahydro-2′,2″-Dioxodispiro[3′H-Indole-3′,2-Pyrrolizine-3,3″[3″H]-Indoline]-4-Carboxylates Derivatives.

The formation of cycloadduct follows the same pathway discussed by us previ-ously, but the stereoselectivity in the product formation is completely different, which can be explained by plausible mechanism shown in Scheme 3.9. This cycloaddition is regioselective with the addition of the electron-rich carbon of the dipole to the β-car-bon of dipolarophiles and stereoselective affording only one diastereomer exclusively despite more than one stereocentre being present in the cycloadducts. In the cycload-duct, both amide carbonyl groups are in trans-relationship with respect to each other; this is presumably ascribable to the preferred spatial arrangement of the dipolarophile and the azomethine ylide in cycloaddition, which would minimize the unfavorable ste-ric repulsion of the bulky groups of **E** isomerized Knoevenagel adduct (dipolarophile) with the generated azomethine ylide.

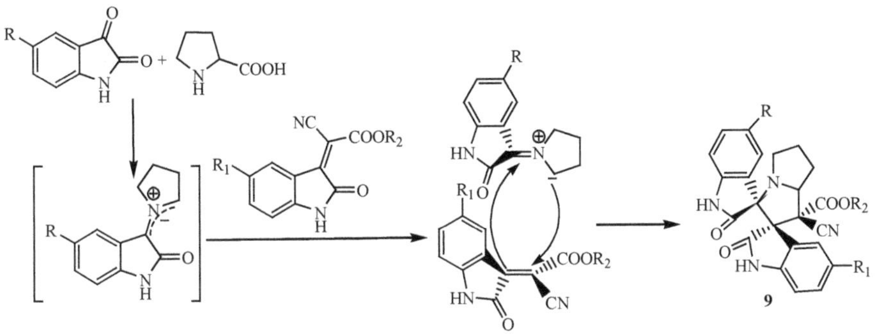

SCHEME 3.9 Plausible Mechanism for the Synthesis of Ethyl $5^{2''}$-Chloro-4-Cyano-3,4,4a,5,6,7-Hexahydro-$2',2^{2''}$-Dioxodispiro[$3'H$-Indole-$3',2$-Pyrrolizine-$3,3^{2''}$[$3^{2''}H$]-Indoline]-4-Cyano-3,4,4a,5,6,7-Hexahydro-$2',2^{2''}$-Dioxodispiro[$3'H$-Indole-$3',2$-Pyrrolizine-$3,3^{2''}$[$3^{2''}H$]-Indoline]-4-Carboxylates Derivatives.

We also reported our new findings for the 1,3-dipolar cycloaddition of 2-oxo-($2H$)-acenaphthylen-1-ylidene-malononitrile as dipolarophiles with the azomethine ylides generated in situ by N-substituted isatin and sarcosine to furnish novel dispiropyrrolidine oxindoles (**10**) in dry toluene with stirring for 6–7 h under refluxing conditions (Scheme 3.10).[28] In addition, we have successfully developed the regioselective bioactive dispiropyrrolidine oxindole (**10**) framework containing two cyano groups, which diversifies the existing 1,3-dipolar cycloaddition of azomethine ylides.

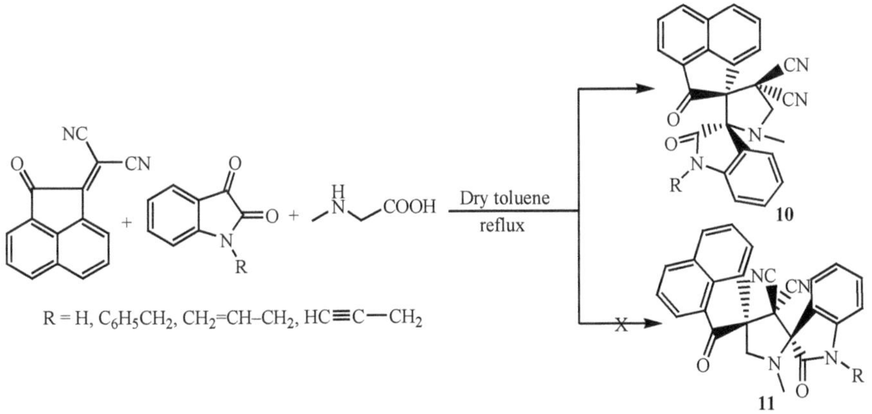

R = H, $C_6H_5CH_2$, CH_2=CH–CH_2, HC≡C—CH_2

SCHEME 3.10 Synthesis of Dispiropyrrolidine Oxindoles Derivatives.

The formation of regioisomer could be explained as follows: decarboxylative condensation of the isatin with sarcosine gives the azomethine ylide, which then undergoes 1,3-dipolar cycloaddition reaction with the dipolarophile regioselectively as shown in Scheme 3.11 (path **A**). The regioselectivity in the product formation can be explained by considering the SOI of the orbital of the carbonyl group of dipolarophile with those of the ylide as shown in Scheme 3.11. Accordingly, the observed regioisomer via path **A** is more favorable because of the SOI that is not possible in path **B** ruling out the formation of stereoisomer **11**.

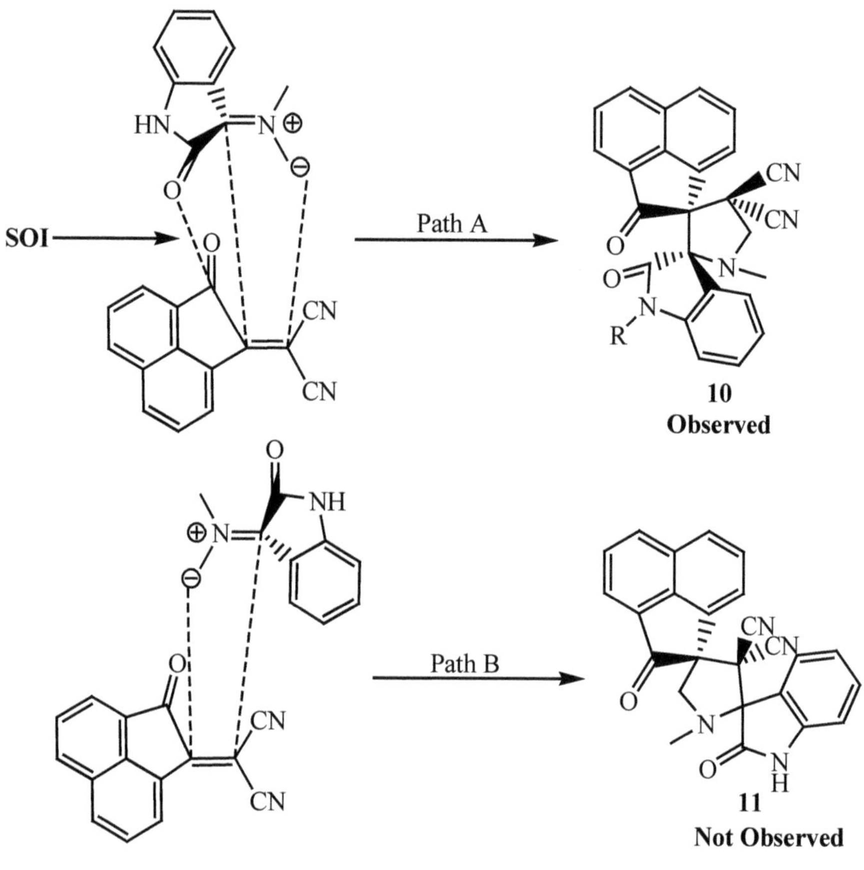

SCHEME 3.11   Mechanism for the Synthesis of Dispiropyrrolidine Oxindoles Derivatives.

Encouraged by this success, our research group extended this reaction to other dipolarophiles such as 2-fluoren-9-ylidene-malononitrile under similar conditions to furnish the respective dispiropyrrolidine oxindole derivatives (**12**) as single product with good yields (Scheme 3.12).

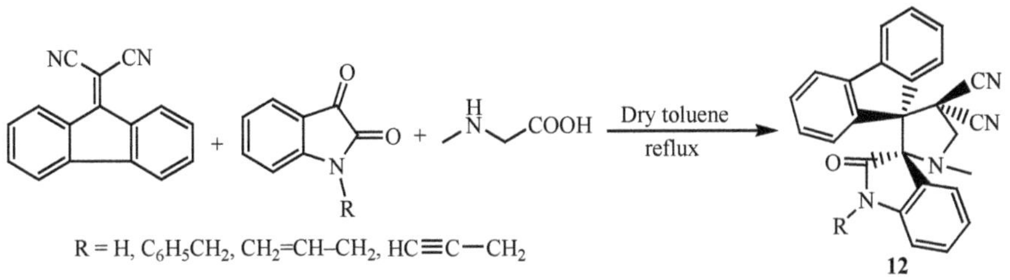

SCHEME 3.12 Synthesis of Dispiropyrrolidine Oxindoles Derivatives.

A simple and highly selective synthetic protocol of dispiro heterocycles containing three pharmacophoric moieties such as piperidinone, 1,3-indanedione, and pyrrolidine in a same molecular framework (**13**) has been afforded by the means of three-component reaction of ninhydrin, sarcosine, and 1-benzyl/methyl-3,5-bis[(E)-arylidene]-piperidin-4-one using task-specific 1,1,3,3-tetramethylguanidine acetate [TMG][Ac] ionic liquid in excellent yield (85–92%) with high degree of chemo-, regio-, and stereoselectivity (Scheme 3.13).[29] The present protocol proceeds in a chemoselective fashion, as only one of the two olefinic double bonds of the dipolarophiles took part in the reaction. This reaction is also regioselective with the addition of the electron-rich carbon of the dipole to the β-carbon of dipolarophiles and stereoselective affording only one diastereomer in excellent yields. In the present case, there was no sign of the formation of stereoisomer **14**.

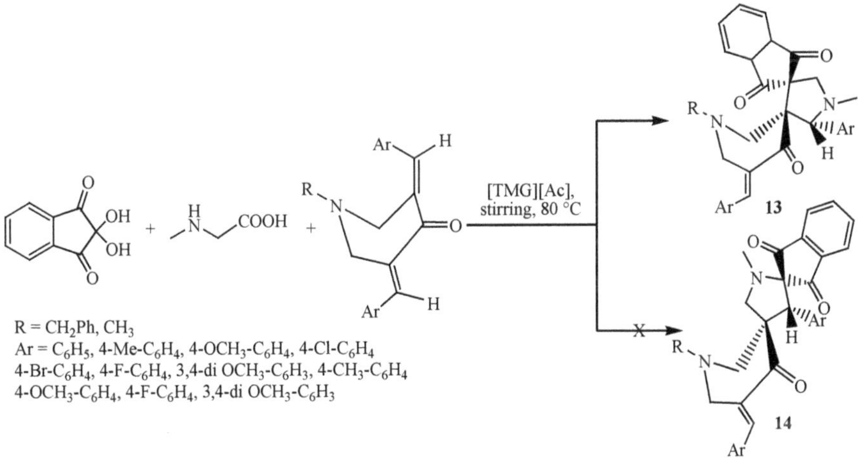

SCHEME 3.13 Synthesis of Dispiro Hetercyclic Derivatives.

In order to determine the effect of solvent on this reaction, we carried out the reaction under different organic solvents such as methanol, ethanol, toluene, dioxane, and acetonitrile, as well as different ionic liquids such as [bmim]BF$_4$ and [bmim]Cl, but the reaction was found to give comparatively lower yield of the products in all these solvents (Table 3.4). After optimization, we observed that [TMG][Ac] was found to be the best solvent in terms of higher yield and shorter reaction times. A single regioisomer was isolated in all cases. No trace of the other regioisomer **16** was found even after prolonged reaction times.

TABLE 3.4 Different Reaction Conditions for the Synthesis of Dispiropyrrolidines (13)

Entry	Solvent	Temperature (°C)	Time (h)	Yield (%)
1	Methanol	Reflux	6	82
2	Ethanol	Reflux	6	78
3	Toluene	Reflux	8	56
4	Dioxane	Reflux	7	42
5	Acetonitrile	Reflux	6	64
6	[bmim]BF$_4$	Reflux	5	89
7	[bmim]Cl	80	5	86
8	[TMG][Ac]	80	3	92

One-pot three-component 1,3-dipolar cycloaddition reaction of substituted isatin, sarcosine, and 1-methyl-3,5-bis[(E)-arylidene]piperidin-4-one in methanol was selectively utilized by us affording dispiro[3H-indole-3,2'-pyrrolidine-3',3"-piperidine]-2(1H),4"-dione derivatives (**15**) (Scheme 3.14).[30] Further, synthesized compounds were subjected to in vitro antimicrobial activity against various bacteria and fungi, and antitubercular activity was carried out against *Mycobacterium tuberculosis* H$_{37}$Rv strain.

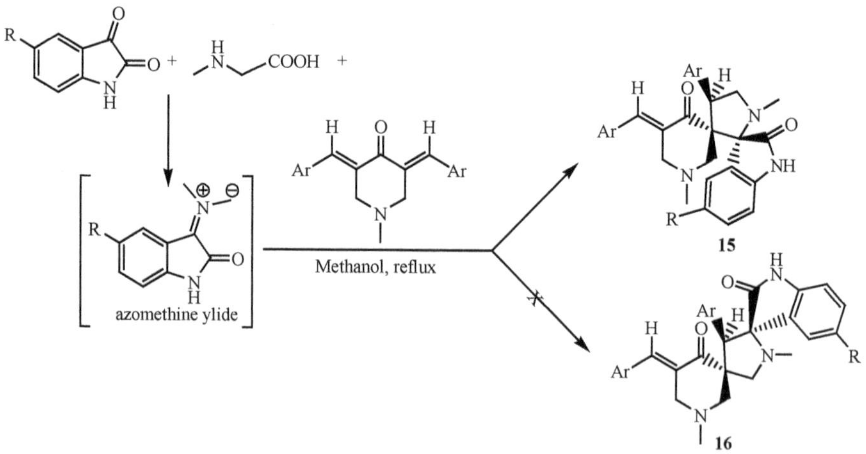

SCHEME 3.14 Synthesis Dispiro[3H-Indole-3,2'-Pyrrolidine-3',3$^{2"}$-Piperidine]-2(1H),4$^{2"}$-Dione Derivatives.

3.4 SYNTHESIS OF TRISPIRO HETEROCYCLES

Fluorinated alcohols are well known as a polar solvent[31] having low nucleophilicity[32] and have been the subject of considerable interest since their introduction as "green" solvents for reactions. Besides their usefulness as powerful reaction media, fluorinated alcohols have been well recognized as efficient catalysts and successfully applied in many organic reactions.[33]

We accomplished an efficient protocol using 2,2,2-trifluoroethanol (TFE) as a promoter, recoverable, greener, environmentally benign solvent for the highly profi-cient regio- and stereoselective synthesis of novel 1-N-methyl-spiro[2,3']oxindole-spiro[3,9″]-7″-arylmethylidene-1,4-dioxa-spiro[4″,5″]decan-4-aryl-pyrrolidine-8″,2′-diones (17) via MCR of 7,9-bis[(E)arylidene]-1,4-dioxa-spiro[4,5]decane-8-ones, sarcosine, and substituted isatins (Scheme 3.15).[34] Priority has been given to this method due to environmental friendliness, higher atom economy, shorter reaction time, and convenient operation. The simple product isolation procedure combined with ease of recovery and reuse of this novel reaction medium adds to the contribution toward the development of green and waste-free chemical process.

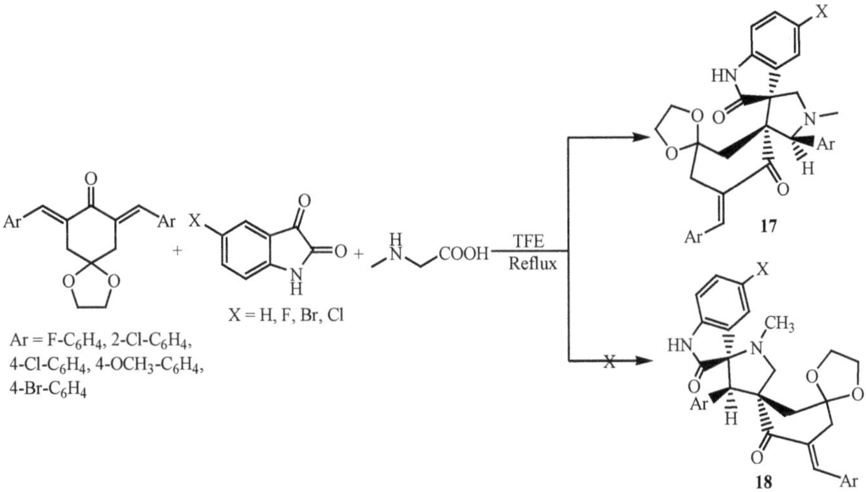

SCHEME   3.15 Synthesis of Novel 1-N-Methyl-Spiro[2,3']oxindole-Spiro[3,9″]-7″-Arylmethylidene-1,4-Dioxa-Spiro[4″,5″]decan-4-Aryl-Pyrrolidine-8″,2′-Diones Derivatives.

Optimization for the reaction conditions of this cycloaddition reaction using dif-ferent solvents such as ethanol, methanol, acetonitrile, 1,4-dioxane, THF, and 2,2,2-tri-fluoroethanol was studied. The results are summarized in Table 3.5. As evident from Table 3.5, the best results were obtained using 2,2,2-trifluoroethanol to yield products as a single regioisomer in high yield in shorter reaction time. Reaction in other solvents did not give satisfactory yield of the product even after longer reaction time.

TABLE 3.5 Preparation of 1-*N*-methyl-spiro[2,3']oxindole-spiro[3,9"]-7"-arylmethylidene-1,4-dioxa-spiro[4",5"]decan-4-aryl-pyrrolidine-8",2'-diones (17) in Different Solvents for Optimization of Reaction Conditions

Entry	Solvent	Temperature (°C)	Time (h)	Yield (%)
1	Ethanol	Reflux	8	78
2	Methanol	Reflux	6	75
3	Acetonitrile	Reflux	5	69
4	1,4-Dioxane	Reflux	6	71
5	Tetrahydrofuran	Reflux	7	68
6	TFE	Reflux	30 min	93

A plausible mechanism for the formation of the cycloadducts is proposed in Scheme 3.16. It is known that the Bronsted acidity ($pK_a = 12.4$) and strong ionizing power of 2,2,2-trifluoroethanol play unique behavior in organic transformations. The present cycloaddition reaction of isatin with sarcosine leads to the "in situ" formation of an azomethine ylide. Subsequent 1,3-dipolar cycloaddition reaction of dipolarophile and azomethine ylide afforded trispiropyrrolidine derivatives. The hydrogen atom of TFE is electron-deficient and can form hydrogen bonds with carbonyl groups of both isatin and dipolarophile, thereby catalyzing reaction. Further, the polar transition state of the reaction could be stabilized well by high ionizing solvent TFE. This catalysis presumably expedites the reaction in TFE relative to other solvents ruling out the possibility of formation of the other stereoisomer **18**.

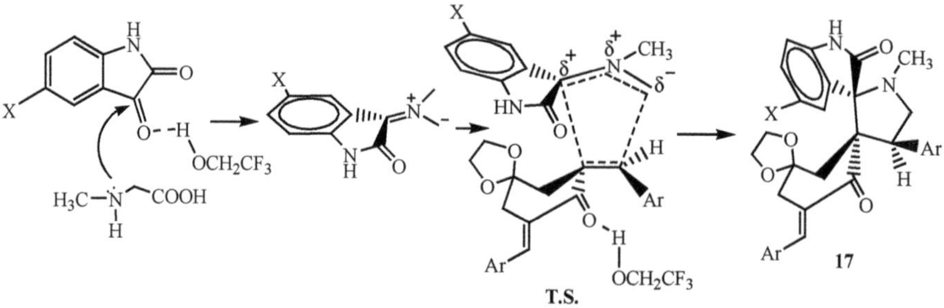

SCHEME 3.16 Plausible Mechanism for the Formation of 1-*N*-Methyl-Spiro[2,3']oxindole-Spiro[3,9"]-7"-Arylmethylidene-1,4-Dioxa-Spiro[4",5"]decan-4-Aryl-Pyrrolidine-8",2'-Diones Derivatives.

Encouraged by diverse medicinal importance of pyrrolothiazole derivatives, a novel class of spiro[4,3']oxindole-spiro[5,9"]-7"-arylmethylidene-1,4-dioxaspiro[4",5"]

decan-6-aryl-thiapyrroli zidine-8″,2′-diones were also synthesized under the optimized reaction conditions (Scheme 3.17) replacing sarcosine by thiaproline.

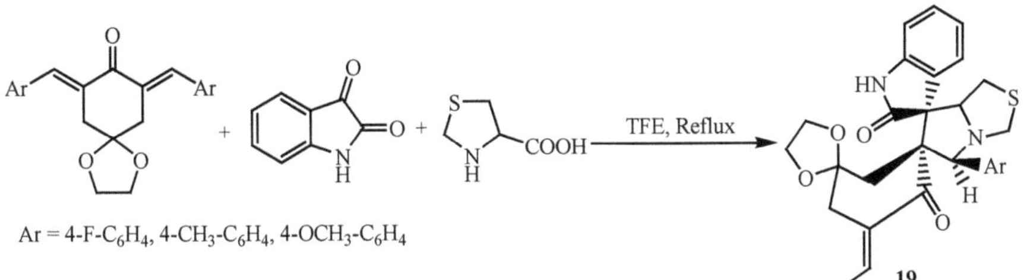

Ar = 4-F-C$_6$H$_4$, 4-CH$_3$-C$_6$H$_4$, 4-OCH$_3$-C$_6$H$_4$

SCHEME **3.17** Synthesis of Spiro[4,3′]oxindole-Spiro[5,9″]-7″-Arylmethylidene-1,4-Dioxaspiro[4″,5″]decan-6-Aryl-Thiapyrroli Zzidine-8″,2′-Diones Derivatives.

A new regio- and diastereoselective 1,3-dipolar cycloaddition reaction of 7,9-bis[(E)-arylidene]-1,4-dioxa-spiro[4,5]decane-8-ones, sarcosine/1,3-thiazolane-4-carboxylic acid, and acenapthequinone in 2,2,2-trifluoroethanol as solvent has also been developed by us for the synthesis of novel trispiropyrrolidine and thiapyrrolizidines, which include in their structures a 1,3-dioxalane moiety (Scheme 3.18).[35] This method has the advantages of good yield, mild reaction condition, low cost, and simplicity in process and handling. 2,2,2-Trifluoroethanol can be easily separated from product and recovered in excellent purity for direct use.

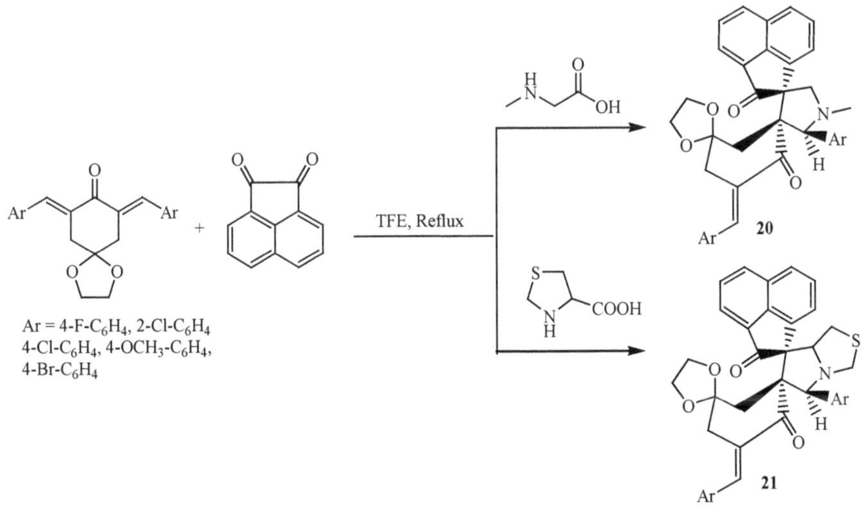

Ar = 4-F-C$_6$H$_4$, 2-Cl-C$_6$H$_4$
4-Cl-C$_6$H$_4$, 4-OCH$_3$-C$_6$H$_4$,
4-Br-C$_6$H$_4$

SCHEME 3.18 Synthesis of Trispiropyrrolidine and Thiapyrrolizidines Derivatives.

3.5 CONCLUSION

Indeed, it is well known in recent years that among the various methodologies known in synthetic organic chemistry, 1,3-dipolar cycloaddition reaction occupies the top slot for the synthesis of five-membered heterocycles and their analogs. This area is clearly expanding in several aspects and will reveal spectacular applications in the near future, ranging from the synthesis of spiro heterocyclic compounds. The development in this research area and understanding of stereochemical mechanistic path provided the chemists a platform to devise and synthesize a desired molecule of pharmacological interest.

As this review discusses the advancements in methodology, clear mechanistic path, stereochemistry involved, and synthetic applications of 1,3-dipolar cycloaddition reactions, it might be very useful for researchers and academicians to understand 1,3-dipolar cycloaddition reactions.

Further, the regio- and stereochemical outcome of the cycloadducts was determined unambiguously by single crystal X-ray analyses of the compounds **1, 3, 4, 5, 7, 9, 10, 12, 13, 15, 17,** and **20** having the CCDC numbers as **840556, 840557, 840558, 844357, 838902, 856837, 827499, 833828, 873496, 883122, 874668,** and **874669,** respectively.

ACKNOWLEDGMENTS

Financial assistance from the CSIR, New Delhi, is gratefully acknowledged.

KEYWORDS

- **Selectivity**
- **spiro heterocyclic compounds**
- **1,3-dipolar cycloaddition reactions**
- **multicomponent reactions (MCRs)**

REFERENCES

1. (a) Domınguez, G.; Perezcastells, J. Recent advances in [2+2+2] cycloaddition reactions. *Chem. Soc. Rev.* 2011, 40, 3430–3444; (b) Anderson, E. D.; Boger, D. L. Inverse electron demand diels alder reactions of 1,2,3-triazines: pronounced substituent effects on reactivity and cycloaddition scope. *J. Am. Chem. Soc.* 2011, 133, 12285–12292; (c) Lopez, F.; Mascarenas, J. L. Recent developments in gold-catalyzed cycloaddition reactions. *Beilstein J. Org. Chem.* 2011, 7, 1075–1094; (d) Moulay, S.; Touati, A. Cycloaddition reactions in aqueous systems: a two-decade trend endeavour. *C. R. Chimie.* 2010, 13, 1474–1511; (e) Xu, D. Q.; Xia, A. B.; Luo, S. P.; Tang, J.; Zhang, S.; Jiang, J. R.; Xu, Z. Y. In situ enamine activation in aqueous salt solutions: highly efficient asymmetric organocatalytic diels–alder

reaction of cyclohexenones with nitroolefins. *Angew. Chem. Int. Ed.* 2009, 48, 3821–3824; (f) Youcef, R. A.; Dos Santos, M.; Roussel, S.; Baltaze, J. P.; Lubin-Germain, N.; Uziel, J. Huisgen cycloaddition reaction of C-alkynyl ribosides under micellar catalysis: synthesis of ribavirin analogues. *J. Org. Chem.* 2009, 74, 4318–4323; (g) Meldal, M.; Tornoe, C. W. Cu-catalyzed azide-alkyne cycloaddition. *Chem. Rev.* 2008, 108, 2952–3015.

2. (a) Erhard, T.; Ehrlich, G.; Metz, P. A. Total synthesis of (±)-codeine by 1,3-dipolar cycloaddition. *Angew. Chem. Int. Ed.* 2011, 50, 3892–3894; (b) Chen, D.; Wang, Z.; Li, J.; Yang, Z.; Lin, L.; Liu, X.; Feng, X. Catalytic asymmetric 1,3-dipolar cycloaddition of nitrones to alkylidene malonates: highly enantioselective synthesis of multisubstituted isoxazolidines. *Chem. Eur. J.* 2011, 17, 5226–5229; (c) Kissane, M.; Maguire, A. R. Asymmetric 1,3-dipolar cycloadditions of acrylamides. *Chem. Soc. Rev.* 2010, 39, 845–883; (d) Stanley, L. M.; Sibi, M. P. Enantioselective copper-catalyzed 1,3-dipolar cycloadditions. *Chem. Rev.* 2008, 108, 2887–2902; (e) Kamata, K.; Nakagawa, Y.; Yamaguchi, K.; Mizuno, N. 1,3-dipolar cycloaddition of organic azides to alkynes by a dicopper-substituted silicotungstate. *J. Am. Chem. Soc.* 2008, 130, 15304–15310; (f) Pellissier, H. Asymmetric 1,3-dipolar cycloadditions. *Tetrahedron.* 2007, 63, 3235–3285.

3. (a) Dauban, P.; Malik, G. A. Masked 1,3-dipole revealed from aziridines. *Angew. Chem. Int. Ed.* 2009, 48, 9026–9029; (b) Gothelf, K. V.; Jorgensen, K. A. Asymmetric 1,3-dipolar cycloaddition reactions. *Chem. Rev.* 1998, 98, 863–910.

4. (a) Huisgen, R. 1,3-dipolar cycloadditions. Past and future. *Angew. Chem. Int. Ed. Engl.* 1963, 2, 565–598; (b) Huisgen, R. Kinetics and mechanism of 1,3-dipolar cycloadditions. *Angew. Chem. Int. Ed. Engl.* 1963, 2, 633–645; (c) Huisgen, R. Cycloadditions -definition, classification, and characterization. *Angew. Chem. Int. Ed. Engl.* 1968, 7, 321–328; (d) Woodward, R. B.; Hoffmann, R. The conservation of orbital symmetry. *Angew. Chem. Int. Ed. Engl.* 1969, 8, 781–853.

5. (a) Padwa, A.; Trost, B. M.; Flemming, I. Comprehensive organic synthesis. *Pergamon Press.* 1991, 4, 1069–1109; (b) Wade, P. A.; Trost, B. M.; Flemming, I. Comprehensive organic synthesis. *Pergamon Press.* 1991, 4, 1111–1168; (c) Kissane, M.; Maguire, A. R. Asymmetric 1,3-dipolar cycloadditions of acrylamides. *Chem. Soc. Rev.* 2010, 39, 845–883.

6. (a) Korotaev, V. Y.; Barkov, A. Y.; Moshkin, V. S.; Matochkina, E. G.; Kodess, M. I.; Sosnovskikh, V. Y. Highly diastereoselective 1,3-dipolar cycloaddition of nonstabilized azomethine ylides to 3-nitro-2-trihalomethyl-2H-chromenes: synthesis of 1-benzopyrano[3,4-c]pyrrolidines. *Tetrahedron.* 2013, 69, 8602–8608; (b) Tsuge, O.; Kanemasa, S. Recent advances in azomethine ylide chemistry. *Adv. Heterocycl. Chem.* 1989, 45, 231–349; (c) Pandey, G.; Banerjee, P.; Gadre, S. R. Construction of enantiopure pyrrolidine ring system *via* asymmetric [3+2]-cycloaddition of azomethine ylides. *Chem. Rev.* 2006, 106, 4484–4517; (d) Najera, C.; Sansano, J. M. Azomethine ylides in organic synthesis. *Curr. Org. Chem.* 2003, 7, 1105–1150; (e) Coldham, I.; Hufton, R. Intramolecular dipolar cycloaddition reactions of azomethine ylides. *Chem. Rev.* 2005, 105, 2765–2810.

7. Huisgen, R.; Padwa, A. 1,3-dipolar cycloaddition chemistry. *Wiley.* 1984, 1, 1–176.

8. Deyrup, J. A.; Szabo, W. A. Deprotonation of ternary iminium salts. *J. Org. Chem.* 1975, 40, 2048–2052.

9. (a) Joucla, M.; Mortier, J. 2,5-unsubstituted pyrrolidines from formaldehyde and amino acids through in *situ* azomethine-ylide 1,3-dipolar cycloaddition to alkenes. *J. Chem. Soc. Chem. Commun.* 1985, 1566–1567; (b) Joucla, M.; Mortier, J.; Hamelin, J. Pyrrolidines from α-amino-acids derivatives. *Tetrahedron Lett.* 1985, 26, 2775–2778

10. (a) Grigg, R.; Thianpatanagul, S. Decarboxylative transamination. mechanism and applications to the synthesis of heterocyclic compounds. *J. Chem. Soc. Chem. Commun.* 1984, 180–181; (b) Grigg, R.; Aly, M. F.; Sridharan, V.; Thianpatanagul, S. Decarboxylative transamination. A new route to spirocyclic and bridgehead-nitrogen compounds. relevance to α-amino acid decarboxylases. *J. Chem. Soc. Chem. Commun.* 1984, 182–183.

11. Vedeja, E.; West, F. G. Ylides by the desilylation of α-silyl onium salts. *Chem Rev.* 1986, 86, 941–955.

12. (a) Pandey, G.; Banerjee, P.; Gadre, S. R. Construction of enantiopure pyrrolidine ring system via asymmetric [3+2]-cycloaddition of azomethine ylides. *Chem. Rev.* 2006, 106, 4484–4517; (b). Braun, K. R.; Freysoldt, T. H. E.; Wierschem, F. 1,3-dipolar cycloaddition on solid supports: nitrone

approach towards isoxazolidines and isoxazolines and subsequent transformations. *Chem. Soc. Rev.* 2005, 34, 507–516; (c) Pellissier, H. Asymmetric 1,3-dipolar cycloadditions. *Tetrahedron.* 2007, 63, 3235–3285; (d) Hong, B. C.; Liu, K. L.; Tsai, C. W.; Liao, J. H. Proline-mediated dimerization of cinnamaldehydes *via* 1,3-dipolar cycloaddition reaction with azomethine ylides. A rapid access to highly functionalized hexahydro-1h-pyrrolizine. *Tetrahedron Lett.* 2008, 49, 5480–5483; (e) James, B. C.; Sunderhaus, D.; Martin, S. F. Concise total synthesis of (±)-pseudotabersonine *via* double ring-closing metathesis strategy. *Org. Lett.* 2010, 12, 3622–3625.

13. Recent review (a) Trost, B. M.; Brennan, M. K. Asymmetric syntheses of oxindole and indole spiro-cyclic alkaloid natural product. *Synthesis.* 2009, 18, 3003–3025; (b) Galliford, C. V.; Scheidt, K. A. Pyrrolidinyl-spirooxindole natural products as inspirations for the development of potential thera-peutic agents. *Angew. Chem. Int. Ed.* 2007, 46, 8748–8758; (c) Marti, C.; Carreira, E. M. Construction of Spiro[pyrrolidine-3,3'-oxindoles]–recent applications to the synthesis of oxindole alkaloids. *Eur. J. Org. Chem.* 2003, 63, 2209–2219.

14. Okita, T.; Isobe, M. Synthesis of the pentacyclic intermediate for dynemicin A and unusual formation of spiro-oxindole ring. *Tetrahedron.* 1994, 50, 11143–11152.

15. (a) Abou-Gharbia, M. A.; Doukas, P. H. Synthesis of tricyclic arylspiro compounds as potential an-tileukemic and anticonvulsant agents. *Heterocycles.* 1979, 12, 637–640; (b) Lundahl, K.; Schut, J.; Schlatmann, J. L. M. A.; Paerels, G. B.; Peters, A. Synthesis and antiviral activities of adamantane spiro compounds. 1. adamantane and analogous Spiro-3'-pyrrolidines. *J. Med. Chem.* 1972, 15, 129–132; (c) Kornet, M. J.; Thio, A. P. Oxindole-3-spiropyrrolidines and -piperidines. Synthesis and local anes-thetic activity. *J. Med. Chem.* 1976, 19, 892–898.

16. (a) Miyake, Y. F.; Yakushijin, K.; Horne, D. A. Preparation and synthetic applications of 2-halo-tryptamines: synthesis of elacomine and isoelacomine. *Org. Lett.* 2004, 6, 711–713; (b) Cui, C. B.; Kakeya, H.; Okada, G.; Onose, R.; Osada, H. Novel Mammalian cell cycle inhibitors, tryprostatins A, B and other diketopiperazines produced by *Aspergillus fumigatus*. I. Taxonomy, fermentation, isola-tion and biological properties. *J. Antibiot.* 1996, 49, 527–533.

17. (a) Reddy, V. J.; Douglas, C. J. Highly enantioselective Intramolecular cyanoamidation: (+)-horsfi-line, (−)-coerulescine, and (−)-esermethole. *Org. Lett.* 2010, 12, 952–955; (b) Jossang, A.; Jossang, P.; Hadi, H. A.; Sevenet, T.; Bodo, B. Horsfiline, an oxindole alkaloid from horsfieldia superb. *J. Org. Chem.* 1991, 56, 6527–6530.

18. (a) Edmondson, S.; Danishefsky, S. J.; Sepp-Lorenzino, L.; Rosen, N. Total synthesis of spirotrypros-tatin A, leading to the discovery of some biologically promising analogues. *J. Am. Chem. Soc.* 1999, 121, 2147–2155; (b) Usui, T.; Kondoh, M.; Cui, C. B.; Mayumi, T.; Osada, H. Tryprostatin A, a specific and novel inhibitor of microtubule assembly. *Biochem. J.* 1998, 333, 543–548.

19. (a) Chou, C. H.; Gong, C. L.; Chao, C. C.; Lin, C. H.; Kwan, C. Y.; Hsieh, C. L.; Leung, Y. M. Rhyn-chophylline from *Uncaria rhynchophylla* functionally turns delayed rectifiers into A-Type K⁺ channels. *J. Nat. Prod.* 2009, 72, 830–834; (b) Endo, K.; Oshima, Y.; Kikuchi, H.; Koshihara, Y.; Hikino, H. Hypotensive principles of *Uncaria hooks*[1]. *Planta Med.* 1983, 49, 188–190.

20. Cioc, R. C.; Ruijter, E.; Orru, R. V. A. Multicomponent reactions: advanced tools for sustainable organic synthesis. *Green Chem.* 2014, 16, 2958–2975.

21. Gu, Y. Multicomponent reactions in unconventional solvents: state of the art. *Green Chem.* 2012, 14, 2091–2128.

22. Dandia, A.; Jain, A. K.; Laxkar, A. K.; Bhati, D. S. A. Highly efficient protocol for the regio- and stereo-selective synthesis of spiro pyrrolidine and pyrrolizidine derivatives by multicomponent reaction. *Tet-rahedron Lett.* 2013, 54, 3180–3184.

23. Dandia, A.; Jain, A. K.; Laxkar, A. K. Ethyl lactate as a promising bio based green solvent for the syn-thesis of spiro-oxindole derivatives *via* 1,3-dipolar cycloaddition reaction. *Tetrahedron Lett.* 2013, 54, 3929–3932.

24. Bhaskar, G.; Arun, Y.; Balachandran, C.; Saikumar, C.; Perumal, P. T. Synthesis of novel Spirooxindole derivatives by one pot multicomponent reaction and their antimicrobial activity. *Eur. J. Med. Chem.* 2012, 51, 79–91.

25. Dandia, A.; Jain, A. K.; Laxkar, A. K.; Bhati, D. S. Synthesis and stereochemical iinvestigation of highly functionalized novel dispirobisoxindole derivatives *via* [3+2] cycloaddition reaction in ionic liquid. *Tetrahedron.* 2013, 69, 2062–2069.

26. Lakshmi, N. V.; Thirumurugan, P.; Perumal, P. T. An expedient approach for the synthesis of dispiro-pyrrolidine bisoxindoles, spiropyrrolidine oxindoles and spiroindane-1,3-diones through 1,3-dipolar cycloaddition reactions. *Tetrahedron Lett.* 2010, 51, 1064–1068.

27. Shanmugam, P.; Viswambharan, B.; Selvakumar, K.; Madhavan, S. A. Facile and effcient synthesis of highly functionalized 3,3'-dispiropyrrolidine- and 3,3'-dispiropyrrolizidine bisoxindoles *via* [3+2] cycloaddition. *Tetrahedron Lett.* 2008, 49, 2611–2615.

28. Dandia, A.; Jain, A. K.; Bhati, D. S. Direct construction of novel dispiro heterocycles through 1,3-dipolar cycloaddition of azomethine ylides. *Tetrahedron Lett.* 2011, 52, 5333–5337.

29. Dandia, A.; Jain, A. K.; Sharma, S. An efficient and highly selective approach for the construction of novel dispiro heterocycles in guanidine-based task-specific [TMG][Ac] ionic liquid. *Tetrahedron Lett.* 2012, 53, 5859–5863.

30. Dandia, A.; Jain, A. K.; Laxkar, A. K. Synthesis and biological evaluation of highly functionalized dispiro heterocycles. *RSC Adv.* 2013, 3, 8422–8430.

31. Reichardt, C. Solvatochromic dyes as solvent polarity indicators. *Chem. Rev.* 1994, 94, 2319–2358.

32. (a) Minegishi, S. S.; Kobayashi, S.; Mayr, H. Solvent nucleophilicity. *J. Am. Chem. Soc.* 2004, 126, 5174–5181; (b) Denegri, B.; Minegishi, S.; Kronja, O.; Mayr, H. SN1 reactions with inverse rate profiles. *Angew. Chem. Int. Ed.* 2004, 43, 2302–2305; (c) Hofmann, M.; Hampel, N.; Kanzian, T.; Mayr, H. Electrophilic alkylations in neutral aqueous or alcoholic solutions. *Angew. Chem. Int. Ed.* 2004, 43, 5402–5405; (d) Westermaier, M.; Mayr, H. Electrophilic allylations and benzylations of indoles in neutral aqueous or alcoholic solutions. *Org. Lett.* 2006, 8, 4791–4794.

33. (a) Richard, J. P.; Toteva, M. M.; Amyes, T. L. What is the stabilizing interaction with nucleophilic solvents in the transition state for solvolysis of tertiary derivatives: nucleophilic solvent participation or nucleophilic solvation. *Org. Lett.* 2001, 3, 2225–2228; (b) Abe, H.; Amii, H.; Uneyama, K. Pd-catalyzed asymmetric hydrogenation of α-fluorinated iminoesters in fluorinated alcohol: a new and catalytic enantioselective synthesis of fluoro α-amino acid derivatives. *Org. Lett.* 2001, 3, 313–315; (c) Schadt, F. L.; Bentley, T. W.; Schleyer, P. V. R. The SN2-SN1 spectrum. 2. Quantitative treatments of nucleophilic solvent assistance. A scale of solvent nucleophilicities. *J. Am. Chem. Soc.* 1976, 98, 7667–7674; (d) Habusha, U.; Rozental, E.; Hoz, S. Could ionic γ-elimination be concerted: clocking the internal displacement across a cyclobutane ring. *J. Am. Chem. Soc.* 2002, 124, 15006–15011; (e) Heydari, A.; Khaksar, S.; Tajbakhsh, M. Trifluoroethanol as a metal-free, homogeneous and recyclable medium for the efficient one-pot synthesis of α-amino nitriles and α-amino phosphonates. *Tetrahedron Lett.* 2009, 50, 77–80; (f) Philippe, C.; Milcent, T.; Crousse, B.; Bonnet-Delpon, D. Non lewis acid catalysed epoxide ring opening with amino acid esters. *Org. Biomol. Chem.* 2009, 7, 2026–2028.

34. Dandia, A.; Singh, R.; Joshi, J.; Kumari, S. An eco-compatible synthesis of medicinally important novel class of trispiroheterocyclic framework using 2,2,2-trifluoroethanol as a reusable medium. *J. Fluorine Chem.* 2013, 156, 283–289.

35. Dandia, A.; Singh, R.; Joshi, J.; Kumari, S. A Green regio- and diastereoselective synthesis of novel trispiroheterocycles in 2,2,2-trifluoroethanol. *Eur. Chem. Bull.* 2013, 2(9), 683–686.

CHAPTER 4

Recent Trends in Plasma Chemistry and Spectroscopy Diagnostics

Ram Prakash*

Plasma Devices Technology Team, Microwave Tubes Division, CSIR-Central Electronics Engineering Research Institute (CEERI), Pilani 333031, Rajasthan, India
*Email: ramprakash@ceeri.ernet.in, rplavania@yahoo.com

CONTENTS

ABSTRACT

Plasma chemistry is most important subject for plasma-processing systems and applications. Basically, it is controlled by highly active species or by the higher temperature properties of the plasmas. In this chapter, clarity about plasma as physics or chemistry has been clearly brought out. Also, chemistry of nonthermal plasma applications specially for dielectric barrier discharge-based VUV/UV excimer light sources has been discussed. In the chemically active plasma discharges, in situ diagnostics are not possible due to contamination issues. The passive optical emission spectroscopy diagnostic is a better solution. The recent trends of plasma spectroscopy diagnostics in concurrence with the plasma chemistry understanding has been presented.

4.1 INTRODUCTION

Plasma chemistry plays an important role in most plasma-processing systems and applications. The industrial plasma applications range from nonthermal to thermal plasmas. In nonthermal plasma applications, the highly active species of plasma play an important role, whereas in thermal plasma applications, the higher temperature properties of the plasma control the chemistry.[1] In fact, plasma has become an important element and has touched many aspects of our lives. For example, people are well aware of plasma TV, plasma thrusters, and fluorescent lamps, as well as popular-culture concepts such as plasma guns and plasma shields (Ref. [2] and references therein). Cell phones, computers, and other modern electronic devices are also manufactured using plasma-enabled chemical processing (Ref. [2] and references therein). Most of the synthetic fibers used in photomaterials, clothing, and advanced packaging materials are also treated using plasmas.[3] A significant amount of clean water in the world is purified using ozone-plasma technology.[4] Many different tools and special surfaces are plasma-coated to protect and provide them with new extraordinary surface properties. The developments in plasma chemistry are enabling tremendous growth in a variety of applications for environmental remediation, manufacturing, therapeutic, preventive medicine, etc.[5,6] The motivation for this chapter is to bring out the foundational understanding of some of the physical and chemical aspects associated with both thermal and nonthermal plasmas. Efforts have also been made to discuss the recent trends in plasma chemistry and spectroscopy diagnostics.

Council of Scientific and Industrial Research-Central Electronics Engineering Research Institute is currently working on the nonthermal applications of plasma by way of developing dielectric barrier discharge (DBD)-based VUV/UV excimer light sources[7,8] for medical and industrial applications. In the excimer light sources, spontaneous radiations that employ nonequilibrium radiation of excimer and exciplex molecules play an important role.[9] The excimers are weakly bound excited states of molecules that do not have a stable ground state. The main advantage of excimer light sources is that they provide high-intensity narrow-band molecular radiation in the absence of a strong molecular bond in the ground state, which eliminates radiation absorption in the plasma. There are many methods to generate excimers, such as corona, high-energy electron beam, X-ray, protons, heavy ions, synchrotron radiation, microwave discharges. However, the DBD-based plasmas provide one of the most efficient ways to produce the necessary precursors for excimer formation due to its ability to produce high energetic electrons at high working pressures (≥ 200 mbar). High pressure further promotes the suitable conditions for excimer formation by three-body reactions in the plasma. The excimer formation in the DBD is well favored by high collision rates at elevated pressure and efficient excitation or ionization of precursor species. A brief review of the plasma chemistry is brought out in the chapter for the VUV/UV excimer formation.

We know well that in the chemically active plasma discharges, in situ diagnostics are not possible due to contamination issues but the plasma spectroscopy plays an important role in understanding the chemically active plasmas.[10,11] In fact, the line radiation contains information on the nature of the plasma and can be utilized passively to understand the plasma environment spectroscopically. However, interpretation of the spectroscopic measurements is not straightforward. The delicacy lies in the complex nature of the plasmas and subsequently the manifold atomic and molecular processes. In order to infer the useful parameters from the measured intensities of lines, proper calculations of the population densities of excited states are required, which is possible by means of collisional–radiative (CR) model analysis.[11] The recent trends of plasma spectroscopy diagnostics in concurrence with the plasma chemistry understanding have also been briefly discussed.

A brief description of the plasma is given in Sections 4.2 and 4.3. Section 4.4 is related to the discharge sources for plasma chemistry. Section 4.5 contains brief information about the major components of chemically active plasmas. Section 4.6 deals with the atomic chemistry and spectroscopy diagnostics, which also include understanding about corona, local thermodynamic equilibrium (LTE), and CR models that are used for quantitative plasma spectroscopic diagnostics in most chemically active plasmas. The chapter is concluded in Section 4.7.

4.2 WHETHER PLASMA IS PHYSICS OR CHEMISTRYβ

To begin with, let us understand the definitions of physics and chemistry. Physics is the branch of science where we study energies, whereas chemistry is the branch of science where we study matters. Next question is then, what is difference between matter and energy. This difference is tabulated in Table 4.1.

TABLE 4.1 General Difference Between Matter and Energy

Matter	Energy
It has weight	It has no weight
It occupy space	It does not occupy any space
One can identify by sense organs	One cannot identify by sense organs (e.g., we cannot see light rather light shows us, here light is an energy)
One cannot do work	One can do work

It is quite interesting that the energy has unique property to do work. If we provide energy to matter it starts working, otherwise it is dead. The matter without energy is chemistry, whereas the matter with energy shows chemical energy and becomes physics. However, since we cannot clearly identify energy by sense organs, physics is mostly an understanding, whereas chemistry is mostly sense. To clearly comprehend whether plasma is physics or chemistry, let us try to learn about the states of matter. Based on the energy, character, and particle motion, there are presently six states of matters known, which are listed in Table 4.2.

TABLE 4.2 Characteristics of Different States of Matter

	Zeroth	First	Second	Third	Fourth	Fifth
Name	Bose–Einstein	Solid	Liquid	Gas	Plasma	Filament
Energy	~Zero	<Crystal	<Attraction	>Attraction	$<10^6$ eV	$<10^{26}$ eV
Character	Nonthermodynamical	Thermodynamical	Thermodynamical	Thermodynamical	Thermal	Nonthermodynamical
Particle motion	No (heat) motion	Particles move in all three dimensions				Motion in only one direction

Four states of matter, that is, solid, liquid, gas, and plasma, are well known. These all are the thermal states. Their particles have heat motion in all the three possible dimensions. The energy of the heat motion produces these four thermal states of matter. In solid, liquid, and gas, thermodynamical characters are valid and chemistry is well defined. However, in plasma state, the thermal effects lead to an equilibrium state along with the charge particles having high energies. So, due to associated added energy in the plasma state, one can say that it is both physics and chemistry. If we look only these four states of matter, still there are a few more questions that need an answer.

How many states of matter are possibleβ

Is there another state of matter which is more energetic than plasmaβ

How can we explain the cosmic rays up to particle energies $\sim 10^{21}$ eVβ

Filaments of the solar corona seem to be very hot without clear heating process, whyβ

Why are filaments and jets so very thinβ

One of the states of matter is the Bose–Einstein condensate. It has zero thermal energy (see Table 4.2), and it is proximity to the 0°K, which is why it is named as zeroth state of matter. This is a nonthermodynamical state due to the associated temperature. Considering Bose–Einstein condensate state as "fifth state of matter" is clearly incorrect because heated plasma will never be transformed to this state. The enthalpy of thermal states is shown in Figure 4.1. In this figure, the level of disorder has been used to determine the state numbering in the matter.

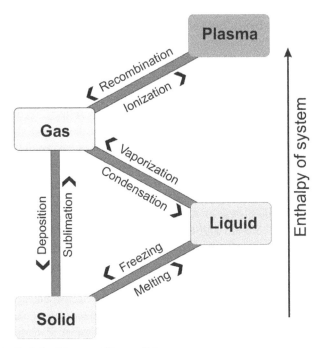

FIGURE 4.1   Enthalpy Diagram of States of Matter.

Taking this into consideration, the filaments should then belong to the "fifth state of matter." Sparks, electron beams in TV, lightnings, ion jets in the spacecrafts, electrons in accelerator machines, ions in the future fusion reactors, solar corona, flares, jets of young stars, and jets at black holes and at neutron stars belong to this most energetic state of matter. Solar flare particles have energy $\sim 10^{10}$ eV but the hottest plasma in supernova has only ~10^5 eV energy.[12] Plasma particles above 10^5 eV energy cannot exist because hypernova cannot exist. Filaments are the largest bodies of the Universe and can have energy <10^{26} eV. Filament particles above 10^{26} eV cannot exist because the largest neutron stars of three sunmasses have energy only ~10^{26} eV. This further indicates that with the existing understanding the above-mentioned six states of matter are only possible. These filaments and jets have an exact circular cross-section, can oscillate, and are produced electrically from plasma. However, one cannot compress the filaments into plasma state. Particles move only in one direction, that is, without the thermal zig-zag. Interestingly, particles in the most energetic state of matter do not emit heat due to their flight along straight lines. These particle motions in filaments are already named recently: "nonthermal motions".[13] Hence, the fifth state of matter is also a nonthermal state. We can say that zeroth and fifth states of matter are nonthermal states of matter and are not part of any chemistry. On the other hand, solid, liquid, and gas are purely chemistry based on the reaction kinetics and thermodynamical properties (bulk behavior), whereas plasma due to reaction kinetics and thermal state (energy behavior) can be said both physics and chemistry.

4.3   WHAT IS PLASMA?

Plasma is known as fourth state of matter. Plasmas are conductive assemblies of charged particles, neutral particles, and fields that exhibit collective effects.[14] The plasma state can exist in any solid, liquid, or gas phases. However, it is to be noted that the plasma state differs with the above states due to its very nature of having additionally long-range electromagnetic (EM) interaction forces. To understand as a layman, let us take a piece of ice, which is a solid. Now if we heat the ice, it will get converted into a liquid, and if we heat it further, it will get converted into gas. What will happen if we heat the gas β The gas molecules break into atoms, and then atoms into electrons and ions. When this ionization becomes large enough, then the charged particles start behaving collectively. As there is a saying "either crowd decides or God decide", this collective approach of charge particles leads to an aggressive state of matter, which is the fourth state of matter, called plasma. So mere ionization in any gas, solid, and liquid does not make plasma state. The charge particles must exhibit collective behavior for plasma state to exist. Basically, collective behavior of particles influences a far particle by long-range EM interaction forces. This happens due to large number of charge particles. To understand the basic nature of crowd (or collective) behavior, let us take an example that there is a crowd of people standing near the road due to some incidence and another person is passing nearby. It is most obvious that this person feels a long-range force toward the crowd due to his own query and will be attracted toward the crowd. Similarly, when there is crowd of charge particles in any solid, liquid, and gas state, it starts influencing a far charge particle of opposite polarity by long-range Coulomb force and this far particle feels a force by the crowd of charge particles. Due to flow of charge particles, there are local currents and apparently magnetic fields. Hence, there come long-range EM interaction forces in the plasma state. The examples of plasmas in gases, liquids, and solids are given in Table 4.3.

TABLE 4.3 Examples of Liquid, Solid, and Gas Phase Plasmas

Sr. No.	Phase	Examples
1.	Gas-phase plasma	H_2, O_2, He, Ar, Cl_2, D + T, SF_6 Plasmas, etc.
		Used in many technological applications
2.	Liquid-phase plasma	Sodium in ammonia, electrolytes, etc.
3.	Solid-phase plasma	Free electrons and holes in semi-conductors, highly correlated dusty plasmas, etc.

Plasmas are described by many characteristics, such as temperature, degree of ionization, and density, the magnitude of which, and approximations of the model describing them, gives rise to plasmas that may be classified in different ways. A few classifications are listed below.

4.3.1 Ideal and Nonideal Plasmas

At low densities, a low-temperature partially ionized plasma can be regarded as a mixture of ideal gases of electrons, atoms, and ions. The particles travel at thermal velocities, mainly along straight paths, and collide with each other only occasionally. With an increase in density, the mean distances between the particles decrease and the particles start spending even more time interacting with each other, that is, in the fields of surrounding particles. So, an ideal plasma is one in which Coulomb collisions are negligible, otherwise the plasma is nonideal. In fact, condition for the ideal plasma existence is that the average kinetic energy of an electron should substantially exceed the average Coulomb energy needed for an ion to bind the electron.

$$\frac{3}{2}kT_e \gg \frac{e^2}{4\pi\varepsilon_0 d^2}.$$

The terms used in this expression have their usual meanings. This expression even can be further used in terms of Debye shielding of the particles and ultimately to understand the collective behaviors of the plasmas.[14]

When mean energy of interparticle interaction becomes comparable with the mean kinetic energy of thermal motion, the plasma becomes nonideal.[15]

4.3.2 Cold, Warm, and Hot Plasmas

A plasma is composed of approximately same number of electrons and ions (i.e., quasi-neutral). However, it is the temperature of electrons, ions, and gas that determines the nature of plasmas. In low-pressure gas discharge, the collision rate between electrons and gas molecules is not frequent enough for nonthermal equilibrium to exist between the energy of the electrons and the gas molecules. So the high-energy particles are mostly composed of electrons, while the energy of the gas molecules is around room temperature. Under these circumstances, we have $T_e \gg T_i \gg T_g$, where T_e, T_i, and T_g are the temperatures of the electron, ion, and gas molecules, respectively. This type of plasma is called as "cold plasma". In cold plasma, the degree of ionization is below 10^{-4}.[16]

In a high-pressure gas discharge, the collision between electrons and gas molecules occurs frequently. This causes thermal equilibrium between the electrons and gas molecules. We have $T_e \sim T_i \sim T_g$. We call this type of plasma "hot plasma". There do exist subnomenclatures for these plasmas and are listed below.

4.3.2.1 Hot Plasma (Thermal Plasma)

A hot plasma is one that approaches a state of LTE. A hot plasma is also called as thermal plasma.[17] Atmospheric arcs, sparks, flames, etc., can produce such plasmas.

4.3.2.2 Cold Plasma (Nonthermal Plasma)

A cold plasma is one in which the thermal motion of the ions can be ignored. Consequently, there is no pressure force, the magnetic force can be ignored, and only the electric force is considered to act on the particles.[18] Examples of cold plasmas include the Earth's ionopshere (about 1000°K compared to the Earth's ring current temperature of about 108°K), low discharge in a fluorescent tube, etc.

4.3.3 Plasma Ionization

The degree of ionization of a plasma is the proportion of charged particles to the total number of particles including neutrals and ions and is defined as $\alpha = n^+/(n + n^+)$, where n is the number of neutrals and n^+ is the number of charged particles.

In a plasma where the degree of ionization is high, charged particle collisions dominate. In plasmas with a low degree of ionization, collisions between charged particles and neutrals dominate. The degree of ionization, which determines when a gas becomes a plasma, will vary between different types of plasmas. It may be as little as 10^{-6}.

Example: Let us take a low-temperature, low-density, weakly ionized plasma for industrial applications. By low temperature, we mean "cold" plasma with a gas temperature normally ranging from 300°K and 600°K, by low density, we mean plasmas with neutral gas number densities of approximately 10^{13}–10^{16} molecules cm^{-3} (pressure between ~0.1 and 10^3 Pa), and by weakly ionized, we mean degree of ionization lies between 10^{-6} and 10^{-1}.[19] At very low ionization level (<10^{-3}), neutral collision dominates. On the other hand, for a few percent ionization, the Coulomb collisions dominate over collisions with neutrals in any plasma.[20]

When the input energy to the plasma increases gradually, the degree of ionization jumps suddenly from a fraction of one percent to full ionization. Under certain conditions, the border between a fully ionized and a weakly ionized plasma is very sharp.[21]

There are many nomenclatures like collisional plasmas, noncollisional plasma, neutral plasmas, non-neutral plasma, high-density plasma, low-density plasma, magnetic plasma, nonmagnetic plasma, high-energy density plasmas, dusty plasmas, grain plasmas, active and passive plasmas, etc., quite popular in plasma community. The classifications are based on dominance of the specific property of plasma, which is reflected by itself from the name.

4.4 DISCHARGE SOURCES FOR PLASMA CHEMISTRY

Plasma chemistry is clearly organized with the plasma, and a plasma source, which in most laboratory conditions is a gas discharge, represents the physical and engineering basis of the plasma chemistry. For simplicity, an electric discharge can be viewed as two electrodes inserted into a glass tube and connected to a power supply. The tube can be filled with various gases and/or evacuated. As the voltage applied across the two electrodes increases, the current suddenly increases sharply at a certain voltage required for sufficiently intensive electron avalanches. If the pressure is low, on the order of a few Torr, and the external circuit has a large resistance to prohibit a large current, a glow discharge develops. This is a low-current, high-voltage discharge widely used to generate nonthermal plasma. A similar discharge is known by everyone as a plasma source in fluorescent lamps. The glow discharges can be considered a major example of low-pressure, nonthermal plasma sources.[22]

A nonthermal corona discharge occurs at high pressures (i.e., ~atmospheric pressure) only in the regions of sharply nonuniform electric fields.[22] The field near one or both electrodes must be stronger than in the rest of the gas. This occurs near sharp points, edges, or small-diameter wires, which tend to be low-power plasma sources limited by the onset of electrical breakdown of the gas. However, it is possible to avoid this restriction through the use of pulse power supplies. Electron temperature in the corona exceeds 1 eV, whereas the gas remains at room temperature. The corona discharges are, in particular, widely applied in the treatment of polymer materials: most synthetic fabrics applied to make clothing have been treated before dyeing in corona-like discharges to provide sufficient adhesion.[23] The corona discharge can be considered a major example of an atmospheric-pressure nonthermal plasma source.

If the pressure is high, on the order of an atmosphere, and the external circuit resistance is low, a thermal arc discharge can be generated between two electrodes.[22]

Thermal arcs usually carry large currents, greater than 1 A at voltages of the order of tens of volts. Furthermore, they release large amounts of thermal energy at very high temperatures often exceeding 10^{40} K. The arcs are often coupled with a gas flow to form high-temperature plasma jets. The arc discharges are well known not only to scientists and engineers but also to the general public because of their wide applications in welding devices.[1] The arc discharge can be considered a major example of thermal plasma sources.

Between other electric discharges, the nonequilibrium, low-pressure radio frequency (RF) discharges play the key roles in sophisticated etching and deposition processes of modern microelectronics, as well as in treatment of polymer materials.[1,22] Between "nontraditional" but very practically interesting discharges, we can point out that there are the nonthermal, high-voltage, and atmospheric-pressure, DBDs.[7,8] The floating electrode DBD can also be used to the human body as a second electrode without damaging the living tissues. Such a discharge obviously provides very interesting opportunities for direct plasma applications in biology and medicine.[24] More detailed information on the discharge sources for plasma chemistry can be found in special books focused on plasma physics and engineering.[1,25]

4.5 MAJOR COMPONENTS OF CHEMICALLY ACTIVE PLASMAS

The chemically active plasmas are multicomponent systems, which are highly reactive due to large concentrations of charged particles (electrons, negative and positive ions), excited atoms and molecules (electronic and vibrational excitation make a major contribution), active atoms and radicals, and UV photons.[26] Each component of the chemically active plasma plays its own specific role in plasma-chemical kinetics. Electrons are usually first to receive the energy from an electric field and then distribute it between other plasma components and specific degrees of freedom of the system. Changing parameters of the electron gas (density, temperature, electron energy distribution function) often permit control and optimization of plasma-chemical processes. Ions are charged but are heavy particles, which make a significant contribution to plasma-chemical kinetics either due to their high energy (as in the case of sputtering and reactive ion etching) or due to their ability to suppress activation barriers of chemical reactions. This second feature of plasma ions results in the so-called ion or plasma catalysis, which is particularly essential in plasma-assisted ignition and flame stabilization, hydrogen production, fuel conversion, exhaust gas cleaning, and even in the direct plasma treatment of living tissues.[27]

The vibrational excitation of molecules also often makes a major contribution to plasma chemical kinetics because the plasma electrons with energies ~1 eV primarily transfer most of the energy in commonly used gases such as N_2, CO, CO_2, H_2, and continuously in the same way into vibrational excitation.[26] Stimulation of plasma-chemical processes through vibrational excitation permits the highest values of energy efficiency to be reached. Electronic excitation of atoms and molecules can also play a significant role, especially when the lifetime of the excited particles is quite long (metastable states). The plasma-generated metastable electronically excites oxygen molecules O_2^* (singlet oxygen) is the best example, which effectively participate in the plasma-stimulated oxidation process in polymer processing, and biological and many medical applications.[26,28]

Plasma-generated photons also play a key role in a wide range of applications, from plasma light sources to UV sterilizations. Since we are working for the UV sterilization of water using plasma-based technology at CSIR-CEERI, it would be useful to discuss the plasma chemistry of DBD-based UV excimer/exciplex sources.[29]

Excimer/exciplex sources are quasi-monochromatic light sources and can operate over a wide range of wavelengths in the UV spectral range, that is, VUV (100–200 nm), UV-C (200–280 nm), UV-B (280–315 nm), and UV-A (315–400 nm).[30] The operation of these sources is based on the formation of excited dimers (excimers) and the following transition from the bound excited excimer state to a weakly bound ground state resulting in an UV-photon radiation. Exciplex means an excited complex. So, excimers/exciplex are diatomic molecules or complexes of molecules that have stable excited electronic states and an unbound or weakly bound ground state. As an example, Xe_2^*, Kr_2^*, and Ar_2^* are excimer molecules, but $XeCl^*$, $KrCl^*$, $XeBr^*$, $ArCl^*$, and Xe_2Cl^* are exciplex molecules.[31] Because the excimer formations are unstable, they disintegrate within a few nanoseconds converting their excitation energy to optical radiation. Because of the excimer molecule nature, the difference between their stable excited state and weakly bound ground state amounts from 3.5 to 10 eV, which is the requisite energy for radiation in the UV spectral range.

Radiation is released when the excimer molecule in upper electronic exited state de-excites to its ground state. The excimer or exciplex molecules are not very stable and rapidly decompose, typically within a few nanoseconds, giving up their excitation energy in the form of a UV photon. The typical Grotrian diagram for excimer/exciplex radiation formation with certain transition possibilities is shown in Figure 4.2.

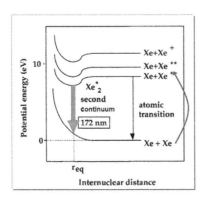
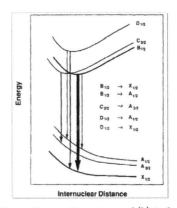

FIGURE 4.2   Typical Sketch Grotrian Diagram of (a) Xenon Excimer Formation and (b) Ordinary Xenon Iodide Exciplex Formation.

Excimer molecule emission happens when

$$Rg_2^* \rightarrow Rg + Rg + h\nu(\text{UV photon})\sqrt{2}$$

Exciplex molecule emission happens when

$$RgX^* \rightarrow Rg + X + h\nu(\text{UV photon})$$

where Rg_2^* is an excimer molecule, RgX^* is an exciplex molecule, Rg is an atom of rare gas, and X is an atom of halogen.

The main principle underlying the operation of excimer source relies on the spontaneous radiative decomposition of excimer states, so the major task is to generate excimer molecules effectively. The important role is played by the electrons in the excimer molecules formation. In order to generate efficiently excimer molecules, the active medium should contain a sufficient concentration of electrons with energies that are high enough to produce the precursors of the excimer molecules, which are mainly excited and ionized rare gas atoms.

Excitation of working gas mixture can lead to form excited and ionized rare gas atoms by the following reactions.

Excitation:

$$Rg + e^- \rightarrow Rg^* + e^- \tag{4.1}$$

Direct ionization:

$$Rg + e^- \rightarrow Rg^+ + 2e^- \tag{4.2}$$

Stepwise ionization:

$$Rg^* + e^- \rightarrow Rg^+ + 2e^-, \tag{4.3}$$

where Rg^* is an excited electronic state of rare gas atom, Rg^+ is an ionized state of rare gas atom, and e^- is an electron. Once active medium accumulates enough quantity of excited rare gas atoms, then the excimer molecules are formed by the following reaction:

$$Rg^* + Rg + M \rightarrow Rg_2^* + M, \tag{4.4}$$

where M is the third particle that can carry away the excess energy. In general, it is a rare gas atom of a working mixture. Since the formation of the excimer molecules is carried out by a three-body reaction, it is advantageous to have pressure high. Higher pressure increases the concentration of atoms and the probability of simultaneous collision of three species that is necessary for excimer formation. However, at the same time, the increasing pressure leads to intensification of excimer molecule quenching (i.e., radiationless decay). Hence, in practice, the optimal pressure of the working mixture in the laboratory is worked out by the experimental ways.

The reaction mechanisms for the formation of exciplex molecules (rare gas halides) are bit complex, in which the ground-state atomic and molecular species, ionic species, and excited atomic species take part.

The formation of exciplex molecules is realized in two main ways. The first one is a three-body ion–ion recombination reaction[32] of the positive rare gas ion and the negative halogen ion as shown in the following reaction:

$$Rg^+ + X^- + M \rightarrow RgX^* + M, \tag{4.5}$$

where M is a third partner which is, in general, the atom or molecule of the working gas mixture. The formation of the negative halogen ion occurs by dissociative attachment reaction when the electron interaction with halogen molecules occurs via the following reaction:

$$X_2 + e^- \rightarrow X + X^-, \tag{4.6}$$

where X is a halogen atom. The second way is a harpooning reaction that is a binary process.[33] In this case, the excited rare gas species transfers their loosely bound electron to the halogen molecule or even the halogen-containing compound may form an electronically excited state of exciplex molecule RgX^* through the following reaction:

$$Rg^* + X_2 \rightarrow RgX^* + X. \tag{4.7}$$

The harpooning reaction is a two-body process so it does not need so high pressure like for three-body reaction. The harpooning reaction made available the exciplex sources to operate at a low pressure of working gas mixture. As a result, in such case, the intensity of excimer molecule quenching is much less than in the excimer sources where excimer molecules are formed by three-body reaction. This has allowed for the achievement of the maximal energy conversion efficiency for the UV radiation. It should be mentioned that high pressure is useful for the domination of the ion–ion recombination reaction and the low pressure is beneficial for the domination of the harpoon reaction. The examples of excimer radiation obtained in DBD-based discharges in our laboratory are shown in Figure 4.3.

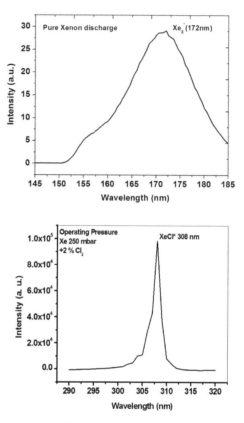

FIGURE 4.3 Radiation Spectrum of the Dielectric Barrier Discharges for (a) Excimer in Pure Xenon Discharge and (b) Exciplex in Mixture of Xenon and Chloride.

It is to be noted that the successful control of plasma can permit chemical processes to be directed in a desired direction and through an optimal mechanism. However, the control of a plasma-chemical system requires detailed understanding of elementary processes and the kinetics of the chemically active plasma. The major fundamentals of plasma physics, elementary processes in plasma, and plasma kinetics could be found in couple of books.[1,25,26] Creation of chemically active species from neutrals by collision with energetic electrons and ions and also the interaction of energetic plasma particles with surfaces play a critical role in deciding the specific type of chemistry. There are three types of chemistry that is important in plasma.

   i)   Atomic-level chemistry
   ii) Molecular-level chemistry
   iii) Surface chemistry

Each of the above plasma chemistry becomes individual subject area of research, and it is difficult to cover in a single chapter. However, an effort has been made to discuss the atomic-level chemistry and spectroscopy diagnostics in Section 4.6. The part of the atomic-level chemistry along with its applications is published elsewhere in greater details.[34]

4.6 ATOMIC CHEMISTRY AND SPECTROSCOPY DIAGNOSTICS

4.6.1 Radiation Processes in Plasmas

The plasma, being an ionized state of matter at larger temperature, emits radiation over a wide range of EM spectrum. The radiations are mainly caused by the collisions that are taking place between particles leading to ionization or recombination and also due to acceleration or deceleration of charged particles. A study of radiation processes in plasma is important because it provides information about plasma parameters. The radiation study is also important from the point of view of energy balance in plasma.

The emitted radiation may exhibit either a continuum or a line spectrum or both. Analysis of continuum intensities, line intensities, and line profile can provide considerable information about plasma parameters such as electron temperature, electron density, ion temperature, plasma motion, etc.[35-39] The radiation frequency range may lie in microwaves, optical, and X-ray regions depending on plasma parameters under consideration.

There are basically three types of radiation processes in unmagnetized plasmas, namely, bound–bound or line radiation, free–bound or continuum radiation, and free–free or Bremsstrahlung radiations. Cyclotron radiation, which occurs in magnetized plasmas, is due to magnetic centripetal acceleration of charge particles as they spiral about the magnetic field lines. Blackbody radiation emitted from plasma in thermodynamic equilibrium is important only in astrophysical plasmas in view of the large size required for plasma to radiate as a black body.

4.6.1.1 Bound–Bound or Line Radiations

The principal characteristic of bound–bound radiation is that the radiation is emitted at discrete frequencies. Whenever plasma contains atoms whose orbital electrons have not been completely stripped, line radiation will appear. Electrons changing their orbits remain bound to the nucleus, before and after the transitions, so this type of transition is called bound–bound transition. The corresponding wavelength extends from IR to far UV. The radiation frequency is characteristic of both the atom or ion and the emitting levels. Bound–bound transition is a more dominant phenomenon in low-temperature or moderate-temperature plasmas. At very high temperatures, plasma can be completely ionized, striping the atoms of all its orbital electrons. Due to their electronic excitation, atoms and ions emit a spectrum of lines such as

$$E_u - E_l = h\nu_{ul} = \frac{hc}{\lambda_{ul}},$$ (4.8)

where u and l are, respectively, upper and lower excited levels between which the transition takes place. This is a spontaneous transition. Since the energy expected to be emitted in time dt is $h\nu_{ul}A_{ul}dt$, so the power radiated at ν_{ul} is $h\nu_{ul}A_{ul}$. The power density of bound–bound radiation is proportional to $P \mu Z^6$, and thus even a very small fraction of impurities of high Z value can be detrimental for energy balance.

In plasma, spectral lines are broadened due to several factors, including collisions and pressure.[37] The Doppler and Stark broadening plays an important role in shaping the spectral lines.[37] Consequently, the study of line profiles offers a powerful tool for plasma diagnostics. In processing plasmas, the spectral study of molecules and radicals may also be required, which is far more complex than those of single atom. The electronic excitation of diatomic molecules gives band systems in the UV, visible, and IR regions. If the electronic state does not change during a transition, IR bands called the vibration–rotation bands may be observed.

4.6.1.2 Free–Bound or Continuum Radiation (Recombination Radiation)

Free electrons in plasma can recombine with ions, or in some cases, be captured by neutral molecules. The energy lost by the electron in these processes may appear as radiation. Here the originally unbound charged particle is captured by another particle, and it emits the radiation, the process is called free–bound radiation.

The excess energy of an electron with velocity radiation according to the relation ϑe is converted into radiation according to the relation

$$m_e \frac{\vartheta_e}{2} + E'_{N^+} - E_{N_j} = h\nu,$$ (4.9)

where E'_{N^+} is the ionization energy corresponding to the reaction $N \rightarrow N^+ + e$ and E_N is the energy of the excited state j to which the electron is trapped. Since free electrons have any value of velocity ϑ_e, its recombination with an ion will result in continuum radiation. A threshold value exists for the wavelength resulting from the trapping of electrons with zero velocity in the excited state E_{N_j}. Hence,

$$\Delta' E_N = h\nu_{min} \quad \text{or} \quad \lambda_{max} = \frac{hc}{\Delta' E_N}$$ (4.10)

where $\Delta'E_N = E^I_{N_I} - E_{N_j}$ because the electrons are distributed over a large spectrum of energies, their recombination into energy level j will give rise to a spectrum between v_{min} and large value of v. Electrons can be trapped into every available energy state of an atom, so the continuous spectrum coincides with the number of available energy levels.

Power density of free–bound emission to the jth level is

$$P_j^{jb} \propto Z^4 \frac{N_e N_i}{\sqrt{T}}. \tag{4.11}$$

So its value increases with number density and high Z value. Generally for processing plasmas, the continuum associated with free–bound transition will be dominant.[36]

4.6.1.3 Free–Free or Bremsstrahlung Radiation

Accelerated charges radiate E.M. energy proportional to square of their acceleration. Bremsstrahlung or breaking radiation is the energy emitted when a free charged particle makes a transition between two states of a continuum in the field of an atom or ion. In this process, a free electron makes a transition to another state of low energy with emission of photon. Since an electron is free before and after the coulomb interaction, Bremsstrahlung is refereed as "free–free" radiation. It is a continuous radiation.

At low energies, ions are massive to suffer the necessary acceleration, and Bremsstrahlung radiation arises almost entirely from electron radiation in electron–ion interaction. The wavelength (maximum) related to temperature is given by the expression

$$\lambda = \frac{hc}{kT} \approx \frac{14,000}{T_e(eV)} \text{Å} \tag{4.12}$$

For high temperature, it usually occurs in X-ray or UV wavelength range.

The radiant energy W emitted by an electron per unit time is proportional to the square of their acceleration,

$$W = \frac{e^2 a^2}{6\pi\varepsilon_0 c^3}, \tag{4.13}$$

where a is the acceleration of the particle and is given by

$$a = \frac{Ze^2}{4\pi\varepsilon_0 mr^2} \tag{4.14}$$

The power radiated per unit volume from plasma in electron–ion interaction with Maxwellian velocity distribution is given by

$$P_b \cong \frac{8\pi^2 Z^2}{3me^2c^3} \left(\frac{e^2}{4\pi\varepsilon_0} \right)^3 \frac{\sqrt{3mkT}}{h} N_e N_i \ \text{W/m}^3 \qquad (4.15)$$

where T_e is in K and N_e and N_i are electron and ion densities in m^{-3}. Usually frequency of Bremsstrahlung radiation is higher than the plasma frequency, and thus Bremsstrahlung radiation propagates as if plasma were not just there. Since P_b scales on N_e, Bremsstrahluug becomes important in high-density plasmas. Even more important is the dependence on Z^2; this shows that even a small percentage of high-Z ions in the plasma can cause a large increase in energy loss by radiation. This is the reason why in fusion plasma, a lot of attention is paid to the reduction of high-Z impurities.

4.6.1.4 Cyclotron Radiation

The radiation occurring when an electron or ion moves within a magnetic field has the fundamental cyclotron frequency of rotation in the magnetic field in which it is trapped. There usually will be higher harmonics of this frequency present but of much weaker intensity. For low electron energies, it occurs as a line at the electron Larmor frequency, whereas for higher electron energies, one finds radiation emitted at harmonics of the electron frequency in addition to the fundamental frequency. Further details could be found elsewhere.[37]

4.6.1.5 Black Body Radiation

Plasma is said to be optically thin when radiation trapping within the plasma can be neglected. Otherwise, it is optically thick. In optically thin plasma, the radiation represents a volume effect, whereas in optically thick plasma, the observed radiation is from surface. If the EM radiation generated within the plasma is absorbed and reemitted many times before reaching the boundaries of plasma, the radiation will come into equilibrium with itself and with plasma particles. In such plasma, concepts of radiations involving single or binary particle interaction can no longer be applied. Instead, plasma is considered to be a medium in thermal equilibrium. In such a case, emission and absorption radiation are balanced and Kirchoff's law applies. A body, which absorbs completely all radiations at all temperatures, is termed as a black body. This equilibrium radiation is called "the black body radiation." The total radiation from the surface of a black body is given by Planck's law and depends only on the temperature of body as,

$$I = \sigma T^4 \ \text{W}/\text{m}^2 \tag{4.16}$$

Plasma having complete thermodynamical equilibrium exists only in stars or during the short interval of a strong explosion. Plasma size should be extremely large so that it can radiate as a black body.

4.6.2 Plasma Models (Corona, LTE, and CR Model)

It is very well known that the impurity spectral lines that one observes are the spontaneous emissions of excited impurity atoms/ions and obviously their intensities are proportional to the excited-state population density N_u, which is generally determined by the local plasma parameters. The measured intensity is generally described by the simple relationship,

$$I_{ul} = \frac{1}{4\pi} \int_{x_1}^{x_2} N_u A_{ul} dx \text{ photons m}^{-2} \text{ s}^{-1} \text{ Sr}^{-1}. \tag{4.17}$$

For plasma, the absolute intensity (or upper level population density N_u) depends not only on the properties of the isolated radiating species but also on the properties of the plasma in the intermediate environment of the radiator. The population of excited states is described by a Boltzmann distribution provided that they are in thermal equilibrium. Since low-pressure plasmas are nonequilibrium plasmas, in them population density does not necessarily follow a Boltzmann distribution. Thus, according to physical situations encountered, the plasma is said to be in LTE or coronal equilibrium or CR equilibrium. Before discussing these theoretical models, let us consider all the processes ultimately responsible for change in upper level population densities.

Population kinetics model:

i) Collisional Processes

Bound–Bound transitions

Collisional de-excitation: $N_S(u) + e \rightarrow N_S(l) + eX_{ul}$ de-excitation rate coefficient

Collisional excitation: $N_S(l) + e \rightarrow N_S(u) + eX_{lu}$ excitation rate coefficient

Free–bound transitions

Collisional ionization: $N_S(u) + e \rightarrow N_i + e + eS(u)$ ionization rate coefficient

Three-body recombination: $N_i + e + e \rightarrow N_S(u) + e\beta(u)$ three-body recombination rate coefficient

ii) Radiative processes

Bound–bound transitions

Spontaneous decay: $N_S(u) \rightarrow N_S(l) + h\nu A_{ul}$ spontaneous transition probability

Photo excitation: $N_S(l) + h\nu \rightarrow N_S(u) + eA_{lu}$ spontaneous absorption probability

Free–bound transitions

Radiative recombination: $N_i + e \rightarrow N_S(u) + h\nu\alpha(u)$ radiative recombination rate coefficient

Photo ionization: $N_S(u) + h\nu \rightarrow N_i + eB(l,u) \; J$ Photo ionization rate coefficient

For nonhydrogenic plasmas, additional processes must be included

Dielectronic recombination: $N_i + e \rightarrow N_S^{**} \rightarrow N_S(u) + h\nu\alpha^D(u)$ Dielectronic recombination rate coefficient and auto-ionization $N_S(u) + h\nu \rightarrow N_S^{**} \rightarrow N_i + eB^D(l,u)$ J^D auto-ionization rate coefficient

The set of equations describing the temporal relaxation of the density $n(u)$ of level u and its CR destruction and production is given by

$$\frac{\partial n(u)}{\partial t} = P(u) - n(u)D(u) ,$$

(4.18)

where $P(u)$ is the so-called production term and $D(u)$ is the destruction factor. The full equation is expressed as

$$\frac{\partial n(u)}{\partial t} = \underbrace{\left(N_e \sum_{l \neq u} n(l)X_{lu} + \sum_{l < u} n(l)A_{lu} + N_e N_i \{\alpha(u) + N_e \beta(u)\} \right)}_{\text{Production term}} - n(u) \underbrace{\left[N_e S(u) + N_e \sum_{l \neq u} X_{ul} + \sum_{l < u} A_{ul} \right]}_{\text{Destruction term}}$$

(4.19)

The solution of this equation is not straightforward. It is also not necessary to solve this equation always. One can take certain valid assumptions and reduce the complexity. In view of this, there are three theoretical models available in plasma spectroscopy for spectral analysis purpose.

i)   LTE model
ii) Coronal model
iii) CR model, which is the more generalized model

Figure 4.4 can easily depict these models.

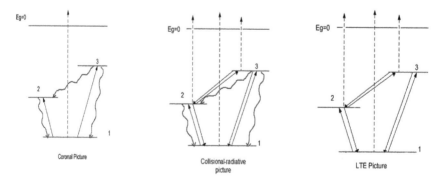

FIGURE 4.4   Comparison of Coronal, Collisional–Radiative Model, and Local Thermodynamic Equilibrium Model for Three-Level Atom (Ground-State and Two Excited States).

In coronal model, an assumption is that all upward transitions are collisional and all downward transitions are radiative. In LTE model, an assumption is that most all processes are collisional, whereas CR model is a general model, which involves solving complete coupled rate equation (4.18). Nevertheless, it is not always necessary to deal with all atomic processes and infinite energy levels simultaneously to solve equation (4.18) under CR model consideration. To qualify these assumptions, the probability for spontaneous decay of an excited atom and electron collision rate plays an important role. For example, radiative decay rate for ground-state hydrogen $A_{ul} \approx 10^8 Z^{-4}$ s^{-1} and electron collision rate are given by $v_e = 2.91 \times 10^{-6} N_e$ (cm^{-3})ln$\Lambda T_e^{-3/2}$(eV) s^{-1}, which is electron density dependent. The simplifications provided by these assumptions are described later.

4.6.2.1 LTE Model

If electron plasma density exceeds $N_e \sim 10^{15}$ cm^{-3}, the collisional rates will increase until radiative processes can be neglected. It means most excited states are depopulated by collisions mainly. In this case, the excited states are in local thermodynamic equilibrium with the ground state. Hence, for high-density plasma region, if any spontaneous decay happens, an electron comes by that time and collides and takes it immediately to the higher energy state. So in the high-density case, only collisional-dominated transitions happen and radiative process becomes negligible. Thus, the distribution of population densities of the electron is determined exclusively by particle collision processes and collisions take place rapidly so that the distribution responds instantaneously to any change in the plasma conditions. In such circumstances, each process is accomplished by its inverse process and these pairs of processes occur at equal rates by the principle of detailed balance.[36] Hence, the distribution of population densities of energy levels of electrons is same as it would be in a system in complete thermodynamical equilibrium. The population distribution is then determined by the statistical mechanical law of equilibrium among energy levels and does not require knowledge of atomic cross-sections for its calculation. Thus, one can describe the population distribution as follows.

Bound–bound case

For bound–bound case, the Boltzmann relation describes the distribution of population levels within a given species,

$$\frac{N_S(u)}{N_S} = \frac{g_S(u)}{B_S(T_e)} \exp\left(-\frac{\chi_S(u)}{kT_e}\right), \tag{4.20}$$

where

$$B_S(T_e) = g_S \exp\left[-\frac{\chi(l)}{kT_e}\right] \tag{4.21}$$

Here $N_S(u)$ is the population density of excited state u of an atom in the ground state, N_S is the ground-state atom density of a species, $g_S(u)$ is the statistical weight of the excited state u, $B_S(T_e)$ is the partition function corresponding to lower level ground-state atom, T_e is the electron temperature, and k is the Boltzmann's constant.

Free–bound case

The relative total populations of successive ionization is given by Saha equation as

$$\frac{N_e N_i}{N_S} = \frac{g_i(u)}{B_S(T_e)} 2\left(\frac{2\pi mkT_e}{h^2}\right)^{3/2} \exp\left(-\frac{\chi_S(\infty)}{kT_e}\right), \tag{4.22}$$

where N_e is the electron density, N_i is the ground-state ion density, and N_S is the ground-state atom density.

Note

i) Although the plasma temperature and density may vary in space and time, the distribution of population densities at any instant and point in space depends entirely on the local values of temperature, density, and chemical composition of the plasma.

ii) The condition of LTE is even valid if the transitions are coming from higher excited energy state u even though the electron density is low enough (below 10^{15} cm^{-3}) because the value of the atomic transition probability decreases as we go up in the excited states. Hence, even for small N_e values, the radiative processes can be easily neglected from the higher-level transitions. However, these transitions produce spectrum in the IR region of the spectrum, which is in general not under the scope of the common spectroscopic studies. Due to this fact, the assumption of LTE is taken valid when $N_e \leq 10^{15}$ cm^{-3} for most plasma studies.

4.6.2.2 Corona Model

When the electron density is low, a rather simple model, which acquires its name from its applicability to the solar corona, may be used. This was a model basically proposed to explain the spectrum of solar corona. In corona model, the assumption is that all upward transitions are collisional and all downward transitions are radiative. In fact, in low-density plasma ($N_e \leq 10^{10}$ cm^{-3}), the probability for spontaneous decay of an excited atom is much higher than for any collisional depopulation process. Hence, it can be easily assumed that

i)   all upward transitions are collisional
ii) all downward transitions are radiative.

Further, these assumptions are also valid even for $N_e \leq 10^{10}$ cm^{-3} provided that the transitions are happening very near to the ground state. Such transition produces spectrum in the extreme UV (EUV) region of the spectrum, which is again not under the scope of the many spectroscopic studies. Hence, the assumption for corona model is taken valid only for low densities for most plasma studies.

We have already seen for low-density case, the following two conditions satisfies

i)   All upward transitions are collisional
√ Collisional excitation: $N_S(l) + e \rightarrow N_S(u) + e$ X_{lu} excitation rate coefficient
× Photo excitation: $N_S(l) + h\nu \rightarrow N_S(u) + e$ A_{lu} (radiative negligible)
which is true when photon density in the system is very less. It means no photo excitation and absorption – all escapes – which is true for low-density optical thin plasma cases.
ii) All downward transitions are radiative
√ Spontaneous decay: $N_S(u) \rightarrow N_S(l) + h\nu$ A_{ul} Spontaneous transition probability
× Collisional de-excitation: $N_S(u) + e \rightarrow N_S(l) + e$ X_{ul} de-excitation rate coefficient (collisional negligible) which is valid when the electron density is very small because for low-density cases X_{ul} becomes negligible.

The net result for bound–bound and free–bound transitions is expressed below.

Bound–bound transition
Since for a detailed balance, the reverse process balances each and every process, the above assumption of corona model says that the spontaneous decay is not balanced by photo-excitation and the collisional excitation is not balanced by collisional de-excitation. But in steady state, the number of atoms being in a particular energy state is constant in time, which means for bound–bound transition in coronal steady-state case, the collisional excitation is simply balanced by spontaneous decay.

$$N_e N_S X_{lu} = N_u \mathring{a} A_{ul}$$

(4.23)

Free–bound transitions
We know that the processes of collisional ionization and three-body recombination are the inverse processes and must take place at equal rates in LTE,

$$N_S + e \Leftrightarrow N_i + e + e$$

However, the ionization rate μN_e and three-body recombination rate μN_e^2. Positive ion may also recombine with electron by radiative process.

$$N_i + e \rightarrow N_S(u) + hn$$

The rate of this process μN_e. At a sufficiently low density, it is more important than the three-body recombination, which is μN_e^2.

Furthermore, for the low-density optical thin plasma cases, the radiation density is very low, and hence, photo-ionization becomes negligible. Then the ionization equilibrium is balanced between radiative recombination and ionization by electron collisions.

$$N_e N_S S = N_e N_i \alpha \tag{4.24}$$

These equations (4.23) and (4.24) are sufficient to describe the spectrum under coronal model case.

4.6.2.3 CR Model

The CR models are 0-dimensional plasma model (and can be extended to 1–2D if impurity transports are important), which are used to calculate atomic-state populations for one or more species, as a function of electron density and temperature.

The model works under the condition that the two types of processes that cause an atom to change its excited state are collision and radiative processes.

This is a general model because for intermediate density case $(10^{10} \text{ cm}^{-3} \leq N_e \geq 10^{15} \text{ cm}^{-3})$, collisional frequency is higher, which leads to a competition between collisional processes and radiative decay processes. Thus, for general model treatment, one has to include all collisional and radiative processes and solve aforesaid equation (4.18) stepwise. How it is done is described below.

As one goes up in higher levels from the ground state, the rate of spontaneous decay/de-excitation decreases quite rapidly. So, in due course, one will end up having certain level above which even in low densities before a spontaneous decay occurs, an electron comes and collides and excite that to a higher state, which is called as LTE. Then one comes to point that the levels above this certain level are in LTE. But how one decides these levels β Here comes the time information. If spontaneous decay time (A_{ul}) for a particular level is comparable to the electron collision frequency (which depends upon the electron density), then the level above that level will not be in the so-called coronal equilibrium.

Then how one will treat these levels in realityβ This treatment is mainly described in the CR model. Most important thing to be noticed is that in the regimes where neither the LTE, nor coronal approximation is valid, the problem is essentially to be solving the complete coupled differential equation describing the population and de-population simultaneously as expressed in equation (4.19). Considerable advances have been made in the last couple of year owing, partially, to the increasing knowledge of cross-sections necessary for the rate coefficients and, partially, to the use of high-speed computers for the calculation of such coupled equations. The CR model is the outcome of this advancement.

Simplifications in CR model are as follows:

i) To avoid dealing with impossibly large summations and equally large number of differential equations, it is necessary to notice one of the most important properties of the CR model. In fact, with increasing quantum number, the level spacing becomes closer, the probability of collisional processes (within bound-bound) becomes greater, while at the same time, the probability of radiative process becomes smaller. Thus, to any desired accuracy, there is always some level above which the effect of the radiative processes may be neglected. So, simply one can assume that the bound–bound transition above this level is like free–bound transition. It is like above this level, the atom goes into next ionizing level. Under these circumstances, a modified form of Saha's equation may be used to evaluate the population densities of these upper levels u. In literature, this is called as equilibrium population densities $\{n_E(u)\}$.

$$N_E(u) = N_S^B(u) = \frac{g_S(u)}{B_S(T_e)} \exp\left(-\frac{\chi_S(u)}{kT_e}\right) N_S \quad \text{From same Boltzmann} \qquad (4.25)$$

$$N_E(u) = N_{Si}^S(u) = \frac{B(T_e)}{2B_i(T_e)}\left(\frac{h^2}{2\pi mkT_e}\right)^{3/2} \exp\left(\frac{\chi_S(\infty) - \chi_S(u)}{kT_e}\right) N_e N_i \quad \text{Modified Saha equation} \qquad (4.26)$$

Equation (4.25) is known as modified Saha equation. It applies in the restricted sense when $u = u_S$. If one normalizes the whole equation (4.19) with equation (4.26), one can compute the value of $N(u)$ easily because the simulations in the rate equation (4.19) are no longer infinitely larger and the number of these equations that need to be considered is reduced to a manageable number.

Now how one will find the value of u_S? The value of $u = u_S$ (above which modified Saha equation may be used) may be found by changing it and checking that the result of the calculation has not changed by more than the required accuracy (within $\pm 10\%$). For $p_S = g$, it will be reduced to LTE limit.

Hence, the first simplification in CR model is that to a desired accuracy there is always some level above which the effect of the radiative processes may be neglected and we can treat these levels under LTE limit. The schematic of energy levels is shown below in Figure 4.5 describing the LTE and Ccronal limits.

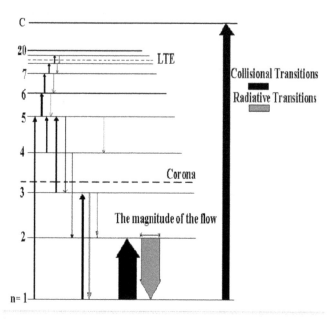

FIGURE 4.5 Schematic of Energy Levels of an Atom.

ii) On examination of rates of various processes, one can find that the relaxation time associated with the ground-level atoms or ions are some orders of magnitude longer than those for transitions among the excited states and the continuum of free electrons. For the excited levels, the relaxation times are the lifetimes for spontaneous radiative decay or shorter. Compared with the relaxation time for the ground-level population, these others may be regarded as instantaneous. It is to be noted that for most of atomic plasmas, this condition holds well. For example, for the lower levels, radiation lifetimes are typically much smaller than even with the transport times (10^{-7} s vs. 10^{-4} s), and for higher excited levels, where radiation lifetimes are larger, the more efficient collisional excitation/de-excitation by electrons usually satisfies the small lifetime requirement.

Hence, the second simplification in CR model is that the dominant populations are ground-state atoms and ions (and also metastables if any), and their sum is constant. Compared with the relaxation times for the ground-level population, these others may be regarded as instantaneous.

$$N_S + N_i = \text{Constant} = N_{S,\text{Total}} \qquad (4.27)$$

Thus, a quasi-steady-state solution of the set of equation (4.19) can be derived as

$$\frac{dN_u}{dt} = 0 \text{ except for } u \neq 1 \qquad (4.28)$$

Now the sets of equation (4.19) reduce to

1) A set of $(u_S - 1)$ simultaneous equations,

$$\sum_{u>1} C_{\text{eff}} n(u) = N_i C_{iu} + N_S C_{Su} \tag{4.29}$$

where C is the function of S, X, α, β, A, T_e, and N_e, etc.

2) Any number of Saha equation for level above u_S

Since N_i (ground-state ion density) and N_S (ground-state atom density) are independent variable, the density of population can be split in atoms and ion ground level by using the said Saha and Boltzmann equations as

$$N_i(u) = r_i'(u)N_S^S(u)$$
$$N_S(u) = r_S'(u)N_S^B(u)$$
$$N(u) = r_i'(u)N_S^S(u) + r_S'(u)N_S^B(u)$$
$$\text{or } N(u) = r_i(u)N_i N_e + r_S(u)N_S \tag{4.30}$$

where r_i' and r_S' are the so-called first-order relative populations, whereas r_i and r_S are the relative population including Saha–Boltzmann coefficient and these values can be found by solving equation (4.29) for $N_i = 0$ and $N_S = 0$, respectively.

3) And equation (4.1) for ground level where we have to retain the time derivative

These equations then give ionization and recombination of a system of ions under consideration. These equations are described in terms of the effective rate coefficients.

$$\frac{dN(1)}{dt} = \alpha_{\text{CR}} N_i N_e - S_{\text{CR}} N_S N_e \tag{4.31}$$

Here the CR ionization and recombination rate coefficients are expressed in terms of population coefficients for $u > 1$ as

$$\alpha_{\text{CR}} = \beta(1)N_e + \alpha(1) + \sum_{u>l} r_S(u)[X_{ul}N_e + A_{ul}] \tag{4.32}$$

$$\alpha_{\text{CR}} = S(1) + \sum_{u>1} X_{1u} - \frac{1}{N_e} \sum_{u>1} r_i(u)[X_{ul}N_e + A_{ul}] \tag{4.33}$$

where $u > 1$ means all excited levels.
when

$$\frac{dN(1)}{dt} = 0 \quad \text{or} \quad \alpha_{\text{CR}} N_i N_e = S_{\text{CR}} N_S N_e \tag{4.34}$$

The plasma is called the ionizing equilibrium plasma. It is important to note that the ionizing coefficient includes r_i and not r_S and vice versa for recombining coefficient and there is no direct dependence of either α_{CR} or S_{CR} on N_i and N_S. This makes it convenient to use these coefficients in plasma transport model where only ground-level atoms, electrons, and ions are considered. The ionization and recombination coefficients can then be used to calculate the transformation flow from ground-level atoms to ions and vise versa.

The solutions of these equations are time-dependent solutions that allow account to be taken of the finite time required for ionization and recombination processes. Due to this fact, this model is also quite useful even for transient plasma cases. The steady-state value can also be deduced in the different regions of models like corona and LTE for given N_e and T_e.

4.6.3 Spectroscopic Equipments

Depending on the plasma environment and chemistry, plasma spectroscopy equipments such as spectrometer/spectrograph, detector, and imaging optics can be chosen. Details of various spectroscopic systems and their components are given in details in a few standard textbooks on plasma spectroscopy.[35–39] A typical spectrometer/spectrograph consists of an entrance slit in the focal plane of a collimating spherical mirror, a grating as a dispersing element, a focusing mirror, which forms an image of the entrance slit in the focal plane, and a exit slit (see Fig. 4.6).

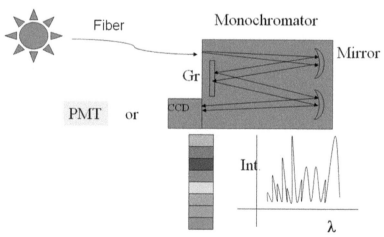

FIGURE 4.6 Schematic Diagram of Spectrometer/Spectrograph.

A detector is mounted on the exit slit. The plasma (emitting source) is either imaged by an optical arrangement on the entrance slit or coupled by fibers to the slit. The choice of grating (lines/mm) decides spectral resolution. Blaze angle of grating determines the useful wavelength range. The focal length of the spectrometer influences the spectral resolution and together with grating defines the aperture and thus the throughput of light. The width of the entrance slit is also of importance for light throughput, which means a large entrance slit gives more intensity.

At the image plane of exit slit either a photomultiplier (PMT) is placed or a charge coupled device (CCD) array is mounted. The width of exit slit or the pixel size of CCD influences the spectral resolution of the system. The overall sensitivity of the system strongly depends on the detector: PMTs with different cathode coatings or CCD arrays with different sensor types (intensified, back-illuminated, etc.).[39] A typical spectrum obtained by CCD-based monochromator is shown in Figure 4.7.

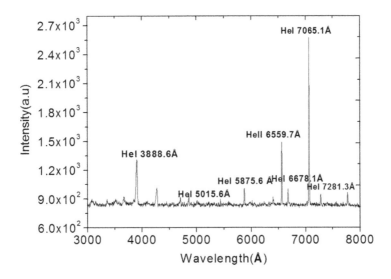

FIGURE 4.7   Typical Spectra from Helium Discharge Plasma.

Spectroscopic systems, which use PMTs, are scanning systems, whereas systems with CCD arrays are capable of recording specific wavelength range. From temporal resolution point of view, usually PMTs are faster, whereas normal CCD arrays are limited by the exposure time and readout time.

For line monitoring, which means following the temporal behavior of a particular emission line, pocket size spectrometers are suitable although they have a poor spectral resolution $\Delta\lambda \approx 1$–2 nm. A spectrometer/spectrograph with a focal length of 0.5–1 m ($\Delta\lambda \approx 40$ pm) having a grating of 1200 lines/mm and a 2-D CCD array gives a good choice for spectral and temporal resolutions. An Echelle spectrometer[35] provides an excellent spectral resolution ($\Delta\lambda \approx 1$–2 pm) by making use of the higher orders of diffraction provided by the special Echelle grating. They are an excellent tool for the measurement of line profiles and line shifts.

The calibration of spectroscopic system is also an important issue.[40] The calibration of wavelength can be done using low-pressure lamps (pencil sources, such as mercury-cadmium lamps). The calibration of intensity can be either relative or absolute calibration. A relative calibration takes into account only the spectral sensitivity of the spectroscopic system along the wavelength axis. For absolute calibration, the light sources are required for which the spectral radiance is known. In visible spectral range, from 350 to 800 nm, tungsten ribbon lamps are used, and down in the UV range 200 nm, the continuum radiation of deuterium lamps are used. In calibration procedure, one must make sure that solid angle is conserved.[40] Nevertheless, some basic principles can be applied even if only relative calibrated systems or systems without calibration are available.

4.6.4 Plasma Spectroscopy Measurements

In plasma spectroscopy measurements, EM radiation emitted from plasma is recorded, spectrally resolved, analyzed, and interpreted in terms of either parameters of the plasma or characteristic parameters of radiating atoms, ions, or molecules. The implementation of these techniques are straightforward since only an optical port in plasma chamber is needed for the radiation to emanate and the apparatus required to resolve the radiation may also be least complicated. However, since the observer always samples radiation along the line of sight, measured values are generally line of sight average values. To get spatially resolved information, special arrangements have to be made or assumption regarding axial symmetry has to be taken. The optical emission spectroscopic methods are based on measuring the intensity of spectral lines, continuum spectrum, or line profile measurements. The plasma parameters can be determined from spectral line widths.

Although spectra are generally easily obtained, their interpretation should be done carefully taking the appropriate aforementioned plasma model depending on physical conditions. It should be emphasized that most diagnostic technique require an optically thin plasma that is very near homogeneous along the line of sight and remains in steady-state for the duration of the observation. Very recently certain calculations of optical thick plasma have also been reported.[41] The accuracy of spectroscopic diagnostic techniques depends critically on the availability of accurate atomic data, especially transition probabilities and broadening parameters.

4.6.4.1 Spectral Identification and Precursor Studies

The emission spectroscopy is an easiest means to identify the particle species (atoms, ions, molecules) in the plasma, provided that particle emits radiation. Since the wavelength is a fingerprint of an element, it is sufficient to have a wavelength-calibrated spectrometer with imaging optics or a fiber. It may be quite useful in identifying the radicals, impurities, etc., and their wavelengths and intensities. As an example, a few identified species and their wavelengths are shown in Figure 4.8.

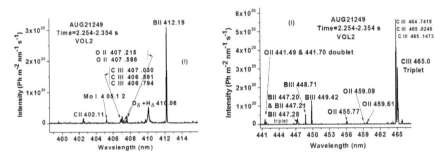

FIGURE 4.8 Identification of Species and Wavelengths in Typical Spectra from Plasma.

4.7 CONCLUSION

Plasma chemistry has become a rapidly growing area of scientific endeavor and holds great promise for practical applications for industrial and medical fields. To understand the plasma chemistry better, there are three types of chemistry, that is, atomic-level chemistry, molecular-level chemistry, and surface chemistry. In atomic-level chemistry, the spectral fingerprints of emitted radiation from different plasmas can provide information about many basic plasma parameters like electron temperature, electron density, ground-state atom density, ground-state ion density, and identification of precursor species by measuring the line intensities or profiles, etc. Spectroscopy diagnostic is nonintrusive and simple to setup. Basic configuration of spectrometer/spectrograph along with theoretical bases has been discussed. In optically thin plasmas, usually these techniques measures average values of plasma parameters along the line of sight. The accuracy of spectroscopic diagnostic technique depends critically on the equilibrium established and availability of accurate atomic data.

ACKNOWLEDGMENTS

Author would like to acknowledge Mr. Niraj Kumar, Dr. Pooja Gulati and Dr. U. N. Pal for useful discussion. The author also thanks Mr. Varun Pathania for preparing some of figure sketches.

KEYWORDS

- Plasma chemistry
- DBD excimer light sources
- plasma spectroscopy diagnostics
- corona model
- LTE-model
- CR-model
- atomic spectroscopy

REFERENCES

1. Roth, J. R. *Industrial Plasma Engineering*; IOP Publishing: London, 2003.
2. John, P. I. *Plasma Sciences and the Creation of Wealth*; Mc Grow Hill Education India: Noida, 2005.
3. Höcker, H. Plasma treatment of textile fibers. *Pure Appl. Chem.* 2002, 74, 423–427.
4. Sato, M.; Ohgiyama, T.; Clements, J. S. Formation of chemical species and their effects on microorganisms used a pulsed high-voltage discharge in water. *IEEE Trans. Ind. Appl.* 1996, 32, 106–112.
5. Bonizzoni, G.; Vassallo, E. Plasma physics and technology; industrial applications. *Vacuum.* 2002, 64, 327–336.
6. Kogelschatz, U. Dielectric-barrier discharges: their history, discharge physics, and industrial applications. *Plasma Chem. Plasma Proc.* 2003, 23, 1–46.
7. Gulati, P.; Pal, U. N.; Prakash, R.; Kumar, M.; Srivastava, V.; Vyas, V. Experimental study of single barrier DBD for the application of water treatment. *IEEE Trans. Plasma Sci.* 2012, 40, 2699–2705.
8. Pal, U. N.; Gulati, P.; Prakash, R.; Srivastava, V.; Konar,S. Analysis of power in an argon filled pulsed dielectric barrier discharge. *Plasma Sci. Tech.* 2013, 15, 635–639.
9. Liu, S.; Neiger, M. Excitation of dielectric barrier discharges by unipolar submicrosecond square pulses. *J. Phys. D: Appl. Phys.* 2001, 34, 1632–1638.
10. Kumar, R.; Prakash, R.; Alphonsa, J; Jain, J.; Pareek, A.; Rayjada, P. A.; Raole, P.M.; Mukherjee, S. Plasma nitriding of AISI 52100 ball bearing steel and effect of the heat-treatment on the nitrided layer. *J. Mat. Sci. Res.* 2012, 1, 11–18.
11. Prakash, R.; Jain, J.; Kumar, V.; Manchanda, R.; Agarwal, B.; Chowdhari, M. B.; Banerjee, S.; Vasu, P. Calibration of a VUV spectrograph by collisional-radiative modelling of a discharge plasma. *J. Phys. B At. Mol. Opt. Phys.* 2010, 43, 144012.
12. Bombaci, I.; Kuo, T. T. S.; Lombardo, U. Temperature and asymmetry dependence of nuclear incompressibility and supernova explosions. *Phys. Rep.* 1994, 242, 165–180.
13. Brown, J. C. The temperature structure of chromospheric flares heated by non-thermal electrons. *Solar Phys.* 1973, 31, 143–169.
14. Chen, F. F. *Introduction to Plasma Physics*, 2nd ed.; Plenum Press: New York, 1984.
15. Fortov, V. E.; Iakubov, I. T. *The Physics of Non-Ideal Plasma*; World Scientific: Singapore, 2000.
16. Wasa, K.; Hayakawa, S. *Handbook of Sputter Deposition Technology: Principles, Technology and Applications.* In *Materials Science and Process Technology Series*; Gary, E., Ed.; William Andrew Inc: Norwich, 1992; p 304.
17. Boulos, M. I.; Fauchais, P.; Pfender, E. *Thermal Plasmas: Fundamentals and Applications*; Springer: Berlin, 1994; p 6.
18. Goossens, M. *An Introduction to Plasma Astrophysics and Magnetohydrodynamics*; Springer: Berlin, 2003; p 25.

19. Christophorou, L. G.; Olthoff, J. K. *Fundamental Electron Interactions with Plasma Processing Gases*; Kluwer Academic Publishers: Norwell, 2004; p 39.

20. Goldston, R. J.; Rutherford, P. H. *Fully and Partially Ionized Plasmas*. In *Introduction to Plasma Physics*; Taylor and Francis Group, LLC: New York, 2000; p 164.

21. Alfvén, H.; Arrhenius, G. *Evolution of the Solar System*, Part C, Plasma and Condensation; NASA Technical Reports SP-345, 1976.

22. Raizer, Y. P.; Kisin, V. I.; Allen, J. E. *Gas Discharge Physics*; Springer: Berlin, 1991.

23. Yan, K.; Hui, H.; Cui, M.; Miao, J.; Wu, X.; Bao, C.; Li, R. Corona-induced non-thermal plasmas: fundamental study and industrial applications. *J. Electrostatics*. 1998, 44, 17–39.

24. Kolb, J. F.; Mohamed, A.-A. H.; Price, R. O.; Swanson, R. J.; Bowman, A.; Chiavarini, R. L.; Stacey, M.; Schoenbach, K. H. Cold atmospheric pressure air plasma jet for medical applications. *Appl. Phys. Lett.* 2008, 92, 241501.

25. Fridman, A.; Kennedy, L. A. *Plasma Physics and Engineering*; Science: Boca Raton, FL, 2004.

26. Fridman, A. *Plasma Chemistry*; Cambridge University Press: Cambridge, 2012.

27. Chen, H. L.; Lee, H. M.; Chen, S. H.; Chao, Y.; Chang, M. B. Review of plasma catalysis on hydrocarbon reforming for hydrogen production-interaction, integration, and prospects. *Appl. Catal. B Environ.* 2008, 85, 1–9.

28. Laroussi, M.; Kong, M. G.; Morfill, G.; Stolz, W. *Plasma Medicine: Applications of Low-Temperature Gas Plasmas in Medicine and Biology*; Cambridge University Press: Cambridge, 2012.

29. Eliasson, B.; Kogelschatz, U. UV excimer radiation from dielectric-barrier discharges. *Appl. Phys. B.* 1988, 46, 299–303.

30. Jun-Ying, Z.; Ian, B. W. UV light-induced deposition of low dielectric constant organic polymer for interlayer dielectrics. *J. Appl. Phys.* 1996, 80, 633.

31. Sosnin, E. A.; Sokolova, I. V.; Tarasenko, V. F. *Development and applications of novel UV and VUV excimer and exciplex lamps for the experiments in photochemistry*. In *Photochemistry Research Progress*; Sanchez, A., Gutierrez, S. J., Eds.; Nova: New York, 2008; pp 225–269.

32. Larsson, M.; Thomas, R. Three-body reaction dynamics in electron-ion dissociative recombination. *Phys. Chem. Chem. Phys.* 2001, 3, 4471–4480.

33. Fajardo, M. E.; Withnall, R.; Feld, J.; Okada, F.; Lawrence, W.; Wiedeman, L.; Apkarian, V. A. Condensed phase laser induced harpoon reactions. *Laser Chem.* 1988, 9, 1–26.

34. Prakash, R. *Plasma and the role of atomic physics in its spectroscopic diagnostic*. In *Atomic and Molecular Physics: Introduction to Advanced Topics*; Srivastava, R., Choubisa, R., Eds.; Narosa Publishing House: New Delhi, 2012.

35. Huddlestone, R. H.; Leonard, S. L. *Plasma Diagnostic Techniques*; Academic Press: New York, 1965.

36. Lochte- Holtgreven, W. *Plasma Diagnostics*; North Holland: Amsterdam, 1968.

37. Hutchinson, I. H. *Principles of plasma diagnostics*; Cambridge University Press: Cambridge, 1987.

38. Ovsyannikov, A. A. *Plasma Diagnostics*; Cambridge International Science Publications: Cambridge, 2000.

39. Kunze, H. J. *Introduction to Plasma Spectroscopy*; Springer: Heidelberg, Dordrecht London New York, 2009.

40. Skoog, D. A.; Holler, F. J.; Nieman, T. A. *Principles of Instrumental Analysis*; Harcout Brace & Conpany: Florida, 1998.

41. Jain, J.; Vyas, G. L.; Kumar, R.; Manchanda, R.; Prakash, R. Opacity Effect on Photon Emissivity Coefficients (PECs) of Neutral Helium Line Emissions and its Impact on Line Ratios. Presented in the 25th National Symposium on Plasma Science and Technology (Plasma-2010) held at IASST, Guwahati, 8–11 Dec, 2010.

CHAPTER 5

Diversity-Oriented Synthesis of Substituted and Fused β-Carbolines from 1-Formyl-9H-β-Carboline Scaffolds

Nisha Devi[1], Ravindra K. Rawal[2], and Virender Singh[1*]

[1]Department of Chemistry, Dr. B R Ambedkar National Institute of Technology, Jalandhar, India
[2]Department of Pharmaceutical Chemistry, Indo-Soviet Friendship College of Pharmacy, Moga, Punjab, India
*Email: singhv@nitj.ac.in/singhvirender010@gmail.com

CONTENTS

ABSTRACT

β-Carboline-based alkaloids are widespread in nature, including plants and animals, possessing a broad spectrum of pharmacological properties showing antimicrobial, antitumor, anxiolytic, anti-HIV, antimalarial, sedative, hypnotic, anticonvulsant, etc. In this context, here, we have compiled the various methodologies that have been explored for the synthesis of 1-formyl-9H-β-carboline. Furthermore, various synthetic applications of this new synthon have been compiled in this chapter, which have been investigated for the generation of various substituted and fused-β-carboline derivatives. It is further envisaged that targeted exploration of rationally designed β-carboline derivatives could afford the new anti-infective drugs.

5.1 INTRODUCTION

The β-carboline-based alkaloids are a large group of natural and synthetic indole alkaloids that possess a common tricyclic pyrido[3,4-b]indole ring skeleton. These molecules can be categorized according to the degree of saturation of their N-containing, six-membered C-ring. Unsaturated members are named as fully aromatic -carbolines, whereas the partially and completely saturated ones are known as dihydro-β-carbolines and tetrahydro-β-carbolines (THβCs), respectively. The β-carboline scaffold represents core unit of several natural compounds and pharmaceutical agents. Compounds containing this subunit are ubiquitously distributed in nature, including plants, foodstuffs, marine organisms, insects, and mammalian including human tissues and body fluids in the form of alkaloids or hormones (Figs. 5.1 and 5.2).[1] β-carboline-based compounds have been particularly known to intercalate into DNA, to inhibit CDK, topoisomerases, and monoamine oxidase, and to interact with benzodiazepine receptors and 5-hydroxy serotonin receptors. In addition to this, these compounds also demonstrate a broad spectrum of pharmacological properties, including anxiolytic, anti-HIV, antimalarial, sedative, hypnotic, anticonvulsant, antitumor, antiviral, and antiparasitic, as well as antimicrobial activities.[2–10] The importance of β-carboline derivatives can be viewed from the fact that two β-carboline-based compounds, tadalafil and abecarnil, are used clinically for the treatment of erectile dysfunction and central nervous system disorders, respectively (Fig. 5.3).[11]

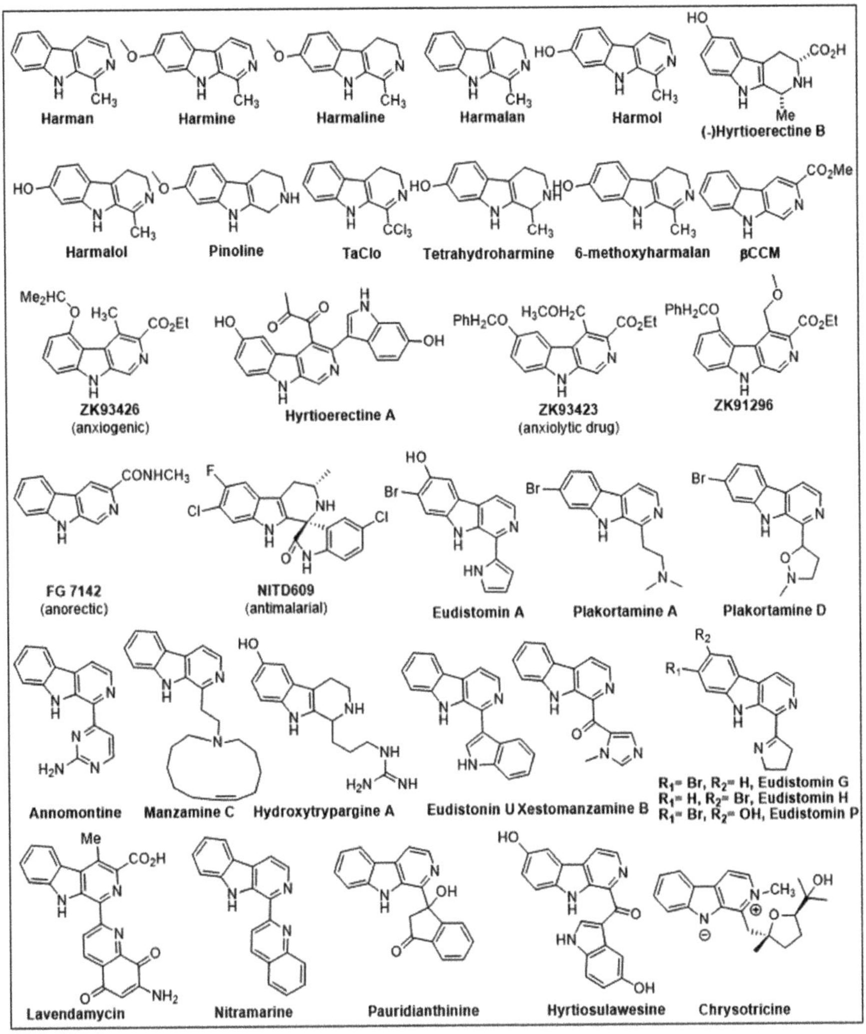

FIGURE 5.1 Few Examples of β-Carboline Substituted Bioactive Natural Products and Synthetic Derivatives.

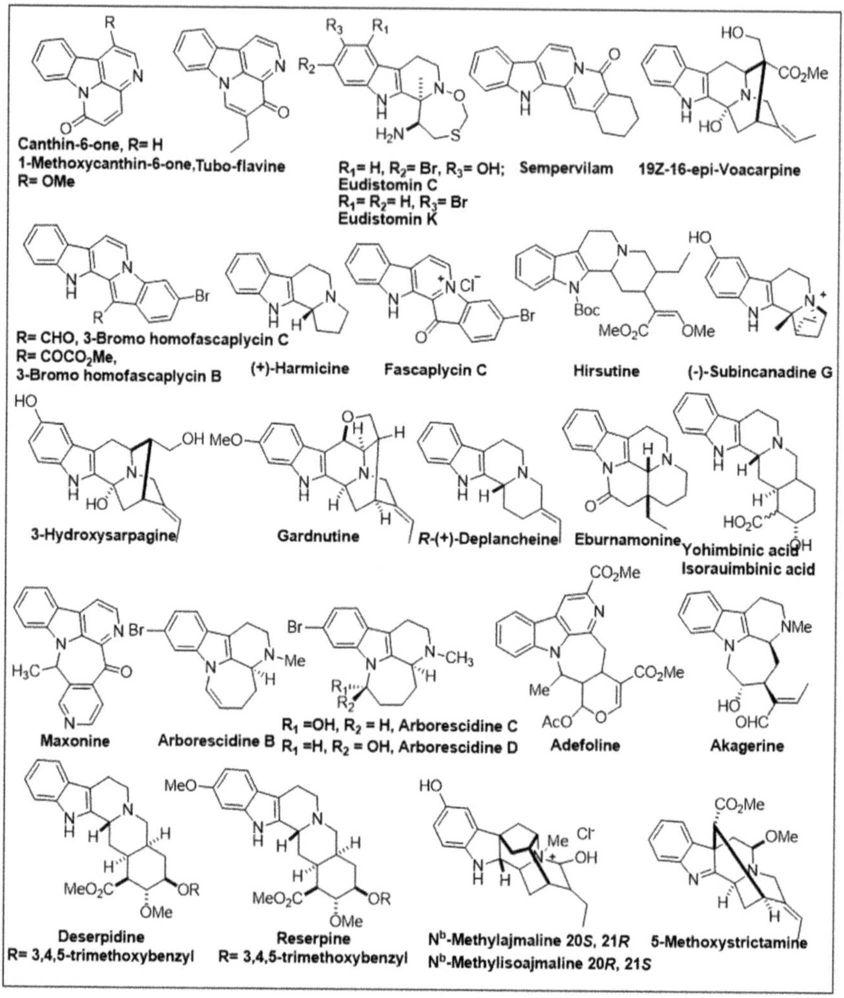

Canthin-6-one, R= H
1-Methoxycanthin-6-one,Tubo-flavine
R= OMe

R_1= H, R_2= Br, R_3= OH; Sempervilam
Eudistomin C
R_1= R_2= H, R_3= Br
Eudistomin K

19Z-16-epi-Voacarpine

R= CHO, 3-Bromo homofascaplycin C
R= COCO$_2$Me,
3-Bromo homofascaplycin B

(+)-Harmicine   Fascaplycin C   Hirsutine   (-)-Subincanadine G

3-Hydroxysarpagine   Gardnutine   R-(+)-Deplancheine   Eburnamonine   Yohimbinic acid
Isorauimbinic acid

Maxonine   Arborescidine B R_1 =OH, R_2 = H, Arborescidine C
R_1 =H, R_2 = OH, Arborescidine D   Adefoline   Akagerine

Deserpidine
R= 3,4,5-trimethoxybenzyl

Reserpine
R= 3,4,5-trimethoxybenzyl

N^b-Methylajmaline 20S, 21R
N^b-Methylisoajmaline 20R, 21S

5-Methoxystrictamine

FIGURE 5.2   Few Examples of Fused β-Carboline Natural Products and Synthetic Derivatives.

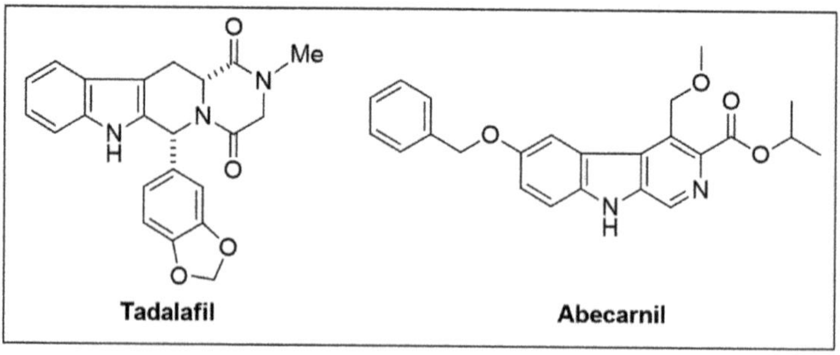

Tadalafil   Abecarnil

FIGURE 5.3   β-Carboline -Based Drugs.

The Pictet–Spengler reaction, since its discovery in 1911 by Ame Pictet and Theodor Spengler, has been the key reaction for generating substituted and fused β-carbolines.[12] Mechanistically, the reaction involves the initial condensation of indole ethylamines or tryptophan with aldehyde or other electrophile to yield the imine, which is followed by the electrophilic substitution at 2-position. After deprotonation, the desired product THβC is formed. The electrophilicity of imine double bond works as the driving force for the cyclization (Fig. 5.4). It is observed that the reaction readily occurs under mild conditions and is temperature- and pH-dependent.

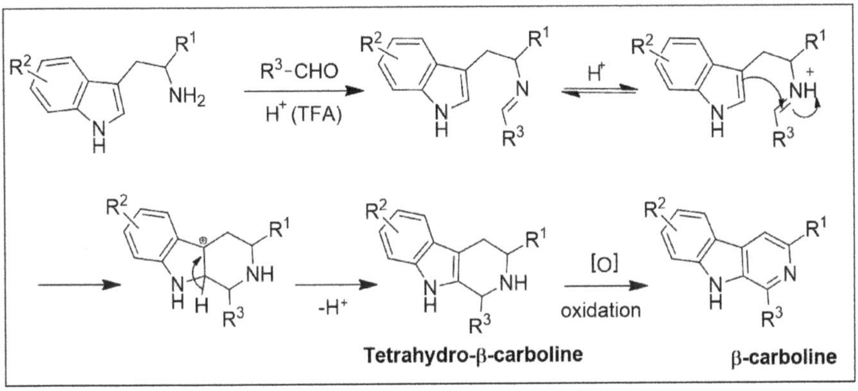

FIGURE 5.4   Pictet–Spengler Condensation between Indolylamines and Aldehyde to Give Simple THβCs Alkaloids. Further Oxidation of THβCs Provides β-Carbolines (βCs).

In general, two strategies have been explored for the synthesis of fused β-carboline compounds. In the first protocol, the THβC core is initially generated via the Pictet–Spengler reaction that is subjected to oxidation followed by further modification through functional group tailoring. The second strategy concerns with installing of different substitutions, first followed by the Pictet–Spengler reaction leading to intramolecular cyclization to afford the THβC. The THβC is then oxidized to generate the desired β-carboline derivative. However, due to immense importance associated with this pharmacophore, alternate and more efficient strategies for generating new β-carbolines are always desired by the chemical community.

In this context, one of the viable alternate strategies could be the generation of a synthon having β-carboline core and equipped with appropriate functional group at a suitable position, which could be synthetically engineered for producing substituted or fused β-carbolines. One of the substrates that has received some attention in this regard is 1-formyl-9H-β-carboline. The presence of an electrophilic site in the form of a formyl functionality in close proximity of the indole NH, which is a nucleophilic site, makes it an attractive template for the synthesis of substituted and C-1–N-9-annulated β-carbolines. Alternatively, intramolecular cyclization could also be achieved between

C-1 and N-2 to generate C-1–N-2-annulated β-carbolines.[13] Therefore, 1-formyl-9H-β-carboline could serve as a useful precursor for diversity-oriented synthesis of substituted and fused β-carbolines by exploring the existing and new methodologies.

A pictorial representation of different possibilities of synthetic diversifications in this scaffold is highlighted in Figure 5.5. Here in this book chapter, we have compiled the literature pertaining to the synthesis and synthetic applications of 1-formyl-9H-β-carboline as a new synthon for generating substituted and fused β-carboline derivatives.

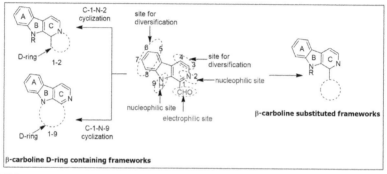

FIGURE 5.5 1-Formyl-9H-β-carboline: A Versatile Synthon for β-Carboline -Containing Systems.

5.2   NATURAL OCCURRENCE OF 1-FORMYL-9H-β-CARBOLINES

The isolation of naturally occurring 1-formyl-9H-β-carboline alkaloid, also known as Kumujian C (**I**), had been reported by different workers from the methanolic extract of root-wood of *Picrasma quassioides*.[14] Kumujancine **II** and **III** are other natural 1-formyl-β-carboline alkaloids isolated from *P. quassioides* and other sources. Subsequently, several synthetic derivatives **IV** and **V** of 1-formyl-β-carboline have also been synthesized and utilized for furnishing new β-carbolines by various researchers (Fig. 5.6).

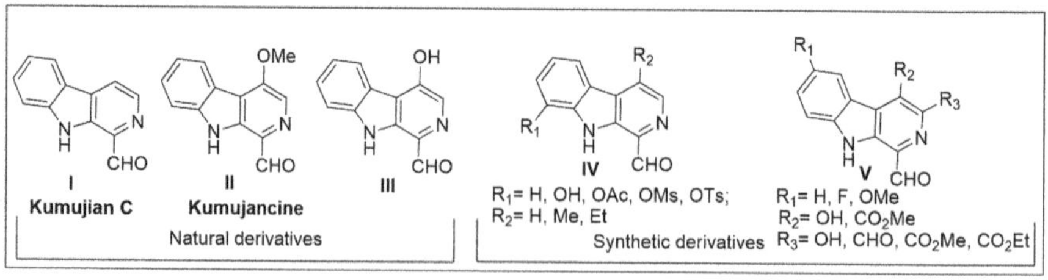

FIGURE 5.6   Synthetic and Natural Derivatives of 1-Formyl-9H-β-Carboline.

5.3 SYNTHESIS OF 1-FORMYL-9H-β-CARBOLINES AND THEIR APPLICATION FOR GENERATING 1-SUBSTITUTED β-CARBOLINE DERIVATIVES

The first synthesis of methyl 1-formyl-9H-pyrido[3,4-b]indole-3-carboxylate (2) was revealed by Gatta and Misiti[15] via SeO$_2$-mediated oxidation of various substituted THβCs. During the reaction of the diastereomeric mixture of 1-methyl,1-benzyl THβC (1) with SeO$_2$ in dioxane, they unexpectedly obtained the methyl 1-formyl-9H-pyrido[3,4-b]indole-3-carboxylate (2) instead of the desired 1-methyl,1-benzyl-3-(methoxycarbonyl)-1,4-dihydro-4-oxo-β-carboline as depicted in Scheme 5.1. It was proposed that the reaction proceeded via oxidation of the benzylic moiety to eliminate the benzaldehyde followed by aromatization of the C-ring, and finally, the oxidation of the C-1-methyl to the formyl group.

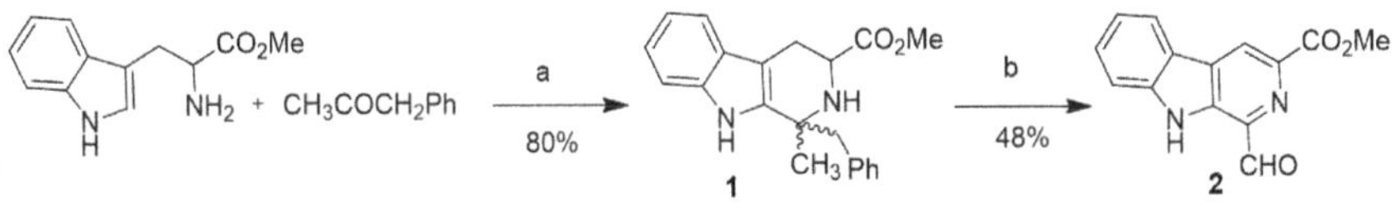

SCHEME 5.1 Synthesis of 1-Formyl-9H-pyrido[3,4-b]indole-3-Carboxylate (2) via SeO$_2$-Mediated Oxidation. Reagents and Conditions: (a) PhMe, Reflux, Dean–Stark, 4 h; (b) SeO$_2$, Dioxane, Reflux, 2 h.

Later, an improved synthesis of methyl 1-formyl-9H-pyrido[3,4-b]indole-3-carboxylate (2) was reported by this group from 1-methyl-3-methoxycarbonyl-β-carboline (3) via oxidation with SeO$_2$ in dioxane, which also served as evidence for their earlier work (Scheme 5.2).[16]

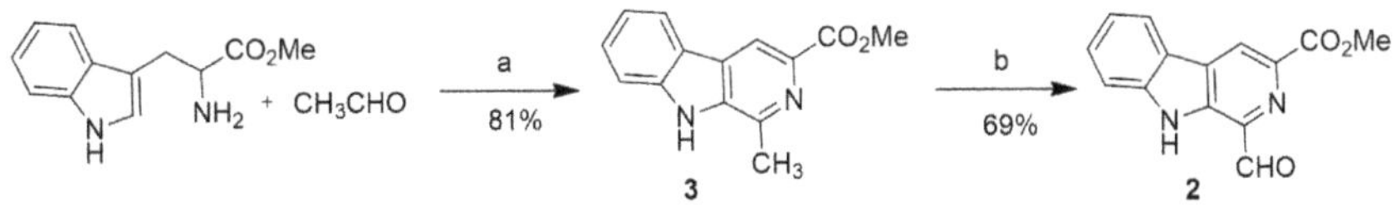

SCHEME 5.2 Synthesis of 1-Formyl-9H-pyrido[3,4-b]Indole-3-Carboxylate (2) from 1-Methyl-3-Methoxycarbonyl-β-Carboline (3) via Oxidation with SeO$_2$. Reagents and Conditions: (a) PhMe, Reflux, Dean–Stark, 20 h; (b) SeO$_2$, Dioxane, reflux, 2 h.

Bracher and Hildebrand[17] studied the reactions of dimetalated β-carboline (4) with different electrophiles and discovered that the reaction with dimethylformamide

(DMF) as an electrophile afforded the 1-formyl-9H-β-carboline (**5**) in 51% yield as depicted in Scheme 5.3.

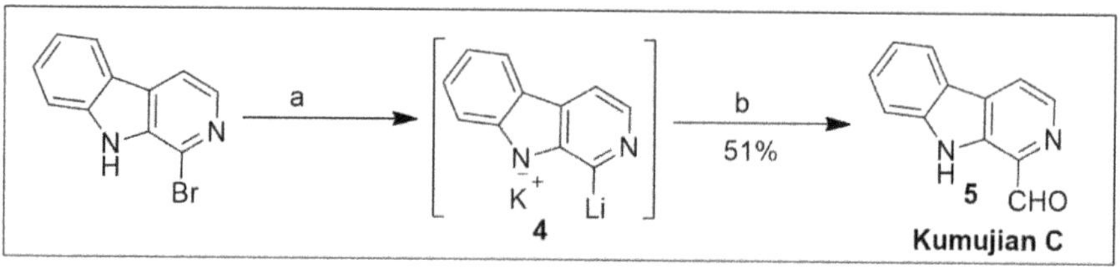

SCHEME 5.3   Synthesis of Kumujian C (**5**). Reagents and Conditions: (a) i) KH, THF, ii) *ter*-BuLi, −78°C, 20 min; (b) DMF, THF, −78°C−rt, 4 h.

Suzuki et al.[18] reported the total synthesis of various naturally occurring 4,8-di-oxygenated β-carboline alkaloids **11** and **12**. The initial synthetic route till β-carboline comprised two key steps of which the first was an improved Fischer-indole synthesis, whereas the second was C-3 selective cyclization of the C-2 substituted indole. Then, 4-methoxy-β-carboline (**6**) was transformed into corresponding N-oxide that afforded the 1-nitrile derivative (**7**) by treatment with diethylphosphoryl cyanide (DEPC) in a modified Reissert–Henze reaction. Compound **7** yielded the 1-formyl derivative (**8**) through acid-mediated oxidation followed by reduction. The Wittig reaction of **8** with methylene triphenylphosphorane afforded the 1-substituted β-carboline **9**, which upon catalytic reduction furnished **10**. Finally, the phenolic O-tosyl group of the β-carboline was removed by Na-naphthalenide and Na-anthracenide to generate picrasidine J (**11**) and picrasidine I (**12**), respectively (Scheme 5.4). The de-

Application of Sharpless oxidation allowed Hibino et al. to accomplish the first enantioselective total synthesis of (+)-oxopropaline D (**23**) and its enantiomer in 13.4% overall yield from **15**.[19] A sequence of nine steps using the common key compound, N-MOM-1-methoxycarbonyl-4-methyl-β-carboline (**15**) that was prepared by thermal electrocyclic reaction of a 1-azahexatriene system (**13**) involving the indole 2,3-bond (Scheme 5.5), led to the formation of (+)-oxopropaline D with 93% ee. The de-protection of MOM in **15** resulted in β-carboline (**16**), which upon reduction gave the aldehyde **17**. Grignard reaction with vinyl magnesium bromide followed by silylation with tert-butyldimethylsilyl (TBDMS)-Cl produced the allyl silyl ether (**18**). Oxidation of **18** with AD-mix-β led to 1,2-diol **19**, which was transformed to acetonide **20**. Removal of the TBDMS group in **20** followed by oxidation of the alcohol **21** furnished the ketone **22**. Finally, a careful deprotection of keto-acetonides **22** with diluted sulfonic acid gave the (+)-oxopropaline D.

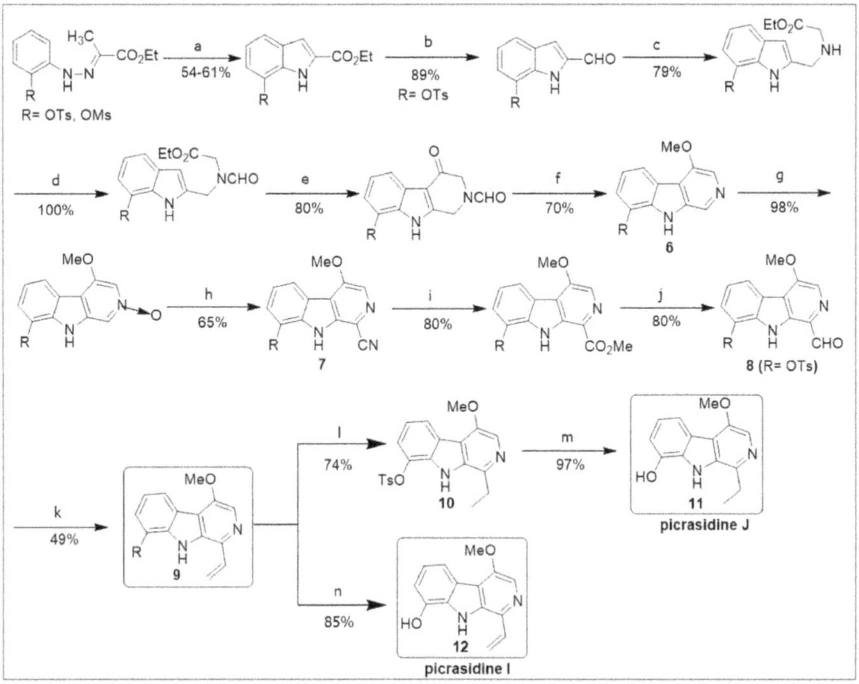

SCHEME 5.4 Synthesis of Picrasidine J and Picrasidine I. Reagents and Conditions: (a) PPA, Heat; (b) i) diisobutylaluminium hydride (DIBAL-H), DCM, −78°C, 5 min, ii) MnO$_2$, DCM, rt, 3 h; (c) NH$_2$CH$_2$COOEt·HCl, NaBH$_3$CN, MeOH, rt, 24 h; (d) HCOOEt, HCOOH, rt, 24 h; (e) MsOH, 55°C, 5 h; (f) (MeO)$_2$CMe$_2$, p-TSA, Chloranil, C$_6$H$_6$, rt, 12 h; (g) m-CPBA, DCM, rt, 24 h; (h) DEPC, Et$_3$N, DCE, 80°C, 2 h; (i) HCl, MeOH, Reflux, 5 h; (j) DIBAL-H, DCM, −50°C; (k) CH$_2$=PPh$_3$, n-BuLi, 0°C, THF, 70°C, 2 h; (l) H$_2$-Pd/C, MeOH, rt; (m) Na-naphthalenide, THF, 0°C, 2 h; (n) Na-anthracenide, THF, 0°C, 1 h.

Takasu et al.[20] reported the synthesis of several β-carboline-based compounds, including the natural products, Kumujancine, 4-methoxy vinyl β-carboline (MVC), creatine, and their corresponding salts. The synthetic strategy involved the Pictet–Spengler reaction of tryptamine with ethyl glyoxalate in ethanol, followed by acylation with acetyl chloride to furnish THβC **24** in 44% yields. Treatment of THβC **24** with 2,3-dichloro-5,6-dicyanobenzoquinone (DDQ) led to an unstable 4-oxocarboline derivative (**25**), which upon reaction with dimethoxypropane in the presence of para toluenesulfonic acid (p-TSA) under azeotropic conditions, followed by oxidative aromatization, produced 1-ethoxycarbonyl-4-methoxy-β-carboline (**26**) along with demethoxy compound (X= H; Kumujian A). Subsequent reduction of ester **26** with diisobutylaluminum hydride (DIBAL-H) produced the aldehyde **5** and Kumujancine (**27**). Finally, Wittig olefination of **27** afforded the first total synthesis of MVC (**28**). The β-carbolinium cations (**29**) were also prepared from the corresponding β-carbolines by simple quaternerization with either alkyl tosylates or alkylhalides (Scheme 5.6). These products were evaluated for in vitro antimalarial activity against *Plasmodium faciparum* and also for cytotoxicity studies. It was observed that quaternary carbolinium salts showed much higher potencies than the corresponding neutral β-carbolines. MVC exhibited EC_{50} 5×10^{-6} M against *P. faciparum* in in vitro assay.

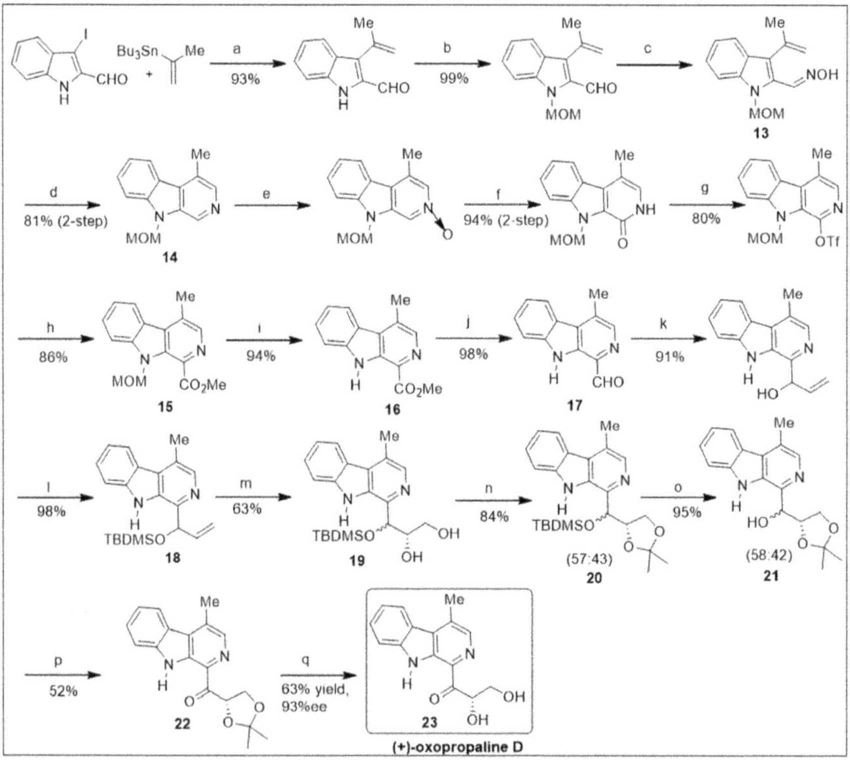

SCHEME 5.5 Synthesis of (+)-oxopropaline D (**23**). Reagents and Conditions: (a) Et$_4$NCl, DMF, PdCl$_2$(PPh$_3$)$_2$, 80 °C, 4 h; (b) NaH, MOMCl, DMF, rt, 16 h; (c) NH$_2$OH·HCl, AcONa, EtOH, Reflux, 15 min; (d) 1,2-(Cl$_2$)C$_6$H$_4$, Reflux, 3 h; (e) *m*-CPBA, CH$_2$Cl$_2$, rt, 5 h; (f) Ac$_2$O, 110 °C, 3 h; (g) Tf$_2$O, Pyridine, CH$_2$Cl$_2$, 0 °C, 3 h; (h) MeOH·CO, Et$_3$N, then Pd(OAc)$_2$, dppf, DMF, 80 °C, 3 h; (i) CF$_3$SO$_3$H, MeOH, CH(OMe)$_3$, MeNO$_2$, 100 °C, 1 h; (j) DIBAL-H, CH$_2$Cl$_2$, −78 °C, 1 h; (k) Vinylmagnesium Bromide, THF, 0 °C, 30 min; (l) TBDMSCl, Imidazole, DMF, 60 °C, 1 h; (m) AD-mix-β, *t*-BuOH, H$_2$O, 0 °C, 1 h; (n) *p*-TSA, Me$_2$CO, Reflux, 24 h; (o) TBAF, THF, rt, 15 min; (p) MnO$_2$, CH$_2$Cl$_2$, rt, 8 h; (q) dil. H$_2$SO$_4$, MeOH, rt, 1 h.

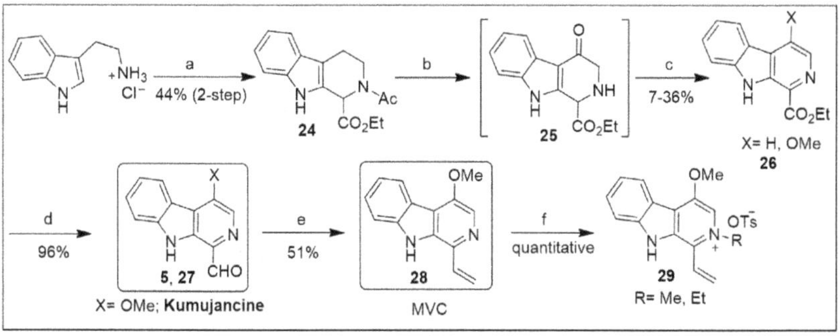

SCHEME 5.6 Synthesis of Kumujancine (**27**), MVC and β-Carbolinium Cations (**29**). Reagents and Conditions: (a) i) CHOCOOEt, EtOH, ii) AcCl, DMAP, DCM; (b) DDQ, THF-H₂O, −78°C–rt; (c) Me₂C(OMe)₂, p-TSA, C₆H₆, p-chloranil, rt; (d) DIBAL-H, DCM, −78°C; (e) Ph₃P=CH₂, THF; (f) ROTs.

A two-step synthesis of MVC (**28**) was later disclosed by Takasu et al.[21] via re-action of Kumujancine (**27**) with MeLi followed by dehydration step as outlined in Scheme 5.7. In a subtle modification to this work, they reported that the oxidation of THβC under the influence of Pd-C and subsequent reduction with DIBAL-H gener-ated 1-formyl-9H-β-carboline (Kumujian C; **5**), which through a similar set of reac-tions afforded 1-vinyl β-carboline (**30**) (Scheme 5.8). Interestingly, all the synthesized compounds and their corresponding salts exhibited good in vitro and in vivo antima-larial activity.

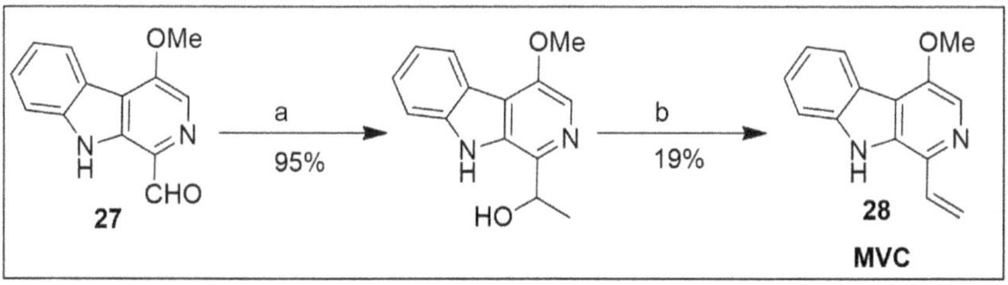

SCHEME 5.7 Two-Step Synthesis of MVC (**28**) from Kumujancine (**27**). Reagents and Conditions: (a) MeLi, DCM, −78°C, 10 min; (b) MsCl, Et₃N, DCM, rt, 8 h.

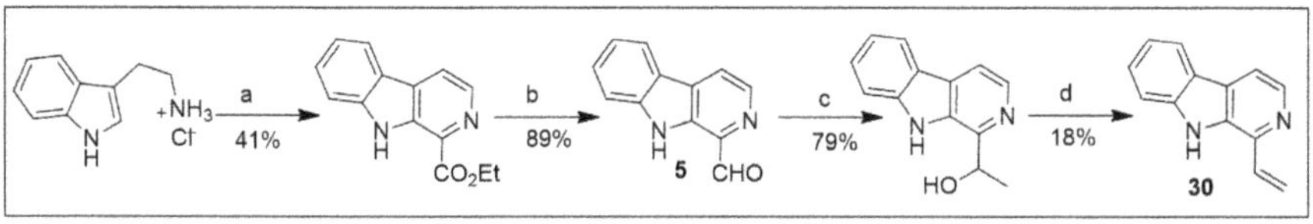

SCHEME 5.8 Synthesis of 1-Vinyl β-Carboline (**30**) from Kumujian C. Reagents and Conditions: (a) i) CHOCO₂Et, EtOH, 0°C–rt, 16 h, ii) Pd-C, xylene, 140°C; (b) DIBAL-H, DCM, −50°C, 10 min, rt, 1 h; (c) MeLi, DCM, −78°C, 10 min; (d) MsCl, Et₃N, DCM, rt, 8 h.

N-acetyl tryptophan (**31**), obtained by the Schotten–Baumann method from commercially available *dl*-tryptophan, was cyclized by a modified Bischler–Napieralski condensation to afford 1-methyl-3,4-dihydro-β-carboline-3-carboxylic acid (**32**) by Bhutani et al.,[22] which was subsequently transformed to methyl ester derivative (**33**) by reacting with thionyl chloride in methanol (Scheme 5.9). The methyl ester was aromatized by employing sodium hydride in dry DMF to yield **3**. Compound **3** underwent oxidation upon treatment with SeO_2 to yield 1-formyl derivative (**2**). The synthetic derivatives were evaluated for anti-HIV activity in human CD4+ T cell line (a cell line for monitoring HIV-1 and HIV-2 infections (CEM-GFP)) infected with HIV-1 $NL_{4.3}$ virus. Interestingly, 1-formyl-β-carboline-3-carbxylic acid methyl ester (**2**) showed inhibition of HIV at $IC_{50} = 2.9$ μM.

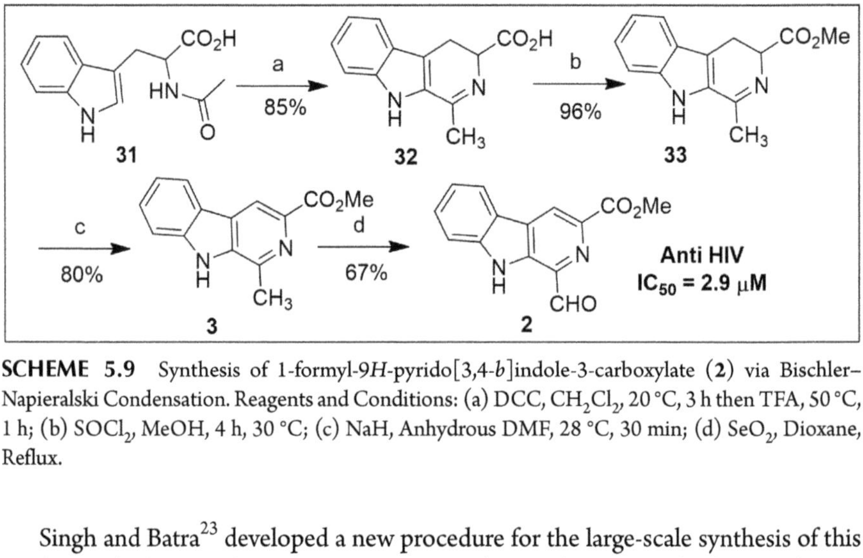

SCHEME 5.9 Synthesis of 1-formyl-9*H*-pyrido[3,4-*b*]indole-3-carboxylate (**2**) via Bischler–Napieralski Condensation. Reagents and Conditions: (a) DCC, CH_2Cl_2, 20 °C, 3 h then TFA, 50 °C, 1 h; (b) $SOCl_2$, MeOH, 4 h, 30 °C; (c) NaH, Anhydrous DMF, 28 °C, 30 min; (d) SeO_2, Dioxane, Reflux.

Singh and Batra[23] developed a new procedure for the large-scale synthesis of this synthon, which was achieved via Pictet–Spengler condensation of substituted tryptophan methyl ester or tryptamine with dimethoxy glyoxal to afford THβC derivatives (**34**) (Scheme 5.10). Subsequent oxidation of C-ring (**34**) with $KMnO_4$ and further demasking of the formyl group (**35**) with $AcOH/H_2O$ generated the desired 1-formyl-9*H*-β-carbolines (**2**, **5**, and **36**) in good yields. The most significant advantage of this three-step synthesis was that at no stage purification was required and the reaction could be scaled-up efficiently up to 10 g.

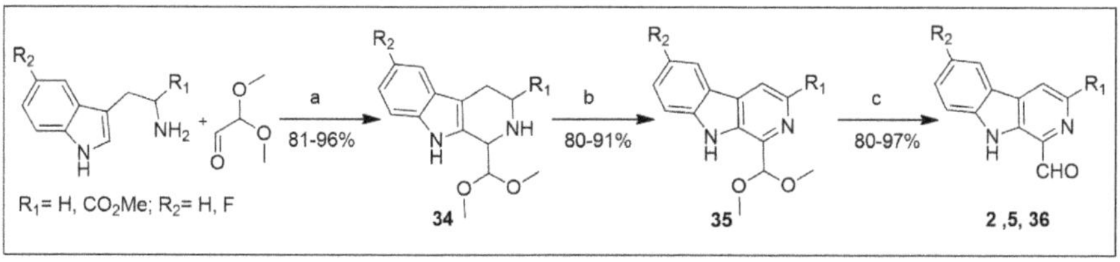

SCHEME 5.10 Three Step Synthesis of 1-Formyl-9H-β-Carbolines (**2, 9, 36**). Reagents and Conditions: (a) 2–5% TFA in DCM, 4 h– 3 Days; (b) $KMnO_4$, dry THF, rt, 6 h– 5 Days; (c) $AcOH:H_2O$ (2:3), 100C, 45 min.

Similar strategy was utilized for the synthesis of another substituted 1-formyl-9H-β-carboline **40** from *rac-threo*-β-methyltryptophan esters (**37**) (Scheme 5.11) via Pictet–Spengler cyclization with dimethoxy acetaledehyde, which led to the formation of diastereomeric mixtures **38**.[24] The diastereomeric mixtures were subjected to $KMnO_4$-mediated oxidation to yield the α-dimethoxymethyl-β-carbolines (**39**), which upon deprotection under acidic conditions afforded 1-formyl-β-carbolines (**40**) in excellent yields.

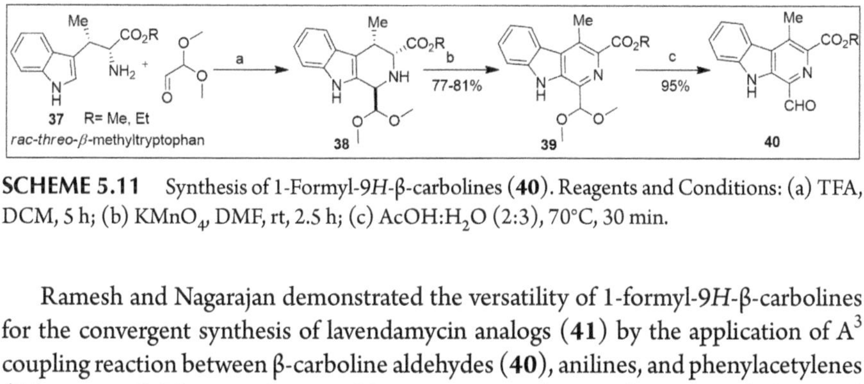

SCHEME 5.11 Synthesis of 1-Formyl-9H-β-carbolines (**40**). Reagents and Conditions: (a) TFA, DCM, 5 h; (b) $KMnO_4$, DMF, rt, 2.5 h; (c) $AcOH:H_2O$ (2:3), 70°C, 30 min.

Ramesh and Nagarajan demonstrated the versatility of 1-formyl-9H-β-carbolines for the convergent synthesis of lavendamycin analogs (**41**) by the application of A^3 coupling reaction between β-carboline aldehydes (**40**), anilines, and phenylacetylenes (Scheme 5.12). The main feature of this strategy was the use of ionic liquids that acted both as catalyst and as an environmentally benign solvent medium. They investigated various ionic liquids like [Bmim][Br], [Bmim][Tfa], [Emim][Tfa], [Bmim][Tsa], [Bmim][PF$_6$], and [Bmim][BF$_4$] for this transformation and revealed that [Bmim][BF$_4$] was superior to other ionic liquids though it provided the products in low yield. Further studies concluded that addition of Lewis acids $(La(OTf)_3,$ 10 mol%) along with ionic liquid could enhance the reaction yield to a significant extent.[25] As per the proposed mechanism, the aldehydes (**40**) and anilines undergo condensation to yield aldimines that are followed by Lewis acid catalyzed coupling reaction with phenylacetylenes and subsequent aromatization lead to afford lavendamycin analogs (**41**).

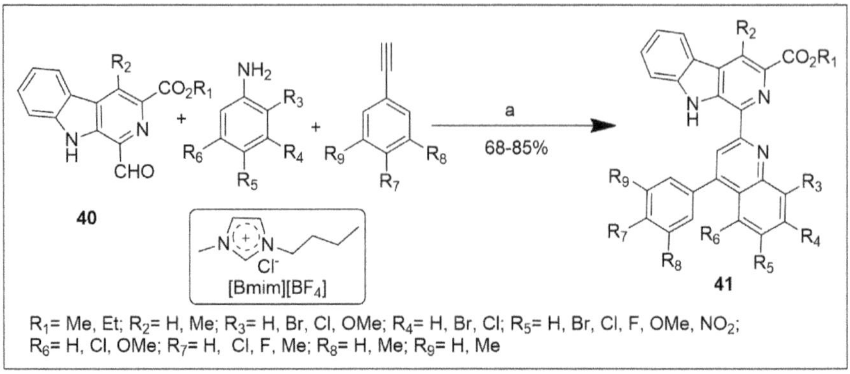

SCHEME 5.12 Synthesis of Lavendamycin Analogs (**41**) by the Application of A^3 Coupling between β-Carboline Aldehydes (**40**), Anilines, and Phenylacetylenes. Reagents and Conditions: (a) La(OTf)$_3$ (10 mol%), [Bmim][BF4], 95–100 °C, 4 h.

In contrast, the reaction of 1-formyl-β-carboline (**40**), aniline, and ethylpropiolate resulted in the formation of α-dihydropyrido-β-carboline (**42**) under the similar reaction conditions as illustrated in Scheme 5.13. The observed different reactivity between phenylacetylene and ethylpropiolate was explained in the proposed mechanism (Fig. 5.7). First, aniline undergoes Michael reaction with ethylpropiolate instead of forming imine with aldehyde **40**, which is followed by consecutive second Michael reaction and finally reaction with aldehyde result in β-carboline substituted dihydropyrido compound **42**.

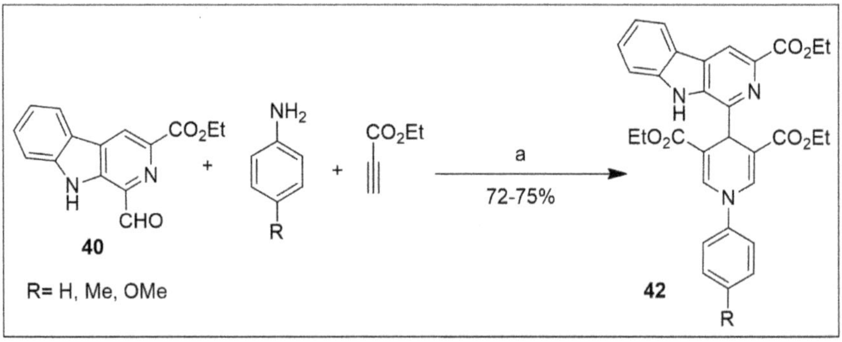

SCHEME 5.13 Synthesis of α-Dihydropyrido-β-Carboline (**42**). Reagents and Conditions: (a) La(OTf)$_3$ (10 mol%), [Bmim][BF4], 95–100 °C, 18 h.

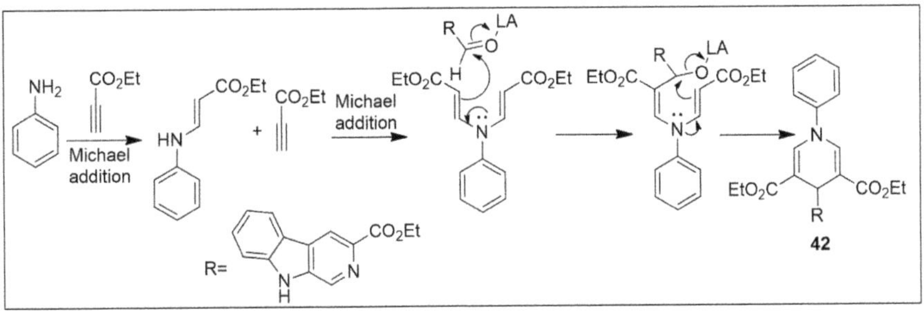

FIGURE 5.7 Mechanism for the Formation of α-Dihydropyrido-β-carboline (42).

The synthesis of β-carboline substituted quinoline derivatives could be achieved via imino Diels–Alder strategy as presented in Scheme 5.14. Several catalysts $(Cu(OTf)_2, Ag(OTf), La(OTf)_3, Sc(OTf)_3,$ and $Yb(OTf)_3)$ were able to trigger the formation of desired products but molecular iodine in CH_3CN was found to be best catalyst for this conversion.[24] Furthermore, it was also observed that the formation of water during Schiff base formation was detrimental to the Povarov reaction. As anticipated, the reaction between isolated aldimine (43) and n-butylvinyl ether under same condition afforded the products in better yield (83–94%).

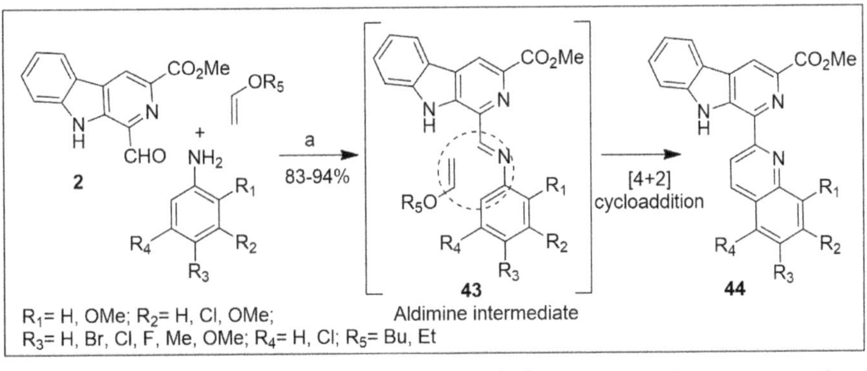

R_1= H, OMe; R_2= H, Cl, OMe;
R_3= H, Br, Cl, F, Me, OMe; R_4= H, Cl; R_5= Bu, Et

SCHEME 5.14 Synthesis of Lavendamycin analogs (44) from 2 through Imine Intermediate. Reagents and Conditions: (a) $MgSO_4$, DCM, reflux, 1 h; then I_2 (10 mol%), THF, reflux, 7–10 h.

These workers also demonstrated the versatility of this approach toward the total synthesis of nitramarine (46). The synthesis was accomplished by de-esterification of 44 using LiOH in $MeOH/H_2O$ (3:1) at room temperature to furnish the acid 45, which was subsequently decarboxylated using the standard conditions to furnish nitramarine (46) in moderate yield (Scheme 5.15).

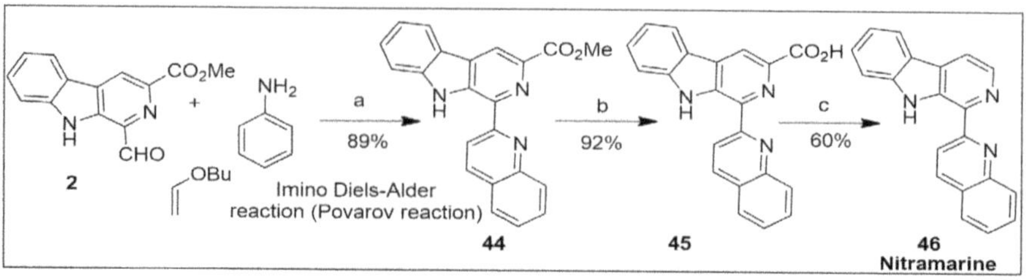

SCHEME 5.15 Synthesis of Nitramine (**46**) from **2**. Reagents and Conditions: (a) I$_2$ (10 mol%), THF, 8 h; (b) LiOH:H$_2$O, MeOH/H$_2$O (3:1), rt, 4 h; (c) CuI/quinoline, 260°C, 1 h.

Ramesh and Nagarajan explored this synthon for the formal synthesis of lavendamycin methyl ester (**48**) by using iodine as an inexpensive catalyst. Povarov reaction of **40**, aniline, and vinyl butyl ether afforded the same intermediate (**47**) that was earlier utilized by Rao et al. for the total synthesis of lavendamycin methyl ester in five steps (Scheme 5.16).

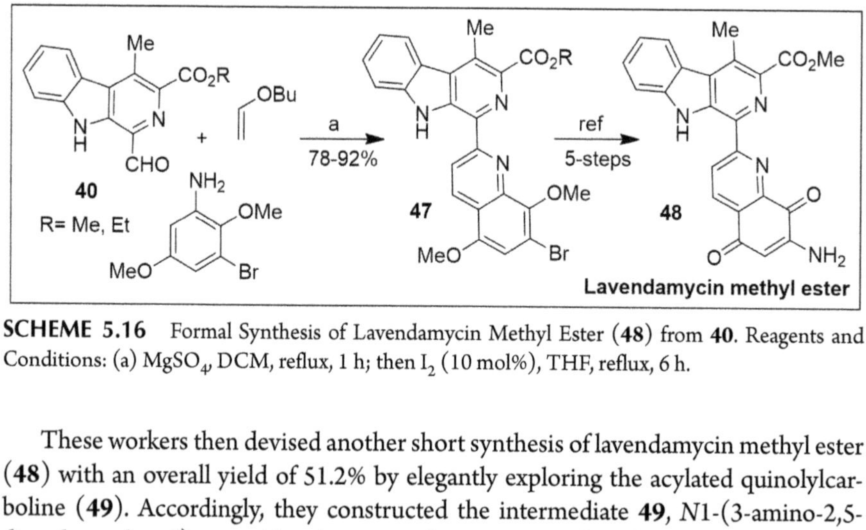

SCHEME 5.16 Formal Synthesis of Lavendamycin Methyl Ester (**48**) from **40**. Reagents and Conditions: (a) MgSO$_4$, DCM, reflux, 1 h; then I$_2$ (10 mol%), THF, reflux, 6 h.

These workers then devised another short synthesis of lavendamycin methyl ester (**48**) with an overall yield of 51.2% by elegantly exploring the acylated quinolylcarboline (**49**). Accordingly, they constructed the intermediate **49**, N1-(3-amino-2,5-dimethoxyphenyl)acetamide, via inverse electron demand Diels–Alder reaction of aldehydes (**40**), aniline, and vinyl ether to afford the expected β-carbolines (**49**) in good yields, and thereby, completing the formal synthesis of lavendamycin methyl ester, as the intermediate **49** could easily be converted into lavendamycin methyl ester in a two-step sequence via intermediary of **50** with excellent yields (Scheme 5.17).

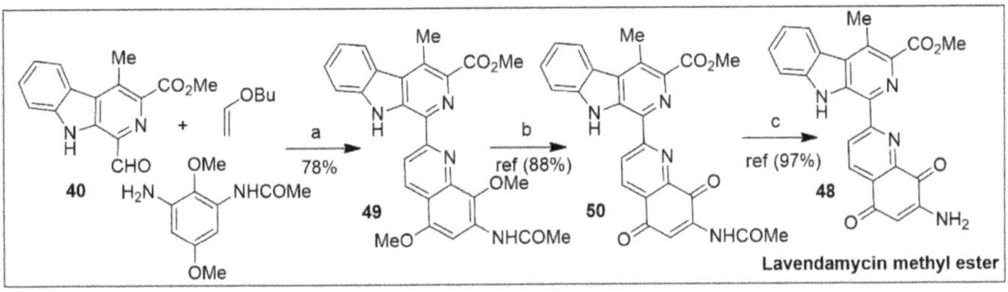

SCHEME 5.17 Three-Step Formal Synthesis of Lavendamycin Methyl Ester (**48**) from **40**. Reagents and Conditions: (a) MgSO$_4$, DCM, Reflux, 1 h; then I$_2$ (10 mol%), THF, Reflux, 10 h; (b) DIB, MeCN/H$_2$O; (c) H$_2$SO$_4$/H$_2$O, 60 °C, 30 min.

5.4 SYNTHETIC STRATEGIES FOR FUSED β-CARBOLINE DERIVATIVES VIA C-1–N-2 CYCLIZATION FROM 1-FORMYL-β-CARBOLINES

Batra et al. disclosed the potential of *N*-alkylated 1-formyl-9*H*-β-carboline (**52**) to accomplish the annulation between C-1 and N-2 to produce indolizinoindole derivatives (Harmicine mimics) as depicted in Schemes 5.18 and 5.19.[26] The synthesis of *N*-alkylated 1-formyl-β-carboline derivatives was achieved via treating the acetal (**35**) with different alkyl halides in the presence of a suitable base to afford (**51**), which under acidic conditions furnished the *N*-alkylated 1-formyl-β-carboline derivatives (**52**). The *N*-alkylated derivative (**52**) upon MBH reaction with various acrylates and cycloalkenones in the presence of DABCO under solvent-free conditions or with 4-dimethyl aminopyridine (DMAP) in aqueous medium afforded the MBH adducts (**53** and **56**). Treating these products with phosphorous tribromide (PBr$_3$) followed by aqueous work-up afforded the indolizinoindole derivatives (**55** and **57**). The reaction was found to proceed through the formation of allyl bromide **54**, which was confirmed by mass spectrometry (HRMS) of crude product. On the other hand, the 1-formyl—carboline originating from tryptamine did not require any activation with PBr$_3$, and cyclized derivatives **58** (R^1=R^2=H) were isolated as the exclusive product during the MBH reaction.

The formation of **55** indicated that the reaction of adduct with PBr$_3$ resulted in the formation of allyl bromide, **54**, which could have undergone a nucleophilic attack by the nitrogen (N-2) of C-ring to form a salt that hydrates and rearranges in the presence of water to afford the isolated product. A plausible mechanism for the formation of **55** has been presented in Figure 5.8.

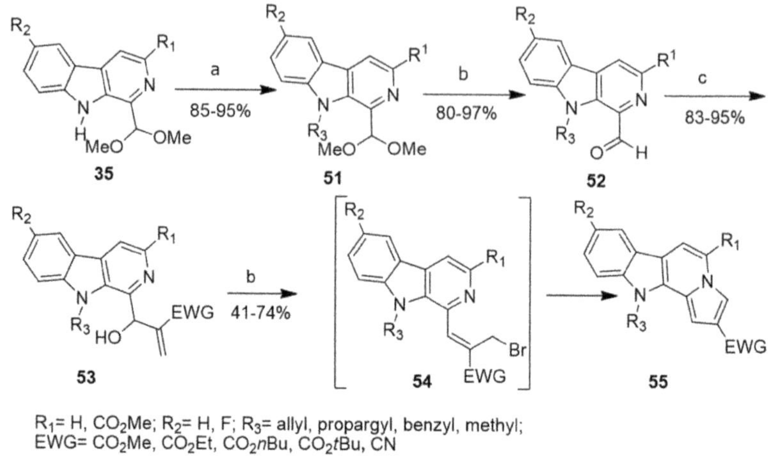

R₁= H, CO₂Me; R₂= H, F; R₃= allyl, propargyl, benzyl, methyl;
EWG= CO₂Me, CO₂Et, CO₂nBu, CO₂tBu, CN

SCHEME 5.18 Synthesis of Indolizinoindole Derivatives (**55**) from MBH Adducts. Reagents and Conditions: (a) R_3-Br (Allyl Bromide, Benzyl Bromide, Propargyl Bromide), Cs_2CO_3, DMF, rt, 1–2 h; or MeI, NaH, DMF, 0 °C–rt, 1 h; (b) AcOH:H_2O (2:3), 100 °C, 45 min –1 h; (c) DABCO, rt, 3 h– 15 days; (d) PBr_3, DCM, 0°C, 30 min–2 h.

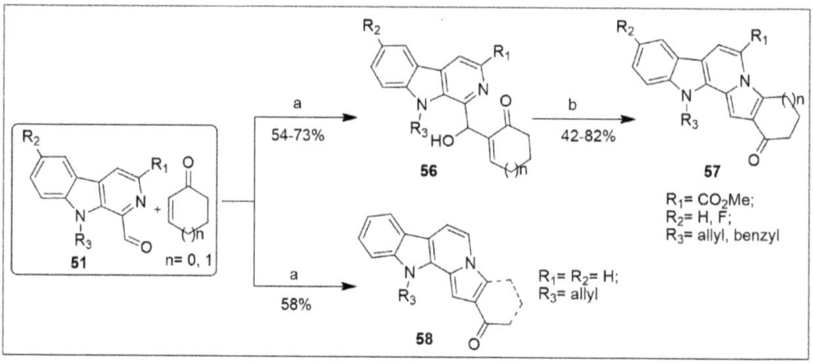

SCHEME 5.19 Synthesis of Indolizinoindole Derivatives (**57–58**) from 1-Formyl -β-Carbolines via MBH Reaction. Reagents and Conditions: (a) DMAP, THF:H_2O (1:1), rt, 2– 4 days; (b) PBr_3, DCM, 0°C, 30 min–1 h.

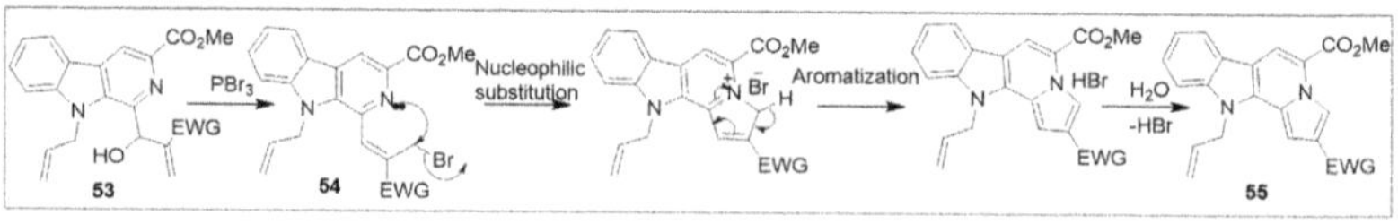

FIGURE 5.8 Plausible Mechanism for the Formation of Indolizinoindole via Allyl Bromide.

The versatility of 1-formyl-β-carbolines was further demonstrated by Batra et al. when they developed a library of 3-aminoindolizino[8,7-b]indoles derivatives (**59**) via Cu-mediated multicomponent route involving coupling/cycloisomerization as illustrated in Scheme 5.20.[27] It was also established that the protocol was amenable to microwave conditions, thereby reducing the reaction time significantly.

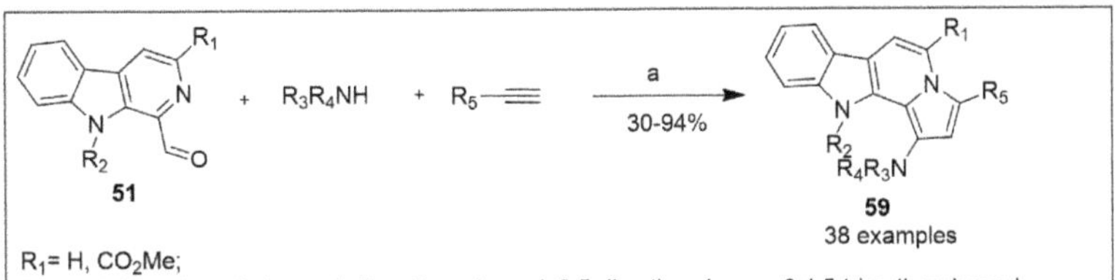

R_1= H, CO$_2$Me;
R_2= allyl, benzyl, methyl,prenyl, 3-methoxy benzyl, 3,5-dimethoxybenzy, 3,4,5-trimethoxybenzyl, 3,4-methylenedioxybenzyl, 4-nitrobenzyl, methoxymethyl;
R_3R_4N= N, N-diethyl amine, N, N-diphenylamine, morpholine, 3-methylpiperidine, thiomorpholine, piperidine, thiomorpholine, pyrrolidine
R_5= phenylacetylen, 4-methylphenyl acetylene, 4-ter-butylphenylacetylene, 3,5-difluorophenylacetylene, hexyne-1, octyne-1, decyne-1, trimethylesilylacetylene

SCHEME 5.20 Synthesis of 3-Aminoindolizino[8,7-b]Indoles Derivatives (**59**). Reagents and Conditions: (a) CuI, PhMe, 85–90°C, 7 h or CuI, PhMe, 90°C, MW, 45 min.

Mechanistically, it was proposed that initially secondary amines reacted with the aldehyde (**51**) leading to the formation of iminium ion with the loss of a water molecule (Fig. 5.9). Subsequently, Cu-coordinated alkyne formed in situ reacted with imine, wherein a nucleophilic attack of pyridyl nitrogen (N-2) on the Cu-coordinated allenyl double bond occurs, resulting in the formation of cationic intermediate (**60**). Thereafter, the secondary amine captures a proton from **60** to furnish the intermediate, **61**, which upon deprotonation yielded the 3-aminoindolizino[8,7-b]indoles derivatives (**59**).

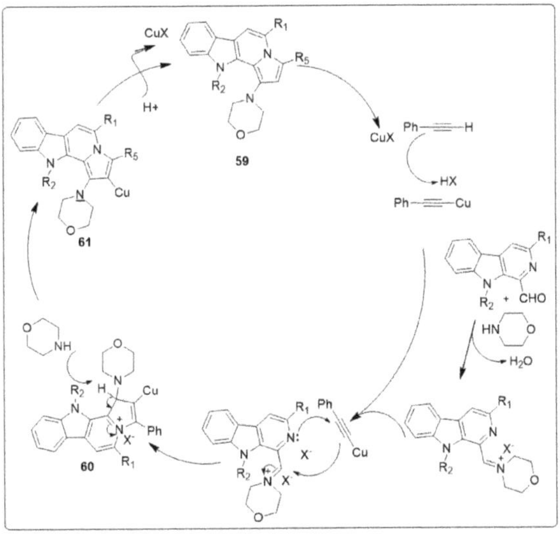

FIGURE 5.9 Plausible Mechanism for the Formation of the Aminoindolizino[8,7-b]indole (**59**).

5.5 SYNTHETIC STRATEGIES FOR FUSED β-CARBOLINE DERIVATIVES VIA C-1–N-9 CYCLIZATION FROM 1-FORMYL-β-CARBOLINES

Gatta et al. reported for the first time the application of 1-formyl-9*H*-β-carboline (**2**) for the synthesis of canthin-6-one and its unknown 5-methyl derivative (**62**) via the reaction with acetic or propionic anhydride in the presence of pyridine under refluxing conditions as illustrated in Scheme 5.21.[16] In a variation to this strategy, they extended the synthetic utility of **2** for the generation of pyrimido[3,4,5-*lm*]pyrido[3,4-*b*]indole derivatives (**63**). The condensation of **2** with various primary amines in methanol followed by reduction of the resulting Schiff's bases with $NaBH_4$ led to the formation of 1-aminomethyl-3-methoxycarbonyl-β-carbolines in good to excellent yields. Finally, cyclization of amine with formaldehyde in the presence of excess of diisopropylethylamine (DIEA) in methanol under heating at reflux yielded pyrimido[3,4,5-*lm*] pyrido[3,4-*b*]indoles (Scheme 5.22).

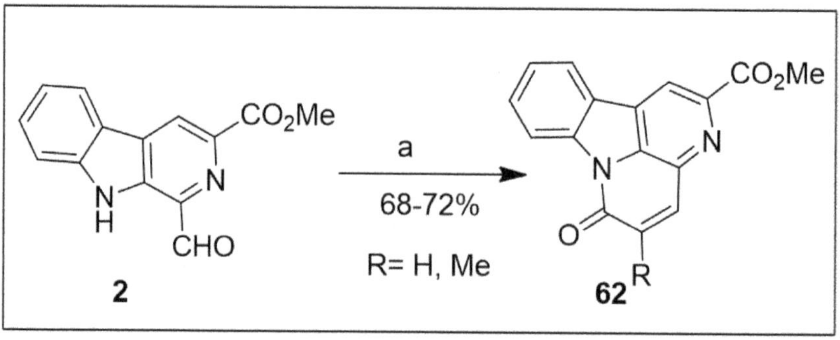
CO2Me
N
N
H
CHO
2
a
68-72%
R= H, Me
CO2Me
N
N
O
R
62

SCHEME 5.21 Synthesis of Canthin-6-one and its Unknown 5-Methyl Derivative (**60**). Reagents and Conditions: (a) $(RCO)_2O$, Pyridine, reflux, 3 h.

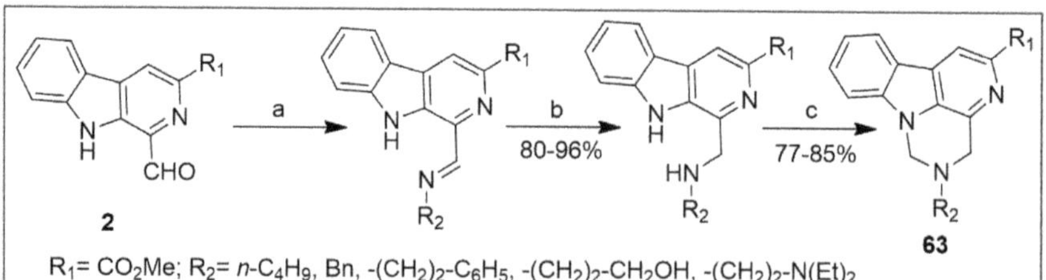
R1
N
N
H
CHO
2
a
R2
b
80-96%
HN
c
77-85%
63
R1= CO2Me; R2= n-C4H9, Bn, -(CH2)2-C6H5, -(CH2)2-CH2OH, -(CH2)2-N(Et)2

SCHEME 5.22 Synthesis of Pyrimido[3,4,5-*lm*]pyrido[3,4-*b*]Indole Derivatives (**63**). Reagents and Conditions: (a) R_2NH_2, MeOH, 0°C, 1 h; (b) $NaBH_4$, MeOH, rt, 2 h; (c) HCHO, DIEA, MeOH, reflux, 1 h.

Takasu et al. explored the synthetic application of **5** for the generation of canthin-6-one (**64**), which was synthesized in a single operation from **5** via acetylation with acetyl chloride, followed by intramolecular aldol condensation in the presence of NaH as depicted in Scheme 5.23.[21]

Condie and Bergman[28] reported the condensation of **2** with ethyl azidoacetate leading to a nonisolable intermediate **65**, which immediately underwent intramolecular cyclization to afford 5-azidocanthin-6-one **66**. Catalytic reduction of **66** resulted in 5-aminocanthin-6-one **67**. Further treatment of **67** with Ac$_2$O under refluxing furnished the corresponding N-acetyl derivative **68** (Scheme 5.24). Similar strategy was also applied to β-carboline-1,3-dicarbaldehyde (**71**) obtained from dichloromethine derivative **69** via sequential hydrolysis, reduction, and oxidation of the intermediate **70** (Scheme 5.25). Compound **71** undergoes selective condensation with dimethyl acetylenedicarboxylate (DMAD) at C-1 formyl group and concomitant cyclization with N-9 furnished a highly functionalized 3-formyl-canthine derivative (**72**).

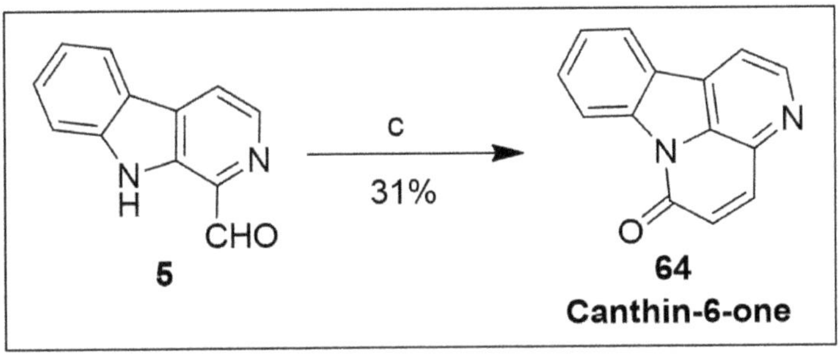

SCHEME 5.23 Synthesis of Canthin-6-one (**64**) from **5**. Reagents and Conditions: (a) AcCl, NaH, THF, 0°C–rt, 5 h.

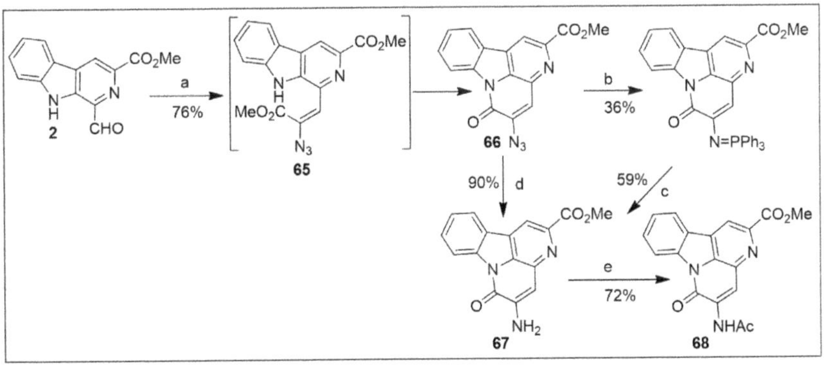

SCHEME 5.24 Synthesis of 5-Aminocanthin-6-One (**68**). Reagents and Conditions: (a) N$_3$CH$_2$CO$_2$Et, NaOEt, MeOH, 0°C, 24 h; (b) PPh$_3$, DCM, rt, 2 h; (c) Chromatographed Silica Gel, MeOH:DCM; (d) H$_2$-Pd/C, EtOAc, 17 h; (e) Ac$_2$O, Reflux, 30 min.

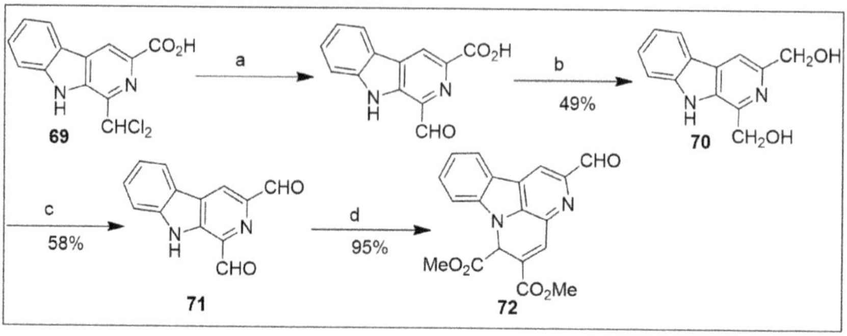

SCHEME 5.25 Synthesis of 3-Formyl-Canthine Derivative (**72**). Reagents and Conditions: (a) HCl, H$_2$O, EtOH, heat, 1 h; (b) LiAlH$_4$, THF, rt, 6 h; (c) MnO$_2$, THF, Reflux, 3.5 h; (d) DMAD, PPh$_3$, DCM, rt, 3.5 h.

In another modified strategy, these workers reported the synthesis of methyl 1-formyl-9H-β-carboline-3-carboxylate derivatives (**2** and **74**), by a four-step procedure from readily available L-tryptophan derivatives.[29] The reaction proceeded through an oxazolone intermediate (**73**) that on treatment with trifluoroacetic acid (TFA) was transformed to β-carboline skeleton **69**. Compound **69** upon esterification with diazomethane followed by hydrolysis of dichloromethine functionality with aq. HCO$_2$H afforded methyl 1-formyl-β-carboline-3-carboxylates (**2** and **74**) as demonstrated in Scheme 5.26. The reactions of **2** and **74** with DMAD and PPh$_3$ led to the formation of canthine derivatives (**75**), while the Wittig reaction with phosphorane (Ph$_3$P=CHCO$_2$Et) furnished a mixture of two products, the 2-(methoxycarbonyl)canthin-6-one derivative (**60**) and uncyclized β-carboline propenoate **76** (Scheme 5.27).

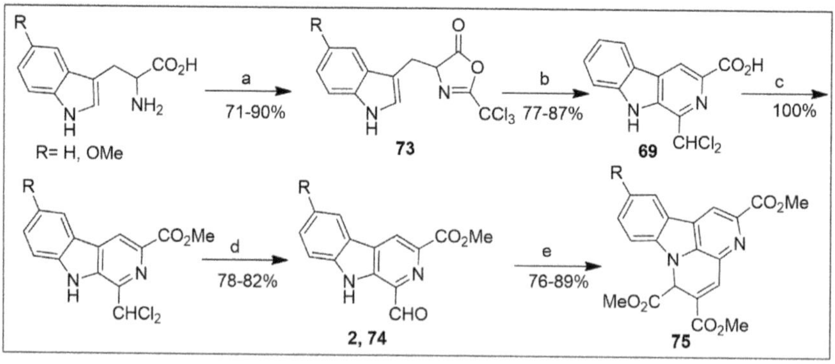

SCHEME 5.26 Synthesis of Canthine Derivatives (**75**). Reagents and Conditions: (a) (Cl$_3$CCO)$_2$O, Et$_2$O, 0°C; (b) TFA, −15°C, rt, 1–24 h; (c) CH$_2$N$_2$, Et$_2$O, 2 days; (d) aq. HCOOH, 3 h; (e) DMAD, PPh$_3$, DCM, rt, 3.5 h.

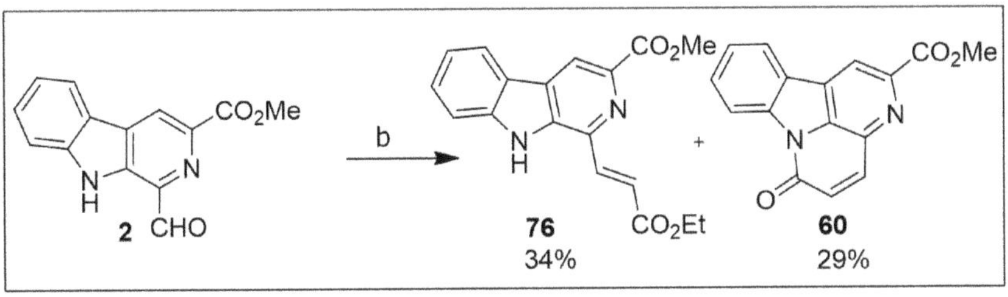

SCHEME 5.27 Synthesis of 2-(methoxycarbonyl)Canthin-6-One Derivative (**60**) and Uncyclized β-Carboline Propenoate **76**. Reagents and Conditions: (a) CH$_2$=CHCH$_2$Br, NaH, DMSO, 2 h; (b) Ph$_3$P=CHCO$_2$Et, PhMe, Reflux, 2 h.

Suzuki et al.[30] reported the synthesis of canthin-6-one derivative from 1-formyl-9H-β-carbolines and its 4-methoxy derivative. The starting aldehydes (**5**, **27**) were prepared in 70–85% yield via a careful reduction of the ester **77** with DIBAL-H. Reaction of **5** and **27** with ethyl acetate (EtOAc) under the influence of t-BuOK in dimethylsulfoxide (DMSO) yielded the 1-methoxy canthin-6-one (**78**) in 39% yield with minor yields of *trans* olefinic compound (**79**). They further investigated the reaction of **5** and ethyl acetate in the presence of lithium hexamethyldisilazide (LiHMDS), which upon quenching with EtOH produced the canthin-6-ones (**64** and **78**) in excellent yields (Scheme 5.28). On the contrary, quenching with NH$_4$Cl afforded the aldol product (**79**) that was transformed to canthin-6-one (**64**) by the use of EtONa (Scheme 5.29).

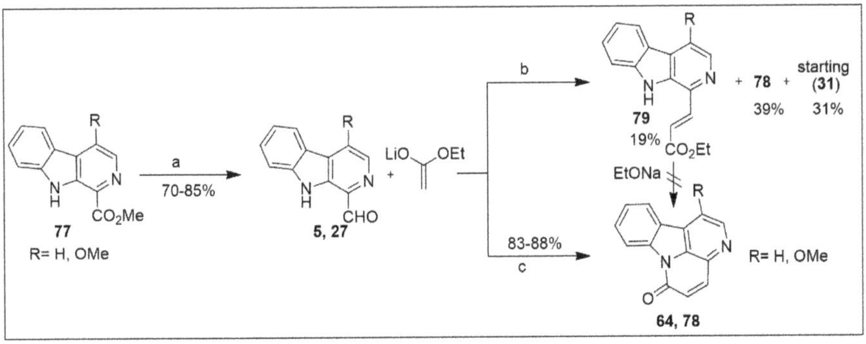

SCHEME 5.28 Synthesis of Canthin-6-Ones (**64** and **78**). Reagents and Conditions: (a) DIBAL-H, DCM, −40°C, 5 min; (b) EtOAc, t-BuOK, DMSO, 100°C, 45 min; (c) LiHMDS, THF, −78°C, then EtOH, rt.

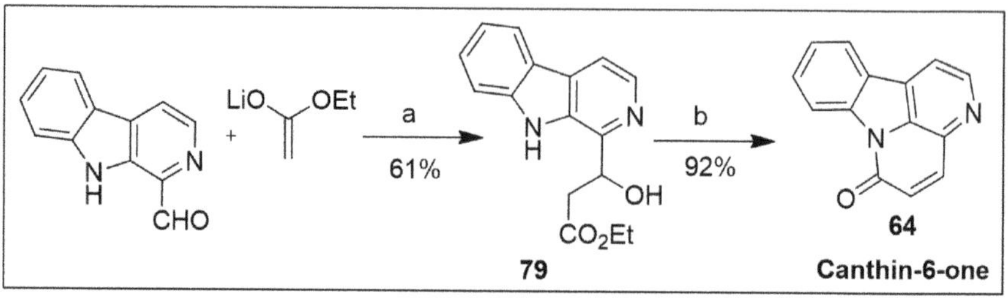

SCHEME 5.29 Synthesis of Cantin-6-One (**64**). Reagents and Conditions: (a) i) LiHMDS, THF, −78°C–rt, 30 min, ii) EtOAc, −78°C, 15 min, aq. NH$_4$Cl; (b) EtONa, EtOH, 0°C, 10 min.

Batra et al. performed the Morita–Baylis–Hillman (MBH) reaction of 1-formyl-9H-β-carbolines (**2** and **5**) with various activated acrylates and observed the formation of the expected MBH adducts (**80**) along with unnatural canthin-6-one derivatives (**81**) in minor yields as shown in Scheme 5.30.[23] However, it was discovered that exclusive formation of either product **80** or **81** could be controlled by modulating the amount of diazabicyclo[2.2.2]octane (DABCO) and the reaction time. It was observed that MBH reaction of **2** and **5** with *tert*-butyl acrylate yielded only the respective adducts **80**. The effect of the amount of DABCO and reaction time on the MBH reaction and isolated yields is presented in Table 5.1.

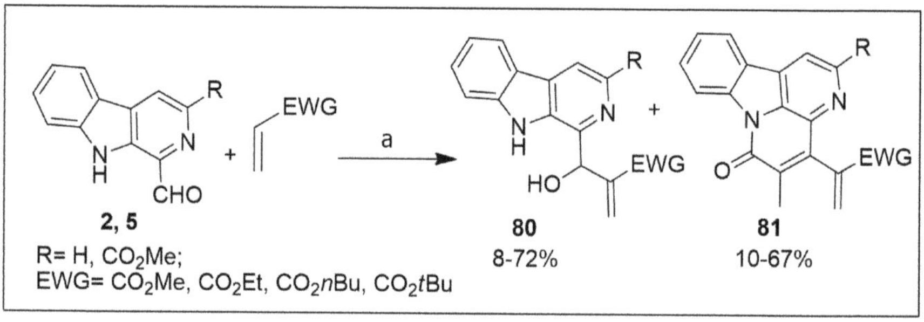

SCHEME 5.30 Synthesis of Unnatural Canthin-6-One Derivatives (**81**) via MBH Reaction. Reagents and Conditions: (a) DABCO, neat, rt, 8–120 h.

Based on the observations, a plausible mechanism for the formation of canthin-6-one skeleton was proposed, which is presented in Figure 5.10. Formation of an amide bond ahead of Michael reaction reflects that maybe the elimination of the DABCO during the first MBH reaction takes place after the formation of the amide bond. After the elimination of DABCO, the resulting species **82** undergoes 1,3-hydrogen shift and tautomerizes to generate a diketone type of intermediate **83**, which further undergoes second MBH reaction to yield the canthin-6-one derivative **81**.

TABLE 5.1 Effect of the Amount of DABCO and Reaction Time on the MBH Reaction and Isolated Yields of the Corresponding Products

Aldehyde	EWG	DABCO [equiv.]	Time [h]	MBH Adduct 80 (%yield)	Canthin-6-one 81 (% yield)
5	CO_2Me	1.5	8	55	10
5	CO_2Et	1.5	13	54	13
5	CO_2Me	5	20	–	59
5	CO_2Et	5	60	–	60
5	CO_2nBu	5	48	38	30
5	CO_2nBu	5	72	–	67
5	CO_2tBu	1.5	80	72	–
2	CO_2Me	1.5	27	60	10
2	CO_2Et	5	72	55	13
2	CO_2Me	5	120	8	57
2	CO_2Et	5	120	55	13
2	CO_2nBu	5	120	55	13
2	CO_2tBu	5	120	55	–

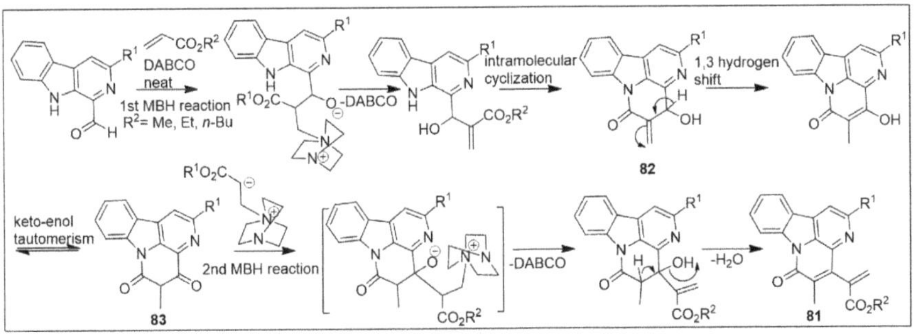

FIGURE 5.10 Proposed Mechanism for the Formation of Canthin-6-one Derivatives.

In an extension to this work, the synthesis of diastereomeric mixture of highly substituted canthine derivatives (**85**) was accomplished via the MBH reaction of **2** and **5** with acrylonitrile followed by a base-mediated intramolecular cyclization of the adducts (**84**) as illustrated in Scheme 5.31.

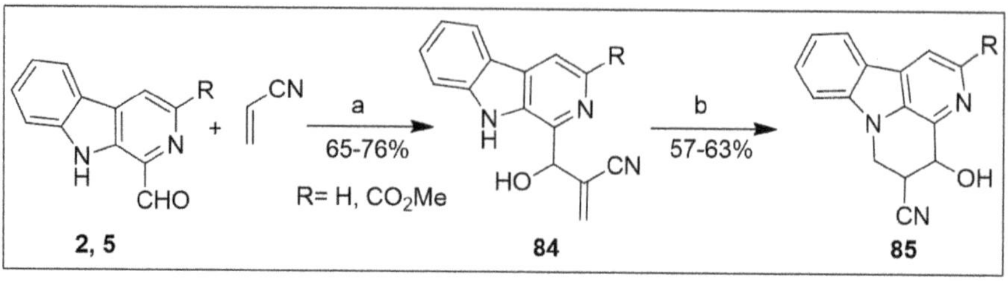

SCHEME 5.31 Synthesis of Substituted Canthine Derivatives (**85**). Reagents and Conditions: (a) DABCO, rt, 3–5 h; (b) K$_2$CO$_3$, DMF, rt, 30 min–3 h.

Further, the synthesis of another novel fused β-carboline aldehyde **86** was achieved via sequential Heck reaction of indole with 2-iodo-nitrobenzene, reduction of the nitro group, Pictet–Spengler condensation of dimethoxyglyoxal, and deprotection of the formyl group. The MBH reaction of **86** with acrylonitrile followed by intramolecular cyclization of MBH adduct (**87**) afforded a new canthine analog **88** as a mixture of diastereomers (Scheme 5.32).

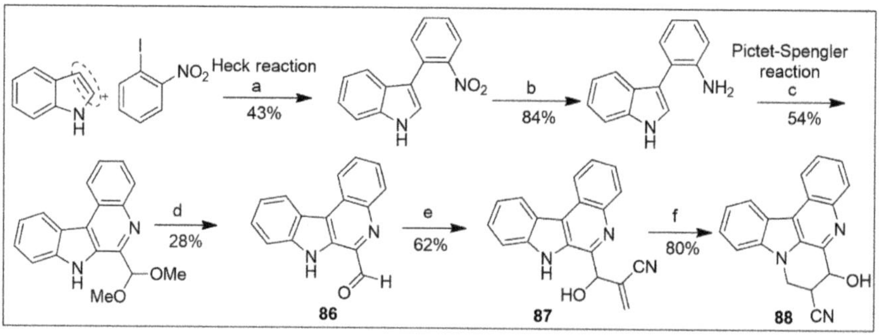

SCHEME 5.32 Synthesis of Fused Canthine Analog (**88**) *via* Intramolecular Cyclization of MBH Adduct (**87**). Reagents and Conditions: (a) Pd(OAc)$_2$, K$_2$CO$_3$, dioxane, N$_2$, 110°C, 24 h; (b) Fe-AcOH, N$_2$, 80°C, 1.5 h; (c) OHCCH(OMe)$_2$, 2% TFA in DCM, rt, 24 h; (d) TFA:H$_2$O (1:1), ACN, 80°C, 5 h; (e) CH$_2$=CHCN, DABCO, rt, 2 h; (f) K$_2$CO$_3$, DMF, rt, 1 h.

Batra et al. also observed that the *N*-prenylated β-carboline aldehydes (**89**) readily undergo intramolecular carbonyl-ene reaction in a diastereoselective fashion to afford the *syn* or *anti* isomer of a new canthine derivative (**90**) in the presence of a suitable catalyst including AcOH, ZnBr$_2$, or Yb(OTf)$_3$ (Scheme 5.33).[31]

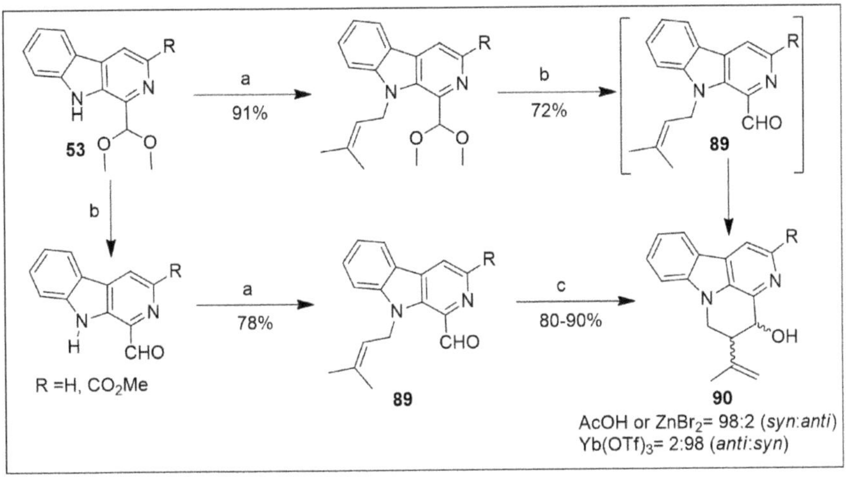

SCHEME 5.33 Synthesis of *syn* or *anti* Isomer of a New Canthine Derivative (**90**). Reagents and Conditions: (a) Prenyl Bromide, Cs$_2$CO$_3$, dry DMF, rt, 45 min; (b) AcOH:H$_2$O (2:3), 100°C, 45 min; (c) ZnBr$_2$, dry C$_6$H$_6$, 80°C, 12 h or Yb(OTf)$_3$, dry MeCN, 80°C, 3 h.

The N-prenylated 1-formyl-9H-β-carbolines (**89**) were also demonstrated to be a viable precursor for the synthesis of dihydroquinoline-fused canthines (**92**) by an intramolecular aza-Diels–Alder reaction (Povarov reaction). The reaction of **89** with substituted anilines in the presence of Yb(OTf)$_3$ afforded a diastereomeric mixture of fused tetrahydroquinolines (**91**) with traces of carbonyl-ene product (**90**). The tetrahydroquinolines (**91**) were then subjected to DDQ-promoted dehydrogenation to afford dihydroquinoline-fused canthines **92** (Scheme 5.34).

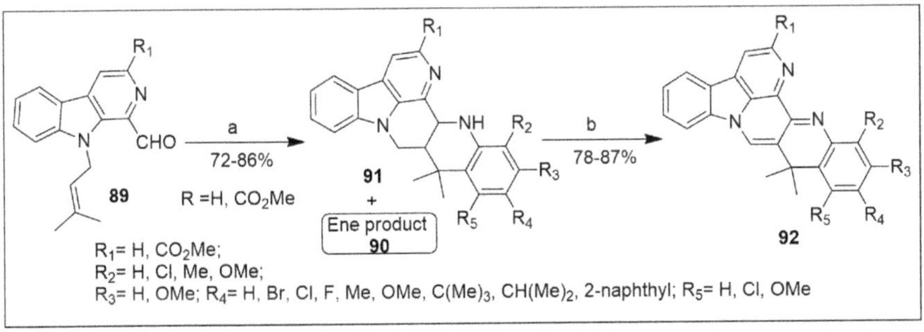

SCHEME 5.34 Synthesis of Dihydroquinoline-Fused Canthines (**92**). Reagents and Conditions: (a) Substituted Anilines, Yb(OTf)$_3$, dry ACN, 80°C, 3 h; (b) DDQ, dry MeCN, rt, 30 min.

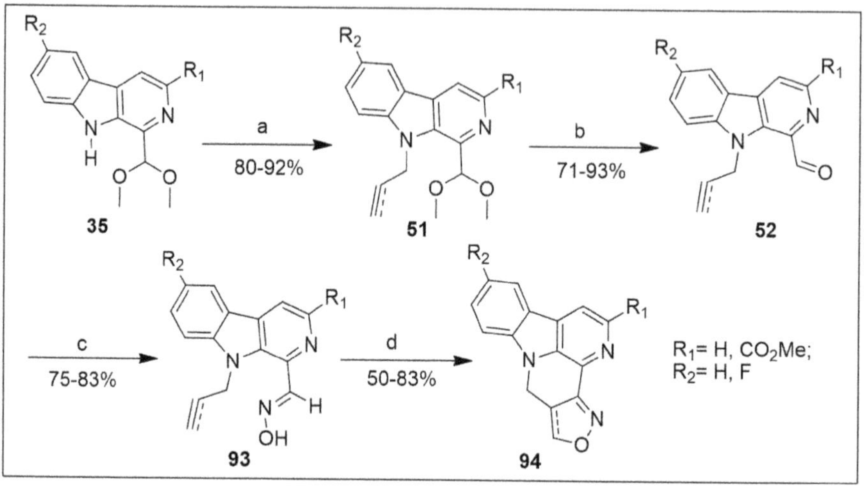

SCHEME 5.35 Synthesis of Novel Isoxazole or Isoxazoline Derivatives (**94**) via 1,3-Dipolar Cycloaddition Reaction. Reagents and Conditions: (a) Allyl Bromide or Propargyl Bromide, Cs$_2$CO$_3$, dry DMF, rt, 3 h; (b) AcOH:H$_2$O (2:3), 100°C, 45 min; (c) NH$_2$OH·HCl, AcONa, MeOH, reflux, 1 h; (d) NaOCl, Et$_3$N, DCM, rt, 3 days.

Substituted 1-formyl-9H-β-carbolines (**52**) were also demonstrated to be viable precursors for the synthesis of a library of new β-carboline-based polycyclic systems via 1,3-dipolar cycloaddition reaction.[32] The dienophile introduced at N–H of B-ring in the form of allyl or propargyl in 1-formyl-9H-β-carbolines (**2, 5, 36**) was achieved through a modified route via initial N-alkylation of acetal **35** with allyl or propargyl bromide in the presence of Cs$_2$CO$_3$ to generate **51** followed by deprotection of formyl group to afford the N-substituted derivatives (**52**). Initially, aldehydes **52** were transformed to oximes **93** via treatment with NH$_2$OH·HCl (Scheme 5.35). Intramolecular 1,3-dipolar cycloaddition of nitrile oxide **93** led to the formation of novel isoxazole or isoxazoline derivatives (**94**).

The synthetic utility of substituted 1-formyl-9H-β-carbolines was further extended to generate more diverse products via cycloaddition strategy. The NH of B-ring in **35** was alkylated with substituted allyl bromides (**a–d**) prepared from the MBH chemistry to prepare highly substituted alkenes **95** (Scheme 5.36). Similar deprotection, oximation, and reaction with NaOCl in the presence of Et$_3$N afforded the highly substituted isoxazoline derivative **96** as a mixture of diastereomers.

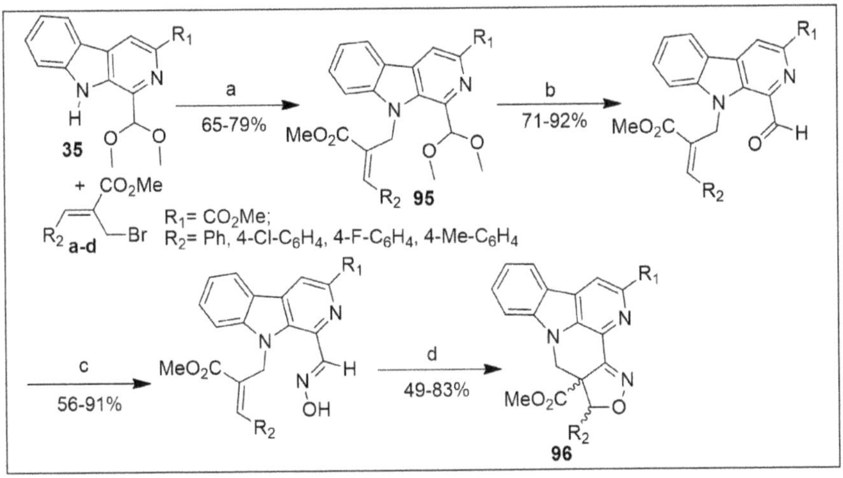

SCHEME 5.36 Synthesis of β-carbolines D-ring Fused Isoxazoline Derivative (**96**). Reagents and Conditions: (a) Cs_2CO_3, dry DMF, rt, 3 h; (b) $AcOH:H_2O$ (2:3), 100°C, 45 min; (c) $NH_2OH \cdot HCl$, AcONa, MeOH, reflux, 1 h; (d) NaOCl, Et_3N, DCM, rt, 3 days.

Further, the azidation of ε,ε-dimethoxy alkyne derivatives **51** with trimethylsilylazide ($TMSN_3$) in the presence of $In(OTf)_3$ under microwave (MW) heating resulted in the formation of new D-ring fused triazole derivative (**97**) (Scheme 5.37). In another variation, preparation of triazole derivative (**98**) devoid of the methoxy substitution in D-ring was formulated by another simplified approach consisting of reduction of formyl group to hydroxymethyl with $NaBH_4$ followed by mesylation of alcohol and subsequent treatment with NaN_3 as outlined in Scheme 5.38.

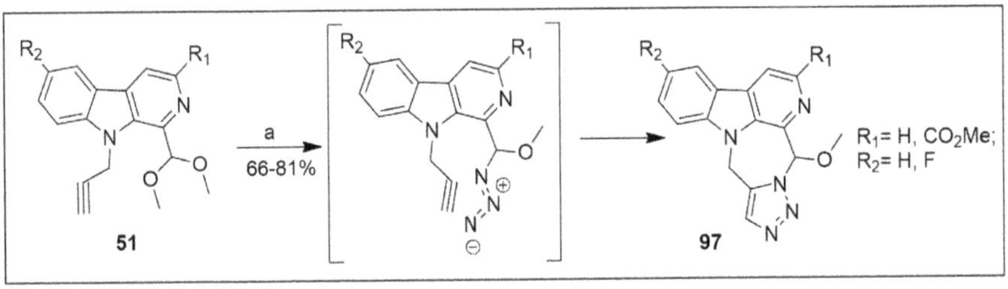

SCHEME 5.37 Synthesis of β-carbolines D-ring Fused Triazole Derivative (**97**). Reagents and Conditions: (a) $TMSN_3$, DCE, MW, 150°C, 10–15 min.

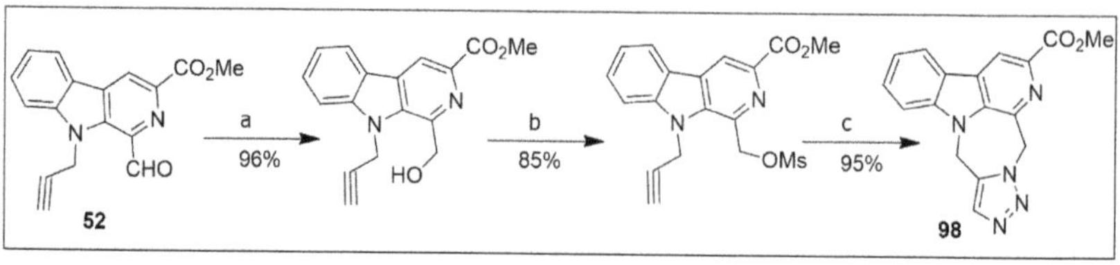

SCHEME 5.38 Synthesis of Fused Triazole Derivative (**98**). Reagents and Conditions: (a) NaBH$_4$, MeOH, rt, 30 min; (b) MsCl, Et$_3$N, DCM, 0°C, 45 min; (c) NaN$_3$, DMF, 90°C, 3 h.

The potential of N-substituted 1-formyl-9H-β-carbolines for generating fused β-carbolines was demonstrated via another intramolecular 1,3-dipolar cycloaddition reaction, wherein the in situ generated unstable azomethine ylide was made to react with alkyne. The reaction of aldehydes **61** with sarcosine afforded the β-carboline D-ring fused pyrroles (**80**) albeit in low yields (Scheme 5.39).

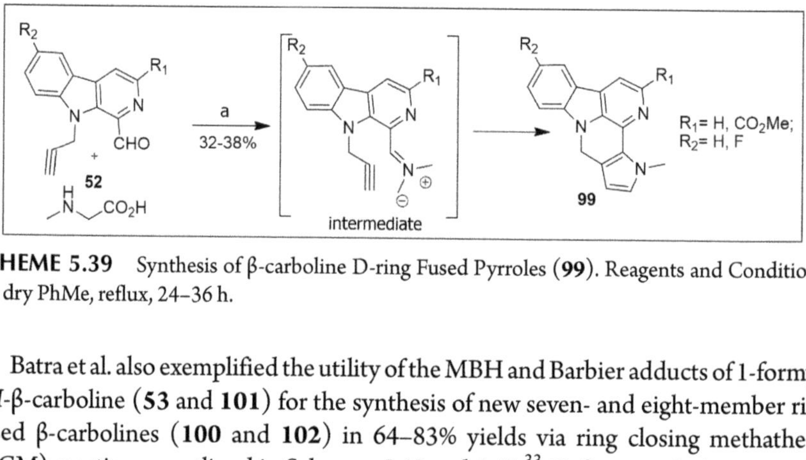

SCHEME 5.39 Synthesis of β-carboline D-ring Fused Pyrroles (**99**). Reagents and Conditions: (a) dry PhMe, reflux, 24–36 h.

Batra et al. also exemplified the utility of the MBH and Barbier adducts of 1-formyl-9H-β-carboline (**53** and **101**) for the synthesis of new seven- and eight-member ring fused β-carbolines (**100** and **102**) in 64–83% yields via ring closing methathesis (RCM) reaction as outlined in Schemes 5.40 and 5.41.[33] Unfortunately, however, the Grignard products **103** failed to undergo the RCM reaction to yield the corresponding products (**104**).

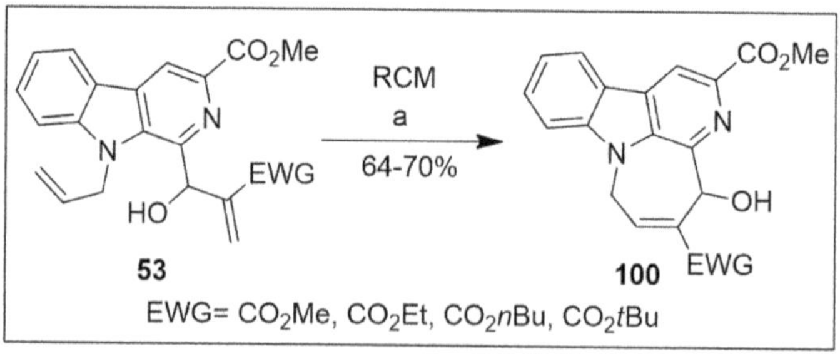

SCHEME 5.40 Synthesis of Seven-Member Ring Fused β-Carbolines (**100**) via RCM Reaction. Reagents and Conditions: (a) Grubb's II Generation Catalyst, dry DCM, reflux, 3–6 h.

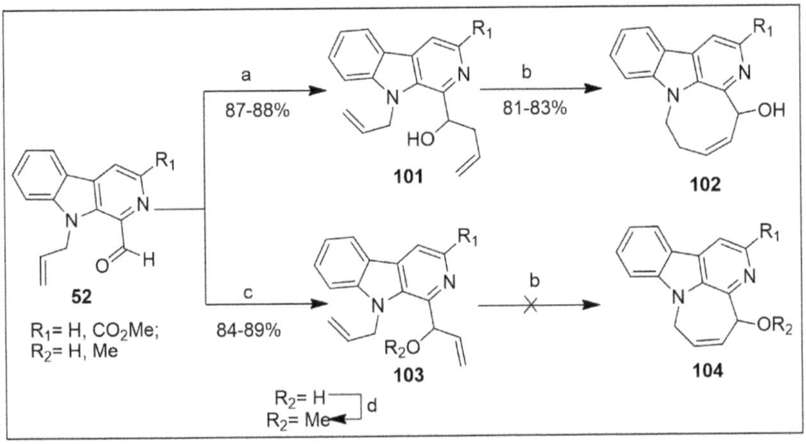

SCHEME 5.41 Synthesis of Eight-Member Ring Fused β-Carbolines (**102**) via RCM Reaction. Reagents and Conditions: (a) Allyl Bromide, THF:H$_2$O (2:1), rt, 3 h; (b) Grubb's II Generation Catalyst, Dry DCM, reflux, 12 h; (c) CH$_2$=CHMgBr, THF, −78°C–rt, 1 h; (d) Ag$_2$O, MeI, rt, 18 h.

More recently, Hutait et al. accomplished the synthesis of a chemical library of lactam-fused-β-carbolines (**106**) via Ugi four-center three-component multicomponent reaction employing 2-(1-formyl-9H-pyrido[3,4-b]indol-9-yl)acetic acid derivatives (**105**) as the bifunctional starting materials. The reaction of **105** with a wide variety of amines and isonitriles yielded the corresponding β-carboline-fused-lactams (**106**) in good yields (Scheme 5.42).[34]

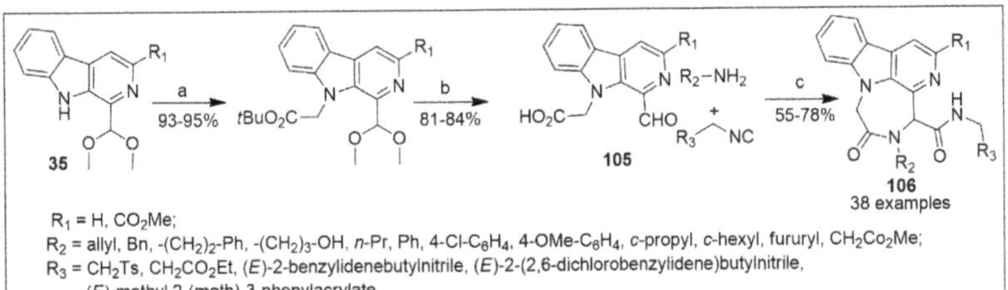

SCHEME 5.42 Synthesis of β-Carboline-Fused-Lactams (**106**) via Ugi Four-Center Three-Component Multicomponent Reaction. Reagents and Conditions: (a) Cs$_2$CO$_3$, dry DMF, rt, 45 min; (b) TFA:H$_2$O (100:1 v/v), 90°C, 1.5 h; (c) MeOH, rt, 3–12 h.

Batra et al. also reported an acid-catalyzed Pomeranz–Fritsch-type reaction between the acetal group at C-1 and the arene unit of the benzyl group at N-9 in 9-substituted benzyl-1-(dimethoxymethyl)-9H-β-carboline (**107**) to generate fused β-carbolines (**109**) that can be readily oxidized to furnish a maxonine-type framework (**110**). Mechanistically, the reaction was proposed to proceed via a protonated aldehyde as the intermediate (**108**), which undergoes attack by the activated arene subunit of the benzyl moiety (Scheme 5.43).[35]

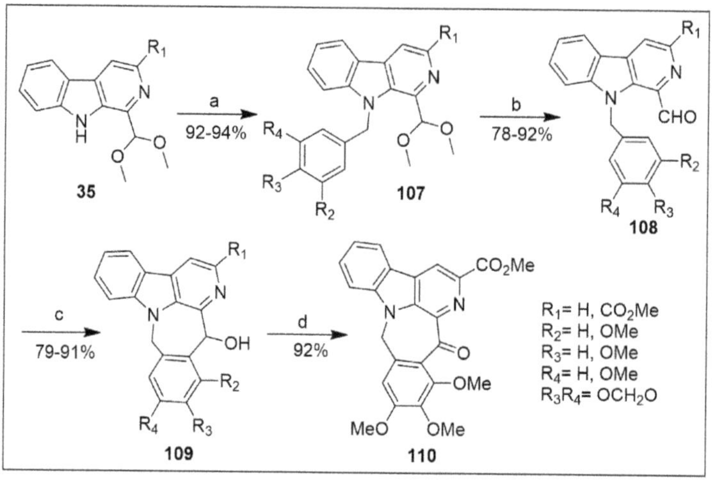

SCHEME 5.43 Synthesis of Fused β-Carbolines (**110**) via Pomeranz–Fritsch-type Reaction. Reagents and Conditions: (a) $CsCO_3$, DMF, rt, 1 h; (b) $AcOH/H_2O$ (2:3), 90°C, 45 min; (c) TFA/ H_2O (100:1), rt, 48–96 h; (d) MnO_2, DCM, rt, 2 h.

They also examined that the MBH adducts (**111**) of 1-formyl-N-substituted benzyl-9H-β-carbolines (**108**) undergo an efficient P_2O_5-mediated intramolecular Friedel–Crafts reaction between the secondary hydroxyl group and the activated phenyl group of the benzyl subunit to yield fused β-carbolines (**112**). Investigations into the scope of the substrates reveal that the success of the methodology relies on the degree of activation of the phenyl ring (Scheme 5.44). For substrates having less-activated phenyl groups, indolizinoindole derivatives (**113**) were isolated in moderate yields only.[32]

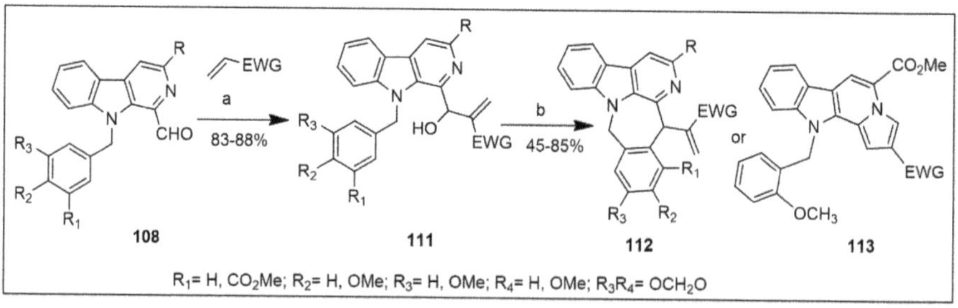

SCHEME 5.44 An Efficient Synthesis of Fused β-Carbolines (**112**) via P_2O_5-Mediated Intramolecular Friedel–Crafts Reaction. Reagents and Conditions: (a) DABCO, rt, 4 h–12 days; (b) P_2O_5, dry DCM, rt–reflux.

5.6 CONCLUSION

It is evident that over the years, a wide range of synthetic methods have been reported for the generation of 1-formyl-9H-β-carboline. However, most of the earlier strategies were limited to small-scale preparation, which is presumed to be one of the major reasons for limited exploration of this substrate for generating β-carboline-based products. However, with the recent efficient strategies developed for the efficient synthesis of this substrate, this prototype has been explored extensively for the construction of novel β-carboline derivatives. It is believed that this substrate has great potential for generating a wide variety of new fused or substituted β-carbolines and should be explored further.

ACKNOWLEDGMENTS

One of the authors, Nisha Devi, acknowledges MHRD New Delhi for financial support in the form of Junior Research Fellowship. The author VS acknowledges the financial support from DST New Delhi (CS-361/2011) and CSIR New Delhi (02(0202)/14/EMR-II) in the form of research grants.

ABBREVIATIONS

Ac	acetyl
AcOH	acetic acid
AcONa	sodium acetate
AIBN	2,2′-azobisisobutyronitrile
aq.	aqueous
Ar	aryl
ATP	adenosine-tri-phosphate
BINAP	2,20-bis(diphenylphosphanyl)-1,10-binaphthyl
Boc	tertiary butyloxycarbonyl
Boc$_2$O	di-tertiary-butyl dicarbonate
BF$_3$·Et$_2$O	boron trifluoride etherate
Bn	benzyl
BTIB	bis(trifluoroacetoxy)iodobenzene
n-Bu	normal butyl
t-But	tertiary butyl
BuLi	butyl lithium
BQCA	benzyl quinolone carboxylic acid
c	cyclo
CAN	ceric(IV) ammonium nitrate
CCl$_4$	carbon tetrachloride
β-CD	β-cyclodextrin

CDI	N,N'-carbonyl diimidazole
CDK	cyclic dependent kinases
CH_3CO_2H	acetic acid
CH_2Cl_2	dichloromethane
$CHCl_3$	chloroform
conc.	concentrated
Cp	cyclopentadienyl
m-CPBA	meta-chloroperoxybenzoic acid
CuBr	cuprous bromide
CuI	copper iodide
DABCO	diazabicyclo[2.2.2]octane
DBU	1,8-dazabicyclo[5.4.0]undec-7-ene
DCC	N,N'-dicyclohexylcarbodiimide
DCE	1,2-dichloroethane
DCM	dichloromethane
DDQ	2,3-dichloro-5,6-dicyanobenzoquinone
de	diastereomeric excess
DEAD	diethyl azodicarboxylate
DEPC	diethylphosphoryl cyanide
DIB	(diacetoxyiodo)benzene
DIEA	diisopropylethylamine
DIBAL-H	diisobutylaluminum hydride
DIC	diisopropylcarbodiimide
DIPEA	diisopropylethylamine
DMAD	dimethyl acetylenedicarboxylate
DMAP	4-dimethyl aminopyridine
DMF	dimethylformamide
DMSO	dimethylsulfoxide
DPP	diphenyl phosphate
DNA	deoxyribonucleic acid
dppp	1,3-bis(diphenylphosphino)propane
dr	diastereomeric ratio
DYKAT	dynamic asymmetric kinetic transformation
E	*Entgegen*
ee	enantiomeric excess
Et	ethyl
Et_3N	triethylamine
Et_2O	diethyl ether
EtOAc	ethyl acetate
EtOH	ethanol
EWG	electron withdrawing group
Fe	iron

h	hour
HBr	hydrobromic acid
HBF_4	hydroflouroboric acid
HCl	hydrochloric acid
$HClO_4$	perchloric acid
HCO_2H	formic acid
Hex	hexyl
HIV	human immune deficiency virus
[Hmim]TFA	1-methylimidazolium trifluoroacetate
HMP	hexamethyl phosphoramide
HMTA	hexamethylenetetramine
HOBT	1-hydroxybenzotriazole
HRMS	high resolution mass spectroscopy
H_2SO_4	sulfuric acid
β-ICD	β-isocupreidine
$InCl_3$	indium chloride
In	indium
K_2CO_3	potassium carbonate
$KMnO_4$	potassium permanganate
LCMS	liquid chromatography coupled with mass spectrometry
LDA	lithium diisopropylamide
$LiAlH_4$	lithium aluminum hydride
LiHMDS	lithium hexamethyldisilazide
MBH	Morita–Baylis–Hillman
MCR	multi component reaction
m-CPBA	meta chloro per benzoic acid
Me	methyl
MeI	methyl iodide
MeOH	methanol
MIC	minimum inhibitory concentration
min	minute
MOM	methoxymethyl
Mont.	montmorillonite
Ms	mesyl
MS	molecular sieves
MVK	methyl vinyl ketone
MW	micro wave
$NaBH_4$	sodium borohydride
NH_4Cl	ammonium chloride
NaH	sodium hydride
NaOAc	sodium acetate
NaOMe	sodium methoxide

NH$_2$OH	hydroxylamine
NMO	*N*-methyl morpholine-*N*-oxide
PBr$_3$	phosphorus tribromide
Pd-C	palladium on carbon
Pd(PPh$_3$)$_4$	tetrakis triphenylphosphine palladium(0)
PEG	polyethylene glycol
Pent	pentyl
pH	potential of hydrogen
Ph	phenyl
Pr	propyl
Py	pyridine
POCl$_3$	phosphorus oxychloride
PPA	polyphosphoric acid
PTC	phase transfer catalyst
p-TSA	*para*-toluenesulfonic acid
RCM	ring-closing metathesis
rt	room temperature
S	sulfur
SFC	solvent-free condition
Sc(OTf)$_3$	scandium triflate
TBAB	tetrabutyl ammonium bromide
TBAF	tetrabutyl ammonium fluoride
TBS	tertiary butyldimethylsilyl
TBDPS	tertiary butyldiphenylsilyl
TBHP	tetrabutylhydroperoxide
TBPA	tris(4-bromophenyl)ammonium
TDAE	tetrakis (dimethyl-1-amino)ethylene
Tf	trifluoromethane sulfonyl
TFA	trifluroacetic acid
TFAE	trifluoroacetaldehyde ethyl hemiacetal
THβC	tetrahydro β-carboline
TfOH	triflic acid or trifluoromethanesulfonic acid
THF	tetrahydrofuran
TIPS	triisopropylsilyl
TMB	2,4,6-trimethoxybenzyl
TMEDA	*N,N,N′,N′*-tetramethylethylenediamine
TMS	trimethylsilane
TMSN$_3$	trimethylsilylazide
TMSCl	trimethylsilylchloride
TMSCN	trimethylsilylcyanide
TMSOTf	trimethylsilyl trifluoromethanesulfonate
(*o*-Tol)$_3$P	tri-ortho-tolyl-phosphine

Ts	tosyl or 4-toluenesulfonyl
p-TSA	or TsOH para-toluenesulfonic acid
TTMSS	tris(trimethylsilyl)silane
UV	ultraviolet
Z	*Zussamen*
$ZnCl_2$	zinc chloride
$ZrCl_4$	zirconium chloride
°C	degree Celsius
μL	microliter
μM	micromolar
π	hydrophobicity constant

KEYWORDS

- β-Carbolines
- Kumujian C
- canthinone
- Pictet—Spengler reaction
- Povarov reaction
- Povarov reaction

REFERENCES

1. (a) Saxton, J. E. Alkaloids of the aspidospermine group. In *The Alkaloids: Chemistry and Biology*; Cordell, G. A., Ed.; Academic Press: San Diego, CA, 1998; Vol. 51, pp 2–197; (b) Mansoor, T. A.; Ramalhete, C.; Molnar, J.; Mulhovo, S.; Ferreira, M. J. U. Tabernines A–C, β-carbolines from the leaves of *Tabernaemontana elegans*. *J. Nat. Prod.* 2009, 72, 1147–1150; (c) Cao, R.; Peng, W.; Wang, Z.; Xu, A. β-Carboline alkaloids: biochemical and pharmacological functions. *Curr. Med. Chem.* 2007, 14, 479–500; (d) Higuchi, K.; Kawasaki, T. Simple indole alkaloids and those with a nonrearranged monoterpenoid unit. *Nat. Prod. Rep.* 2007, 24, 843–868; (e) Kawasaki, T.; Higuchi, K. Simple indole alkaloids and those with a nonrearranged monoterpenoid unit. *Nat. Prod. Rep.* 2005, 22, 761–793; (f) Carbrera, G. M.; Seldes, A. M. A. A β-carboline alkaloid from the soft coral *Lignopsis spongiosum*. *J. Nat. Prod.* 1999, 62, 759–760; (g) Airaksinen, M. M.; Kari, I. β-Carbolines, psychoactive compounds in the mammalian body. Part II: effects. *Med. Biol.* 1981, 59, 190–211; (h) Karin, C.-S. L.; Yang, S.-L.; Roberts, M. F.; Phillipson, J. D. Canthin-6-one alkaloids from cell suspension cultures of *Brucea javanica*. *Phytochemistry.* 1990, 29, 141–143; (i) Gonzalez-Gomez, A.; Domınguez, G.; Perez-Castells, J. Novel chemistry of β-carbolines. Expedient synthesis of polycyclic scaffolds. *Tetrahedron.* 2009, 65, 3378–3391.
2. (a) Allen, J. R.; Holmstedt, B. R. The simple β-carboline alkaloids. *Phytochemistry.* 1979, 19, 1573–1582; (b) Molina, P.; Fresneda, P. M. Iminophosphorane-mediated annelation of a pyridine or pyrimidine ring into an indole ring: synthesis of β-, γ-carbolines and pyrimido[4,5-*b*]indole derivatives. *J. Chem. Soc. Perkin Trans.* 1988, 1, 1819–1822; (c) Bazika, V.; Lang, T.-W.; Pappelbaum, S.; Corday, E.; Ajmalin, a Rauwolfia alkaloid for the treatment of digitoxic arrhythmias. *Am. J. Cardiol.* 1966, 17, 227–231; (d) Wu, Y.; Zhao, M.; Wang, C.; Peng, S. Synthesis and thrombolytic activity of

pseudopeptides related to fibrinogen fragment. *Bioorg. Med. Chem. Lett.* 2002, 12, 2331–2335; (e) Brahmbhatt, K. G.; Ahmed, N.; Sabde, S.; Mitra, D.; Singh, I. P.; Bhutani, K. K. Synthesis and evaluation of beta-carboline derivatives as inhibitors of human immunodeficiency virus. *Bioorg. Med. Chem. Lett.* 2010, 20, 4416–4419.

3. (a) Turner, R. B.; Woodward, R. B. The chemistry of the cinchona alkaloids. In *The Alkaloids*; Manske, R. H. F., Ed.; Academic Press: New York, 1953; Vol. 3, pp 1–63; (b) Uskokovic, M. R.; Grethe, G. The *Cinchona* alkaloids. In *The Alkaloids*; Manske, R. H. F., Ed.; Academic Press: New York, 1973; Vol. 14, pp 181–223; (c) Grethe, G.; Uskokovic, M. R. The Cinchona group. In *The Chemistry of Heterocyclic Compounds*; Sexton, J. E., Ed.; Wiley-Interscience: New York, 1983; Vol. 23, Part 4, p 279; (d) Schwikkard, S.; Heerden, R. V. Antimalarial activity of plant metabolites. *Nat. Prod. Rep.* 2002, 19, 675–692; (e) Steele, J. C. P.; Veitch, N. C.; Kite, G. C.; Simmonds, M. S. J.; Warhurst, D. C. Indole and β-carboline alkaloids from *Geissospermum sericeum*. *J. Nat. Prod.* 2002, 65, 85–88.

4. (a) Takasu, K.; Shimogama, T.; Saiin, C.; Kim, H. S.; Wataya, Y.; Ihara, M. Indole and β-carboline alkaloids from *Geissospermum sericeum*. *Bioorg. Med. Chem. Lett.* 2004, 14, 1689–1694; (b) Boursereau, Y.; Coldham, I. Synthesis and biological studies of 1-amino β-carbolines. *Bioorg. Med. Chem. Lett.* 2004, 14, 5841–5844; (c) Al-Allaf, T. A. K.; Rashan, L. J. Synthesis and cytotoxic evaluation of the first trans-palladium(II) complex with naturally occurring alkaloid harmine. *Eur. J. Med. Chem.* 1998, 33, 817–820; (d) Ishida, J.; Wang, H.-K.; Bastow, K. F.; Hu, C.-Q.; Lee, K.-H. Antitumor agents 201.1 Cytotoxicity of harmine and β-carboline analogs. *Bioorg. Med. Chem. Lett.* 1999, 9, 3319–3324.

5. (a) Braestrup, C.; Nielsen, M.; Olsen, C. E. Urinary and brain β-carboline-3- carboxylates as potent inhibitors of brain benzodiazepine receptors. *Proc. Natl. Acad. Sci. U.S.A.* 1980, 77, 2288–2292; (b) Schlecker, W.; Huth, A.; Ottow, E.; Mulzer, J. Regioselective metalation of 9-methoxymethyl-β-carboline-3- carboxamides with amidomagnesium chlorides. *Synthesis*. 1995, 1225–1227; (c) Dorey, G.; Poissonnet, G.; Potier, M. C.; Carvalho, L. P. D.; Venault, P.; Chapouthier, G.; Rossier, J.; Potier, P.; Dodd, R. H. Synthesis and benzodiazepine receptor affinities of rigid analogs of 3-carboxy-β-carbolines: demonstration that the benzodiazepine receptor recognizes preferentially the *s-cis* conformation of the 3-carboxy group. *J. Med. Chem.* 1989, 32, 1799–1804.

6. (a) Ozawa, M.; Nakada, Y.; Sugimachi, K.; Yabuuchi, F.; Akai, T.; Mizuta, E.; Kuno, S.; Yamaguchi, M. Pharmacological characterization of the novel anxiolytic β-carboline abecarnil in rodents and primates. *Jpn. J. Pharmacol.* 1994, 64, 179–187; (b) Biggio, G.; Concas, A.; Mele, S.; Corda, M. G. Changes in GABAergic transmission induced by stress, anxiogenic and anxiolytic β-carbolines. *Brain Res. Bull.* 1987, 19, 301–308.

7. Molina, P.; Fresnda, P. M.; Gareia-Zafra, S. An iminophosphorane-mediated efficient synthesis of the alkaloid eudistomin U of marine origin. *Tetrahedron Lett.* 1995, 36, 3581–3582.

8. Molina, P.; Fresnda, P. M.; Gareia-Zafra, S.; Almendros, P. Iminophosphorane mediated syntheses of the fascaplysin alkaloid of marine origin and nitramarine. *Tetrahedron Lett.* 1994, 35, 8851–8854.

9. (a) Deveau, A. M.; Labroli, M. A.; Dieckhaus, C. M. The synthesis of amino-acid functionalized β-Carbolines as topoisomerase II inhibitors. *Biorg. Med. Chem. Lett.* 2001, 11, 1251–1255; (b) Pommier, Y.; MacDonald, T. L.; Madalengoitia, J. S. U. S. Patent 5622960, April 22, 1997; (c) Batch, A.; Dodd, R. H. Ortho-directed metalation of 3-carboxy-β-carbolines: use of the SmI(2)-cleavable 9-N-(N',N'-Dimethylsulfamoyl) blocking group for the preparation of 9-N-deprotected 4-amino derivatives via azide introduction or a palladium-catalyzed cross-coupling reaction. *J. Org. Chem.* 1998, 63, 872–877.

10. Burkard, W. P.; Bonetti, E. P.; Haefely, W. The benzodiazepine antagonist Ro 15-1788 reverses the effect of methyl-beta-carboline-3-carboxylate but not of harmaline on cerebellar cGMP and motor performance in mice. *Eur. J. Pharmacol.* 1985, 109, 241–247.

11. (a) Sorbera, L. A.; Martin, L.; Leeson, P. A.; Castaner, J. Treatment of erectile dysfunction - treatment of female sexual dysfunction - phosphodiesterase 5 inhibitor. *Drugs Future.* 2001, 26, 15–19; (b) Daugan, A.; Grondin, P.; Ruault, C.; de Gouville, A.-C. L. M.; Coste, H.; Kirilovsky, J.; Hyafil, F.; Labaudiniere, R. The discovery of tadalafil: a novel and highly selective PDE5 inhibitor. 1: 5,6,11,11a-tetrahydro-1H-imidazo[1',5':1,6]pyrido[3,4-b]indole-1,3(2H)-dione analogues. *J. Med. Chem.* 2003, 46, 4525–4532; (c) Turski, L.; Stephens, D. N.; Jensen, L. H.; Peterson, E. N.;

Meldrum, B. S.; Patel, S.; Hansen, J. B.; Loscher, W.; Schneider, H. H.; Schmiechen, R. Anticonvulsant action of the β-carboline abecarnil: studies in rodents and baboon, Papio papio. *J. Pharmacol. Exp. Ther.* 1990, 253, 344–352; (d) Maw, G. N.; Allerton, C. M.; Gbekor, E.; Million, W. A. Design, synthesis and biological activity of β-carboline-based type-5 phosphodiesterase inhibitors. *Bioorg. Med. Chem. Lett.* 2003, 13, 1425–1428.

12. (a) Pictet, A.; Spengler, T. The formation of isoquinoline derivatives by the action of methylal to phenyl ethylamine, phenylalanine and tyrosine. *Ber. Dtsch. Chem. Ges.* 1911, 44, 2030–2036; (b) Cox, E. D.; Cook, J. The Pictet-Spengler condensation: a new direction for an old reaction. *Chem. Rev.* 1995, 95, 1797–1842; (c) Royer, J.; Bonin, M.; Micouin, L. Chiral heterocycles by iminium ion cyclization. *Chem. Rev.* 2004, 104, 2311–2352.

13. Singh, V.; Batra, S. 1-Formyl-9*H*-β-carboline: a useful scaffold for synthesizing substituted- and fused β-carbolines. *Curr. Org. Syn.* 2012, 9, 513–528.

14. (a) Jordaan, A.; Du Plessis, L. M.; Joynt, V. P. The structure and synthesis of pavettine, an alkaloid from *Pavetta lanceolata. J. S. Afr. Chem. Inst.* 1968, 21, 22–23; (b) Ohmoto, T.; Koike, K. Studies on the constituents of *Picrasma quassioides* BENNET. III. The alkaloidal constituents. *Chem. Pharm. Bull.* 1984, 32, 3579–3583; (c) Yang, J.-S.; Gong, D. Kumujancine and kumujanrine, two new β-carboline alkaloids from *picrasma quassioides. Acta Chim. Sinica.* 1984, 42, 679–683; (d) Matsumura, S.; Enomoto, H.; Aoyasi, Y.; Nomiyama, Y.; Kono, T.; Mastuda, M.; Tanaka, H.9*H*-Pyrido[3,4-*b*]indole derivatives. US 4241064, 1980; (e) Matsumura, S.; Enomoto, H.; Aoyagi, Y.; Nomiyama, Y.; Kono, T.; Matsuda, M.; Tanaka, H. *Grr. Offen.* 1980, 29(41), 449. [*Chem. Abstr.* 1980, 93, p114495rl].

15. Gatta, F.; Misiti, D. Selenium dioxide oxidation of tetrahydro-β-carboline derivatives. *J. Heterocyclic. Chem.* 1987, 24, 1183–1187.

16. (a) Giudice, M. R. D.; Gatta, F.; Settimj, G. New tetracyclic compounds containing the β-carboline moiety. *J. Heterocyclic Chem.* 1990, 27, 967–973; (b) Bennasar, M.-L.; Roca, T.; Monerris, M. Total synthesis of the proposed structures of indole alkaloids lyaline and lyadine. *J. Org. Chem.* 2004, 69, 752–756.

17. Bracher, F.; Hildebrand, D. 1,9-Dimetalated β-carbolines. Versatile building blocks for the total synthesis of alkaloids. *Tetrahedron.* 1994, 50, 12329–12336.

18. Suzuki, H.; Unemoto, M.; Hagiwara, M.; Ohyama, T.; Yokoyama, Y.; Murakami, Y. Synthetic studies on indoles and related compounds. Part 46.1 First total syntheses of 4,8-dioxygenated β-carboline alkaloids. *J. Chem. Soc., Perkin Trans. I.* 1999, Issue 12, 1717–1723.

19. Choshi, T.; Kuwada, T.; Fukui, M.; Matsuya, Y.; Sugino, E.; Hibino, S. Total syntheses of novel cytocidal β-carboline alkaloids, oxopropalines D and G. *Chem. Pharm. Bull.* 2000, 48, 108–113.

20. Takasu, K.; Shimogama, T.; Saiin, C.; Kim, H.-S.; Wataya, Y.; Ihara, M. π-Delocalized β-carbolinium cations as potential antimalarials. *Bioorg. Med. Chem. Lett.* 2004, 14, 1689–1692.

21. Takasu, K.; Shimogama, T.; Saiin, C.; Kim, H.-S.; Wataya, Y.; Reto, B.; Ihara, M. Synthesis and evaluation of β-carbolinium cations as new antimalarial agents based on pi-delocalized lipophilic cation (DLC) hypothesis. *Chem. Pharm. Bull.* 2005, 53, 653–661.

22. Brahmbhatt, K. G.; Ahmed, N.; Sabde, S.; Mitra, D.; Singh, I. P.; Bhutani, K. K. Synthesis and evaluation of β-carboline derivatives as inhibitors of human immunodeficiency virus. *Bioorg. Med. Chem. Lett.* 2010, 20, 15, 4416–4419.

23. Singh, V.; Hutait, S.; Batra, S. Baylis–Hillman reaction of 1-formyl-β-carboline: one-step synthesis of the canthin-6-one framework by an unprecedented cascade cyclization reaction. *Eur. J. Org. Chem.* 2009, 35, 6211–6216.

24. (a) Ramesh, S.; Nagarajan, R. A formal synthesis of lavendamycin methyl ester, nitramarine, and their analogues: A povarov approach. *J. Org. Chem.* 2013, 78, 545–558; (b) Nissen, F.; Detert, H. Total synthesis of lavendamycin by a [2+2+2] cycloaddition. *Eur. J. Org. Chem.* 2011, 2845–2853; (c) Gaddam, V.; Nagarajan, R. A new entry to polycyclic indole derivatives via intramolecular imino Diels–Alder reaction: observation of unexpected reaction. *J. Org. Chem.* 2007, 72, 3573–3576; (d) Gaddam, V.; Nagarajan, R. An efficient, one-pot synthesis of isomeric ellipticine derivatives through intramolecular imino-Diels–Alder reaction. *Org. Lett.* 2008, 10, 1975–1978.

25. Ramesh, S.; Nagarajan, R. A novel route to synthesize lavendamycin analogues through an [A3] coupling reaction. *Tetrahedron.* 2013, 69, 4890–4898.

26. Singh, V.; Hutait, S.; Batra, S. Advancing the Morita–Baylis–Hillman chemistry of 1-formyl-β-carbolines for the synthesis of indolizino-indole derivatives. *Eur. J. Org. Chem.* 2010, 3684–3691.

27. Dighe, S. U.; Hutait, S.; Batra, S. Copper-catalyzed multicomponent coupling/cycloisomerization reaction between substituted 1-formyl-9H-β-carbolines, secondary amines, and substituted alkynes for the synthesis of substituted 3-aminoindolizino [8, 7-*b*] indoles. *ACS Comb. Sci.* 2012, 14, 665–672.

28. Condie, G. C.; Bergman, J. Synthesis of some fused β-carbolines including the first example of the pyrrolo[3,2-*c*]-β-carboline system. *J. Heterocyclic Chem.* 2004, 41, 531–540.

29. Condie, G. C.; Bergman, J. Reactivity of β-carbolines and cyclopenta[*b*]indolones prepared from the intramolecular cyclization of 5(4H)- oxazolones derived from L-tryptophan. *Eur. J. Org. Chem.* 2004, 1286–1297.

30. Suzuki, H.; Adachi, M.; Ebihara, Y.; Gyoutoku, H.; Furuya, H.; Murakami, Y.; Okuno, H. A total synthesis of 1-methoxycanthin-6-one: An efficient one-pot synthesis of the canthin-6-one skeleton from β-carboline-1-carbaldehyde. *Synthesis.* 2005, 1, 28–32.

31. Hutait, S.; Singh, V.; Batra, S. Facile synthesis of dihydroquinoline-fused canthines by intramolecular aza-Diels–Alder reaction. *Eur. J. Org. Chem.* 2010, 6269–6276.

32. Singh, V.; Hutait, S.; Biswas, S.; Batra, S. Versatility of substituted 1-formyl-9H-β-carbolines for the synthesis of new fused β-carbolines via intramolecular 1,3-dipolar cycloaddition. *Eur. J. Org. Chem.* 2010, 531–539.

33. Hutait, S.; Batra, S. *Tetrahedron Lett.* RCM-based approach to 7- and 8-member ring-fused β-carbolines. 2010, 51, 5781–5783.

34. Hutait, S.; Nayak, M.; Penta, A.; Batra, S. Convenient synthesis of a library of lactam-fused β-carbolines via Ugi-reaction. *Synthesis.* 2011, 3, 419–430.

35. Hutait, S.; Biswas, S.; Batra, S. Efficient synthesis of maxonine analogues from N-substituted benzyl-1-formyl-9H-β-carbolines. *Eur. J. Org. Chem.* 2012, 2453–2462.

CHAPTER 6

Plasma Chemistry as a Tool for Eco-Friendly Processing of Cotton Textile

Hemen Dave[1], Lalita Ledwani[1*], and S. K. Nema[2]

[1]Department of Chemistry, Manipal University Jaipur, Jaipur, India
[2]FCIPT, Institute for Plasma Research, Gandhinagar, India
*Email: lalita.ledwani@jaipur.manipal.edu

CONTENTS

ABSTRACT

Cotton is the most used natural fiber in textile industries. Generally, fiber is used as textile material due to its favorable bulk properties although its surface properties are inadequate for processing as per user requirement. In conventional processing, the textile materials undergo wet chemical treatment to get the desire properties; however, consumption of large amount of chemicals and water makes it extremely polluting industry, and thus the textile industries belong to the top 10 polluting industries. Pollution prevention is the prime requirement of the 21st century, and thus there are demands for textile industries for reduction in pollution generation and switch to eco-friendly processing. Non-thermal plasma technology is a versatile technology with proven potential for industrial process enhancement and pollution prevention. Pre-treatment and finishing of textile fabrics using plasma received enormous attention as a solution for the environmental problem of textile industries, and cotton being the most used natural fiber, during last two decades, considerable efforts have been devoted on surface modification of cotton textile with plasma technology. The main purpose of this review is to offer an overview to readers about wide variety of applications of emerging plasma technologies, aiming for reducing environmental impacts of conventional processing of cotton textile. This review offers plasma chemistry as an innovative tool for processing of cotton textile.

6.1 INTRODUCTION

Protection of environment is one of the major concerns before the world today. Pollution of environment is the darker side of industrial development due to which the natural ecosystems are under ever-increasing threat. At the beginning of the 21st century, the mankind has to face multifaceted environmental problems that pose enormous challenges to contemporary science. New and innovative technologies are being sought and existing ones are being manipulated in order to meet the primary demands of environmental protection. One such technology lies in the broad field of plasma science.[1] In recent years, the field of gas discharge plasma applications is rapidly expanding. Plasma technology has been implemented in various applications and now established as a versatile tool for industrial process enhancement. Nonthermal plasmas (NTPs) are very useful in surface modification of textile polymers. Since plasma exposure of polymers enhances their surface properties without alteration of bulk properties, pretreatment and finishing of textile fabrics by plasma received enormous attention as a solution for the environmental problem of textile industries. In the last decades, considerable efforts have been devoted on surface modification of cotton textile with plasma technology to reduce adverse environmental impacts of conventional wet chemical processing. As NTP technology is widely studied for cotton textile modification with enormous amount of potential uses, this chapter on application of NTP treatment of cotton textile gives overviews of different application strategies of plasma treatment on cotton textile reported during the last two decades.

6.2 PLASMA TECHNOLOGY: BRIEF HISTORY AND DEVELOPMENT

Plasma is a partially or fully ionized gas consisting of various particles, such as electrons, ions, atoms, and molecules, and is described as the fourth state of matter. The scientific study of plasma began in 1808 with the development of the steady-state DC arc discharge by Sir Humphry Davy and with the development of the high-voltage DC electrical discharge tube by Michael Faraday and others in the 1830s. Plasma was identified as a fourth state of matter by Sir William Crookes in 1879. The word plasma was coined by Irving Langmuir in 1928 for ionized gas, while studying the unusual magnetic and electric characteristics of super-heated gases.[2] A plasma can be defined as a collection of free charged particles and neutral particles moving in random directions that are, on the average, electrically neutral and enough to make collective electromagnetic effects important for its physical behavior. It is estimated that more than 99% of the known universe is in the plasma state, with the exception of cold celestial bodies and planetary systems (Fig. 6.1). The plasma state can be produced in the laboratory by raising the energy content of matter regardless of the nature of the energy source. To produce and sustain plasma, the easiest way to inject energy into a system in a continuous manner is with electrical energy, and that is the reason why electrical discharges are used to produce the most common man-made plasmas. For the quantitative description of plasma, the term temperature is usually used. Thermal plasma is in a state where almost all its components are at thermal equilibrium. In NTPs, temperature (i.e., kinetic energy) is not in thermal equilibrium and differs substantially between the electrons and the other particles (ions, atoms, and molecules). In this sense, an NTP is also referred to as a "nonequilibrium plasma" or a "cold plasma".[3,4]

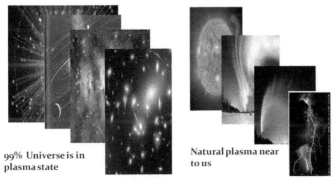

99% Universe is in plasma state

Natural plasma near to us

Figure x.1: Plasma the forth state of matter

FIGURE 6.1 Plasma, the Fourth State of Matter.

Although the scientific study of an electrical discharge in gas phase has a long history (over 200 years), however, it was not until the post-world war years that plasma technology became of major importance as industrial technology.[1] The 19th century was the era for the origin of plasma science when rapid progress was made in electrical discharge physics, and arc and DC electrical discharge plasmas were extensively researched in scientific laboratories. The only widespread application of plasmas during the 19th century was the use of electrical arcs for illumination, a technology which for a time was competitive with gaslights, but by 1900 lost out to the incandescent lamp because of its requirement for high current DC electrical power transmission.[2] At the beginning of the 20th century, the gas discharge branch of Langmuir's plasma physics, a major impetus to plasma physics, has been evolved for controlling the fusion reactions, which began in the major industrialized countries about 1950 and continues to the present day.[2] Also, since World War II, plasma science evolution has enlarged the scope to include plasma chemistry, atomic and molecular physics, surface chemistry and physics, optics, high-temperature physics and chemistry, electrical engineering, and computer science.[5,6] With the rapid development in the understanding of plasma reactions with other materials due to advanced measurement techniques, scientists have successfully developed plasma technologies for various applications in the second half of the 20th century. Since their introduction in the 1960s, the main industrial applications of plasmas have been in the microelectronics industries. In the 1980s, their uses broadened to include many other treatments, especially in the fields of metals and polymers, and other industrial plasma-processing applications have been developed recently.[2]

In recent years, plasma technology is also gaining interest in the field of life sciences, environmental issues, and biomedical applications and indeed has proved to have great promise in the fields of chemistry, biology, physics, and biotechnological and medical sciences.[7-9] At the end of the 20th century, plasma science and its wide field of technological applications became diverse, interdisciplinary fields and are now blooming as novel key research fields in the 21st century,[8,9] which is clearly reflected from various excellent review papers available in scientific literature. From thermal plasma to NTP at low pressure or atmospheric pressure, important fields for industrial application of plasma technology are low-energy light bulbs,[9] thermal plasma processing of materials, metallurgical, metal cutting and welding, powder melting during spraying, in the reduction of oxides,[5,10-12] in manufacturing of glass,[13] ceramic powders, or films,[5,10-12] plasma coating and thin film deposition (aerospace industry, automotive industry, the medical implant industry, the petrochemical industry, the cutting tool industry, photovoltaic),[12,14-20] plasma immersion ion implantation and deposition (microelectronics and metallurgical engineering, biomedical engineering, nanotechnology),[21-24] nanotechnology (nonomaterial synthesis, nanostructuring, processing),[25-33] plasma thrusters,[9,34] plasma catalyst synthesis,[35-37] reforming and chemical synthesis,[38-41] nitrogen fixation,[42,43] hydrogen production,[44] fuel cell technology,[45] and many more.[46] Apart from the application of plasma in the field of biological science (immobilization of biomolecule, sterilization, tissue culture, etc.),[47-52] medical (cancer treatment, wound healing, disinfection issues, skin treatment, biofilm inactivation, teeth disinfection, blood coagulation, surgery, endoscopy, etc.),[53-62] biomedical engineering,[63-72] treatment of polymeric materials (improve adhesion, surface activation, etching),[73-76] and food packaging[77] are greatly explored recently.

6.3 APPLICATION OF PLASMA TECHNOLOGY FOR ECO-FRIENDLY PROCESSING OF TEXTILES

In recent times, applications of plasma technology for environmental protection are of great concern. The invention of silent discharge (also referred to as dielectric barrier discharge [DBD]) by Ernst Werner von Siemens in 1857 was one of the most important landmarks in the history of electrical discharge. Generation of ozone by DBD is the first environmental application of an NTP, and it is still one of the most important applications of NTPs.[4,78] Another important example of NTP application is the electrostatic precipitator (ESP) for dust removal. It was Cottrell who first demonstrated the large-scale application of ESPs for the collection of sulfuric acid mist in 1908, and since then, ESPs have been successfully used for the removal of dust in many industrial fields.[4] Also, during last decades, plasma technology has been greatly explored for environmental pollution cleanup,[79] which includes application of thermal plasma for waste disposal and energy recovery,[80-86] applications of NTPs for wastewater remediation,[87-90] and gaseous pollution control.[4,91-99] Above all plasma treatments of materials are eco-friendly, cost effective, time solving, and more efficient than most other conventional methods used in several industries including polymers (film, textile), metals, semiconductors, etc. Conventionally, industrial materials are manufactured and processed by actions of chemical energies released from raw materials or fuels, and today, these kinds of processes based on chemical energy consumption are still dominant in materials production and processing. But due to ever-increasing need for environmental protection, direct utilization of electrical energy through one carrier-plasma is emerging as a new way to material production/processing, wherever it is feasible in technology and beneficial in economics. The research for new and innovative fields of plasma application resulted in these new approaches solving unmet environmental problems by plasma technology, which is predicted to be the field of maximum growth of future plasma applications.

The advantages of plasma technology are mainly oriented toward the provision of high energy levels at low temperatures by the generation of excited species in an electrical discharge that provides a highly excited medium with no chemical or physical counterpart accessible in the natural environment. This facilitates processing under low-temperature operating conditions and lower residence time compared to conventional methods. Apart from that, the utilization of plasma technology eliminates the need of heat supply and effluent treatment, thereby reducing associated energy cost and thus providing advantages such as lower costs, higher treatment and energy efficiencies, smaller space volume. Accordingly, the interest grows on plasma integration into conventional energy-intensive industrial processing to reduce energy consumption and pollution generation. Textiles and its end-products constitute the world's second largest industries. Textile industries are one of the biggest users of water and complex chemicals and energy and, taken as a whole, are among the top 10 most polluting industries in the world.[100–103] Wet processing is the segment of textile production that involves cleaning, preparation, dyeing, and finishing of textile to provide properties as per user requirement using various thermal, mechanical, and chemical treatments. Most textiles, regardless of their end use, go through wet processing steps. This wet processing of textiles requires huge inputs of energy, water, and chemicals and also results in large volume of chemical laden wastewater that is often difficult to treat due to its complexity and when discharged into environment affects its ecological status by causing various undesirable changes.[100,102]

Positive experience and interest for plasma processing in microelectronics production, automotive industry, biomedical applications, and surface modification of polymers are being slowly transferred to the textile industry though still more at a scientific level.[104,105] NTP treatment of textiles offers plenty of functional, environmental, and economical benefits. Surface properties of polymeric materials that constitute textiles have strong effects on most of their practical applications. Indeed, many properties, such as adhesion, gloss, wettability, permeability, dye ability, printability, surface cleanliness, bonding of different components, biocompatibility, and antistatic behavior, are related more to the surface than to the bulk of the material. This is true whether one is dealing with natural polymers or synthetic polymers. In most cases, a polymer, for instance, is selected primarily because of its favorable bulk properties, such as thermal stability, mechanical strength, solvent resistance, and cost. However, the selected polymer mostly has surface characteristics that are less than optimum for the intended application. Thus, modification of surface properties with plasma can considerably reduce environmental impacts of wet chemical processing.[104,105]

In the past decade, the use of NTPs for selective surface modification of textile has been a rapidly growing research field.[105–110] Thus, plasma processing of textiles opens up new possibilities for industrial applications where the specific advantages of plasma processing provide effective surface modification of textile in environment friendly manner over convention treatment. This technology has other advantages of leaving the bulk characteristics unaffected. For instance, plasma technology provides an attractive means of textile treatment with wide applicability with low quantities of input needed in terms of gases and electricity, making it nonenergy intensive. Plasma technology, being dry technique, does not require water; it involves extremely low quantities of starting materials; the process is realized in gas phase, process duration is low; it provides energy saving and its low temperature avoids textile destruction.[105–111]

In general, the environmental benefits of plasma treatment[104] can be summarized as follows:

- Reduced amount of chemicals needed in conventional processing
- Better exhaustion of chemicals from the bath
- Reduced biochemical oxygen demand (BOD)/chemical oxygen demand (COD) of effluents
- Shortening of the wet processing time
- Decrease in wet processing temperature
- Energy savings

Considering broad field of plasma applications and the limitation to review them all, subsequently in this chapter, experimental results on plasma treatment of cotton textile materials reported during last decades are discussed from environmental standpoint.

6.4 COTTON

Cotton is a soft, fluffy natural vegetable fiber of great economic importance as a raw material for textile. Cotton grows in a boll, or protective capsule, around the seeds of cotton plants of the genus Gossypium of mallow family. There are four commercial species of cotton, the most common of which is *Gossypium hirsutum*. Two of these (*Gossypium arboreum* and *G. herbaceum*) are diploids, and two (*G. hirsutum* and *G. barbadense*) are tetraploids. India is the only country where all the cultivated species and some of their hybrid combinations are commercially grown. Cotton's strength, absorbency, and capacity to be washed and dyed also make it adaptable to a considerable variety of textile products, thus it is the most consumed natural fiber for textile production.[112,113] India is the second largest producer of raw cotton fiber and cellulosic fibers and yarns. The world production of cotton in 2009–2010 is estimated as 25,684 kton, while that of India is 5594 kton, 21.97% of total production.[114]

6.4.1 Structure of Cotton

Cotton is a seed hair fiber taken out from the seeds of cotton plants. After flowering, an elongated capsule or boll is formed in which the cotton fibers grow. Once the fibers have grown completely, the capsule bursts and the fibers come out. A cotton capsule contains about 30 seeds, and each seed hosts around 2000–7000 seed hairs (fibers). The cotton fiber grows as unicellular fiber on the seed coat. The mature cotton fiber forms highly convoluted flat ribbon, varying in width of 12–20 μm. The average fiber length of different kinds of cotton varies from 22 to 50 mm. The mature cotton fiber is basically the cell wall of the cell that is elongated from the seed coat. The cotton fiber is structurally built up into concentric zones and a hollow central core known as the lumen. The mature fiber essentially consists of (from outside to inside) the cuticle, that is, the outermost layer, the primary cell wall, the winding layer, the secondary wall, and the lumen. Figure 6.2 systematically shows the different layers present in the cotton fiber. Cotton contains nearly 88–96% of cellulose and around 10% of noncellulosic substances, which are mainly located in the cuticle and primary wall of the fiber. Typical components in dry mature cotton fibers are cellulose (90–95%), waxes (0.6–1.3%), pectin (0.9–1.2%), protein (0.6–1.3%), ash (1.2%), organic acids (0.8%), and others (1.4%), whereas the chemical components of the outer surface are cellulose (54%), waxes (14%), pectin (9%), protein (%), ash (3%), and others (12%). Thus, most of the noncellulosic impurities are located at the outer surface of cotton fibers.[112,113,115,116]

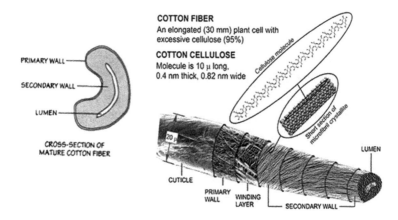

FIGURE 6.2 Schematic Diagram Showing Structure of Cotton Fiber.

The cuticle is the outermost layer of cotton fibers, which consisted of wax and pectin as a thin sheet over the primary wall, and forms grooves on the cotton surface. The primary wall is only 0.5–1 μm thick in mature fiber and comprises noncellulosic materials and amorphous cellulose (50%). The noncellulosic components are hemicelluloses, pectin, proteins, natural colorants, and ions. The thin layer directly adjoining the primary wall is called the winding layer or also known as immediate outer layer of the secondary wall or primary wall and consists of cellulose microfibrils. The secondary wall is mainly made up of crystalline cellulose (92–95%) and concentric layers of densely packed elementary cellulose fibrils (laying parallel), which held together with hydrogen bonds. The lumen forms the center of cotton fibers.[115–117]

The cuticle contains primary alcohols, higher fatty acids, hydrocarbons, aldehydes, glycerides, sterols, acyl components, resins, cutin, and suberin, which are together called waxes. The waxy contents can be divided into two categories: a saponifiable part (nearly 40% of total wax content) and a nonsaponifiable part (which is around 60% of the total wax). Alcohols such as gossypol ($C_{30}H_5OH$), montanyl ($C_{28}H_{57}OH$), and ceryl ($C_{28}H_{53}OH$) are high-molecular weight monohydric alcohols and belong to the category of nonsaponifiable waxes. These n-primary alcohols ($C_{26}-C_{36}$) combined with the fatty acids ($C_{16}-C_{36}$) are the main components of wax from the mature white cotton fiber. Suberin and cutin are insoluble, lipophilic biopolymers also called as bio-polyesters. Suberins and cutins are closely related to each other, the only difference is their chain length and substitution patterns. Suberin from cotton fibers predominately comprises C_{16} and C_{18} compounds. Cutin is the high-molecular weight polyester that comprises various interesterified C_{16} and C_{18} hydroxy and hydroxy–epoxy fatty acids.[115–117]

The noncellulosic components of primary wall are hemicelluloses, pectin, proteins, natural colorants, and ions. The hemicellulose is the name of a heterogeneous group of branched matrix forming polysaccharides. Hemicelluloses bind noncovalently to the surface of cellulose microfibrils in the primary wall. There are several classes of hemicelluloses with an average of 50 glucose units that are linearly $\beta(1\rightarrow4)$-linked to one another. Glycoproteins also account for up to 15% of the primary cell wall mass. Glycoproteins contain a protein backbone with extended rod-like carbohydrates that protrude outwards. The carbohydrates in the glycoproteins account for 65% of the total structure. For a cotton fiber, these rod-shaped extensions are made up of roughly 300 amino acids and abundantly contain hydroxyl-proline (Hyp). The pectin is an acidic polysaccharide and acts as cementing material for the cellulosic network in the primary wall. It is composed of a high proportion of D-galacturonic acid residues, joined together by α (1$\rightarrow$4) linkage. The carboxylic acid groups of some of the galacturonic acid residues are partly esterified with methanol and thus covalently linked with cellulose. The nonesterified or acidic pectin contains many negatively charged galacturonic acid residues. The acidic pectin is organized by crossbridges with calcium ions (Fig. 6.3). The primary wall of a cotton fiber contains different quantities of metal depending on their growing conditions and source. Potassium is the most abundant metal ion in cotton fibers followed by magnesium and calcium.[115,116,118–120]

Cellulose is the main component of cotton fiber. Chemical composition of cellulose is simple; it is a linear polymer of $\beta(1\rightarrow4)$ glucopyranose. Three free hydroxyl groups in C_2, C_3, and C_6 per anhydroglucose unit are available. Each cellulose chain has a nonreducing end group at C_4 and reducing end at C_1. Cellulose in the secondary wall is characterized by a higher degree of polymerization (nearly 5000 units) compared with cellulose in the primary wall. In the secondary wall of a cotton fiber, two cellulose molecules can form a long planar chain of $\beta(1\rightarrow4)$-linked glucose units, forming a sheet called an elementary fibril. Microfibrils of cellulose are crystalline aggregates of approximately 21 elementary cells. Finally, a large number of such micro fibrils that are laid in parallel direction form various layers of the secondary cell wall.[115,116]

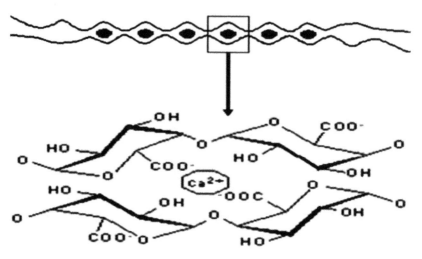

FIGURE 6.3 Egg Box Model Showing Bonding of Nonmethylated Pectins with Ca^{2+} Ions.[120]

6.4.2 Conventional Processing of Cotton Textile

The major portions of noncellulosic impurities are present in the cuticle and primary wall, and these noncellulosic impurities give fiber surface a high hydrophobicity and have adverse effects during wet chemical processing such as bleaching, dyeing, and printing because it interferes with wetting of fiber and penetration of reagents. Therefore, these impurities are removed from cotton textile before its bleaching or dyeing. The typical processing flow diagram of cotton textile is shown in Figure 6.4. Firstly, the desizing process involves the removal of starch from the fabric. After 1950, enzymatic desizing process involving α-amylases has been widely introduced and implemented successfully in the textile industry.[115–117]

However, even today, alkaline scouring of cotton is still the most widespread commercial technique for removing or rupturing the fiber cuticle to make the fiber absorbent for the cotton processing. Scouring of cotton fabric is typically done with a hot solution (90–100°C) of alkali (±1 mol/L) for up to 1 h. NaOH is used in the scouring process to remove the outermost waxy layer during cotton scouring process by saponification and emulsification. In alkali scouring, the primary wall is removed by swelling and hydrolysis with NaOH at elevated temperatures. The scouring process requires large quantities of chemicals, energy, and water and is rather time consuming. Owing to the high sodium hydroxide concentration and its corrosive nature, intensive rinsing is required that leads to high water consumption. The use of high concentrations of sodium hydroxide also requires the neutralization of wastewater, which requires additional acid chemicals. Furthermore, the alkaline effluent requires special handling because of very high BOD and COD values. The energy consumption in cotton scouring is 6.0–7.5 GJ/ton. The challenge now is how to remove this waxy layer efficiently in an environmental friendly way.[101,112,115,116,121,122] Therefore, in the past decades, solvent scouring, enzymatic scouring, and plasma treatment are studied as alternative to alkaline scouring, of which solvent scouring is inherently costly and not environment friendly, and also it only removes wax.[123,124] So, generally, plasma treatment and enzymatic scouring are found to be promising as alternative to alkaline scouring in terms of environmental performance. Enzymatic scouring though widely studied still faces several problems such as long incubation time with enzyme, nonuniformity of treatment, and cost-effectiveness for its application to industrial scale.[125]

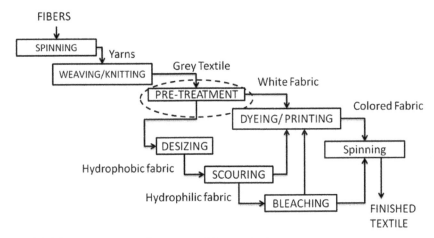

FIGURE 6.4   Flow Diagram of Processing of Cotton Textile.

6.5   PLASMA TREATMENT OF COTTON

6.5.1   Plasma Treatment of Gray Cotton for Removal of Noncellulosic Impurities

In the literature, plasma treatment is reported to improve various properties of gray cotton. Air–chlorine coronas are reported to modify the wax, causing increased wettability, and C-chlorination occurs in about one of every three carbon atoms in the cuticle. Potential practical applications of corona treatment are for improved mercerization, better yarn stretch and setting properties, increased dyeability, and the localized heating and the generation of free radicals, which may make possible heat-induced localized polymerization.[126] The low-temperature oxygen plasma treatment is applied for removal of wax as an alternative to conventional alkaline scouring; it is reported that the oxygen plasma treatment results in almost similar quality in subsequent bleaching, mercerizing, dyeing, and finishing as conventional wet scouring.[127] The low-pressure air plasma treatment is recently reported for improvement in wettability of gray cotton fabric by etching and modification of cuticle layer.[128,129] Also, an effective and cleaner production process for gray cotton fabric pretreatment with an aid of low-pressure nitrogen plasma in combination with a subsequent wild and one-step wet chemical treatment is reported to be more environmentally friendly by shortening the conventional wet chemical treatment process and less consumption of chemicals, water, and energy in comparison with the conventional one-step pretreatment process.[130] The recent development of atmospheric pressure plasma treatment generated a great interest for processing of gray cotton, which predicted to bring manifold potential advantages to textile industry. The treatment of gray cotton with atmospheric pressure plasmas is reported to improve bleaching and dyeing,[131] surface functionality,[132] wettability of gray cotton fabric,[123] wettability and sizing properties of gray cotton yarn,[133,134] wettability and desizing properties of gray cotton fabric,[135] scouring, and dyeing.[136] The oxygen plasma treatment at atmospheric pressure and low-pressure plasma is reported to improve enzymatic scouring of cotton fabric with alkaline pectinase.[137] Influence of temperature on corona discharge treatment is studied, and it is reported that the treatment increased surface oxygen content at a considerably low temperature and then declined when temperature increased. Weight loss rate showed that the treatment was fiercer as treatment temperature increased. The breaking strength and surface adhesion property of the fabric treated with starch sizing increased to a certain extent and then decreased.[138] In order to avoid use of polyvinyl alcohol (PVA), an eco-friendly sizing technology with atmospheric pressure plasma treatment and green sizing recipes has been developed and evaluated with respect to sizing properties and desizing efficiency.[139] Recently, plasma treatment at atmospheric pressure using atmospheric pressure plasma jet (APPJ) apparatus (Atomflo 200-Series manufactured by SurfxTechnologies (California, USA)) is studied for its efficiency for gray cotton processing by comparing it with conventional wet processing. The results obtained from wicking and water drop tests showed that wettability of gray cotton fabrics was greatly improved after plasma treatment and yielded better results than conventional desizing and scouring. The weight reduction of plasma-treated gray cotton fabrics revealed that plasma treatment can help in removing sizing materials and impurities. Similar dyeing results were obtained for samples treated with conventional wet chemical method and plasma-treated samples. This can prove that plasma treatment would be another choice for treating gray cotton fabrics.[140]

However, the use of helium gas to generate atmospheric pressure plasma is the major limitation for industrial application of plasma technology; in this regard, the atmospheric pressure plasma treatment using air DBD seems to be one of the most attractive alternative to conventional alkaline scouring due to its environmental friendliness. In most of the previously reported studies, the plasma treatment of gray cotton fabric is characterized mainly by scanning electron microscopy, X-ray photoelectron spectroscopy, and/or Fourier transformed infrared (FTIR) spectroscopy, and it is concluded that plasma treatment results in the formation of carboxyl groups and free radicals on the surface. Also, considerable changes in surface morphology are reported that are due to etching effect of plasma treatment. Removal of noncellulosic impurities is not clearly evidenced in these previously reported studies. In some studies, changes in wax and pectin content are observed by improved wettability, reduction in weight, solvent extraction, ruthenium red staining, etc.; however, common disadvantage of these techniques is that they are basically qualitative and inaccurate and/or laborious and time consuming.[123,137,141] Improvement in wettability cannot be considered as a strict measure of removal of impurities as both removal of impurities and formation of polar functional groups contribute to this improvement. Reduction in the weight of plasma-treated fabric cannot distinguish between size applied to the fabric and different types of impurities. Also degradation of cellulose by plasma oxidation can contribute to weight reduction. Determination of wax content by solvent extraction is tedious and time consuming, besides its accuracy is questionable.[123,137,141,142] Staining with ruthenium red (ammoniated ruthenium oxychloride) dye is considered as one of the accurate method for the measurement of pectin content. The principle of this test is that amino groups of dye molecules form the covalent bond with carboxyl groups of pectin (Fig. 6.5); thus, spectroscopic measurement of color can be considered as pectin content.[123,143] However, since plasma treatment forms carboxylic groups, accuracy of this method for characterization of plasma-treated fabric remains doubtful.

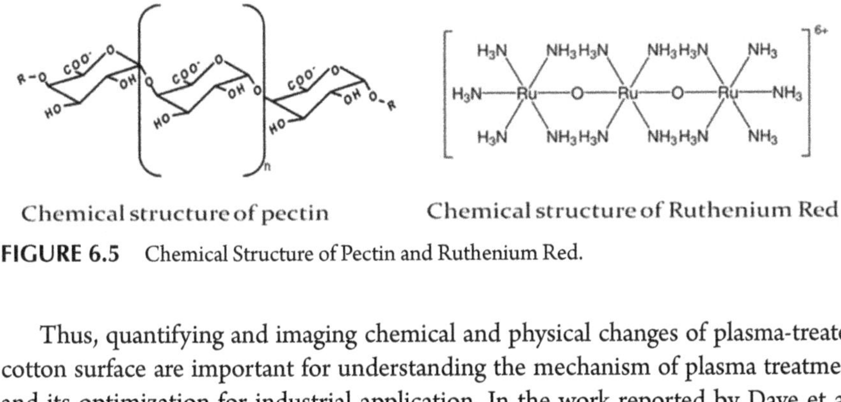

Chemical structure of pectin   Chemical structure of Ruthenium Red

FIGURE 6.5   Chemical Structure of Pectin and Ruthenium Red.

Thus, quantifying and imaging chemical and physical changes of plasma-treated cotton surface are important for understanding the mechanism of plasma treatment and its optimization for industrial application. In the work reported by Dave et al., gray khadi cotton fabric is treated with atmospheric pressure air DBD (treatment time: 1–5 min), and the changes in surface chemical composition occurred during the air DBD plasma treatment are analyzed with attenuated total reflection (ATR)-FTIR spectroscopy method, which can provide information about impurities of cotton fiber in a fast and efficient manner.[144] In Figure 6.6, the selected region of the spectra (from wave number 2600–3000 and 1500–1800 cm^{-1}), which represents peaks related to impurities, is presented. As seen from the figure, intensity of 2956–3463 and 2817–2979 cm^{-1} significantly changed, which is due to mercerizing effect of alkali leading to change in crystal structure of cellulose. Besides, changes observed in the wave number regions of 2600–3000 and 1500–1800 cm^{-1} present peaks related to noncellulosic impurities. Broad C–H stretching peak appeared at 2817–2979 cm^{-1} in the spectra of gray as well as treated fabric. However, in the spectrum of gray fabric, as shown in Figure 6.6, two distinguished peaks corresponding to the symmetric and asymmetric stretching mode of methylene groups in long alkyl chain are clearly visible at 2852.24 and 2917.81 cm^{-1}, respectively. Similar results obtained in case of acid desized cotton fabric. According to Chung et al.,[142] –CH$_2$– groups from cellulose never give separate peaks corresponding to the symmetric and asymmetric stretching mode in ATR spectra of pure cellulose. Also other impurities such as pectin do not have long alkyl chain of methylene groups. Therefore, intensities of these peaks indicate the amount of wax present on the fabric.[142] As seen from Figure 7.6, in the spectra of scoured cotton fabric, these peaks completely disappeared, which indicates that no wax remains on the surface of cotton fiber after alkaline scouring.[144]

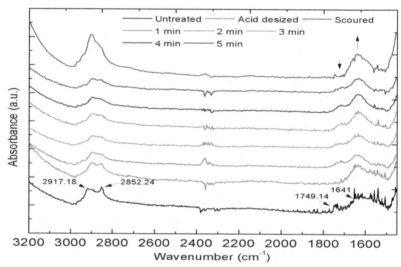

FIGURE 6.6 Comparison of Spectra from Wave Number 2600–3000 and 1500–1800 cm^{-1}.

Similarly, these peaks also disappeared in the spectra of plasma-treated cotton fabric indicating wax removal due to plasma treatment. These peaks are not visible, even when the spectrum of fabric is exposed to plasma for 1 min. The result is also supported by the recent study of Raja et al.,[145] which reports reduction in the intensity of these peaks when scouring of cotton is carried out by bacterial lipase. The spectrum of gray fabric showed small absorbance band of C=O stretch near 1749 cm^{-1}, which indicates free fatty acids and cutin present in wax and free and esterified carboxylic groups of pectin. In the case of scoured and treated fabric, it merely presents as weak absorbance, indicating removal of these impurities.[142] The spectrum of gray fabric displayed a weaker absorption at 1641 cm^{-1}, whereas spectra of scoured and plasma-treated fabric displayed strong absorption at this frequency, which increases with increase in plasma exposure time. The IR band at 1641 cm^{-1} is composed of adsorbed H$_2$O at

1631–1644 cm^{-1} and asymmetric carboxylate stretch at 1600 cm^{-1}, which is difficult to differentiate in ATR-FTIR.[142]

According to Chung et al.,[142] this peak represents impurities such as free fatty acids of wax and pectin that contains carboxylate (COO$^-$) functional groups and can be used for characterization of scouring process. In their work, they reported appearance of a new peak at around 1750 cm^{-1} in ATR-FTIR spectra of gray fabric if it was exposed to the HCl vapor for a few minutes, which they proved from the protonation of carboxylate groups (Fig. 6.7). Also, they demonstrated that scoured fabric is free from these impurities, and thus, no such peak in ATR-FTIR spectrum is induced by HCl vapor exposure.[142] Similar results are reported by Wang et al.[141] for bio-scouring of gray cotton.

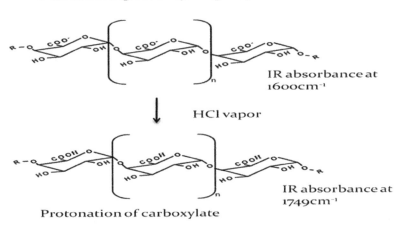

FIGURE 6.7 Change in IR Absorbance after HCl Vapor Exposure by Protonation of Carboxylate Functional Groups of Acidic Pectin.

In the study by Dave et al., when HCl vapor exposure is given to gray, chemically treated and air DBD-treated fabric, absorption intensity at 1641 and 1749 cm^{-1} is varied significantly in ATR-FTIR spectra of HCl vapor-exposed fabrics, which are summarized in Table 6.1. Figure 6.8 represents the comparison of ATR-FTIR spectra obtained before and after HCl vapor exposure for gray cotton, acid desized, and scoured cotton fabric. In case of gray cotton, the band at 1641 cm^{-1} is completely disappeared in ATR-FTIR spectra after exposure to HCl vapor, and a strong absorption band developed at 1749 cm^{-1}. This enhanced absorbance at 1749 cm^{-1} is the attributed protonation of the carboxylate present in wax and pectin. Since gray fabric is super hydrophobic (Fig. 6.9), it can be concluded that no water is adsorbed to the surface. Thus, the peak at 1641 cm^{-1} is from asymmetric carboxylate stretch (COO$^-$) from impurities, which converted in to carboxyl (COOH) through protonation leaving no peak around 1641 cm^{-1} in the spectrum of HCl vapor-exposed gray cotton fabric. Similarly, in the case of acid desized cotton fabric, a strong absorbance band developed after HCl vapor exposure, intensity of which is even higher compared to gray cotton fabric.[144] This can be explained as consequences of desizing, which removes size layer, and thus, now more impurities are exposed to surface. In the case of scoured fabric, no such peak is induced after exposure to HCl vapor, which is in accordance with the results previously reported by Chung et al.[142] due to absence of impurities.

TABLE 6.1 Experimental Peak Obtained in ATR-FTIR Spectra after HCl Vapor Exposure

Samples	Experimental Peak Obtained after HCl Vapor Exposure	
	1641 cm–1	1749 cm–1
Untreated gray cotton	–	Strong peak
Acid desized cotton	–	Strong peak
Scoured cotton	Clear peak	–
1 min plasma-treated cotton	–	Relatively weak peak
2 min plasma-treated cotton	Clear peak	Very weak peak
3 min plasma-treated cotton	Clear peak	Very weak peak
4 min plasma-treated cotton	Clear peak	Clear peak with higher intensity than 1641 cm^{-1}
4 min plasma-treated cotton	Clear peak	Clear peak with higher intensity than 1641 cm^{-1}

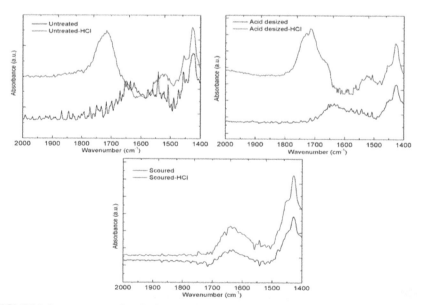

FIGURE 6.8 Experimental Peak Obtained after HCl Vapor Exposure in Case of Gray (Untreated), Acid Desized, and Chemically Scoured Cotton Fabric.

FIGURE 6.9 Image of Water Drop and Water Contact Angle on Gray Cotton Fabric.

Figure 6.10 represents the comparison of ATR-FTIR spectra obtained before and after HCl vapor exposure for cotton fabric treated with air DBD at different time durations. The result for 1 min plasma-treated fabric is similar to that of gray cotton fabric; however, intensity of 1749 cm^{-1} IR bond developed after HCl vapor exposure is lower compared to gray fabric, also a very weak absorbance observed at 1641 cm^{-1}. In the spectra of 2 and 3 min plasma-treated fabric, clear absorbance band appeared at 1641 cm^{-1} after HCl vapor exposure, which is assigned to adsorbed water, while band at 1749 cm^{-1} appeared merely as weak absorbance. It is important to note here that at a first glance, not much difference can be seen in the results of HCl vapor-exposed fabric; however, intensity of 1641 cm^{-1} band is comparatively higher for 3 min plasma-treated fabric to that of 2 min plasma-treated fabric before its exposure to HCl.[144]

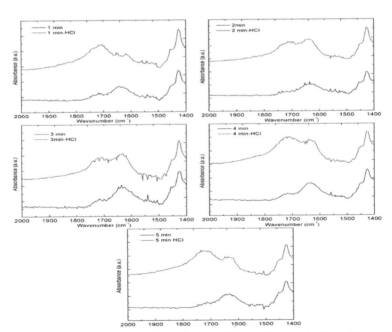

FIGURE 6.10 Experimental Peak Obtained after HCl Vapor Exposure in Case of Cotton Fabric Treated with Air Dielectric Barrier Discharge for 1–5 min.

This indicates that plasma treatment removes noncellulosic impurities by etching and chemical degradation. Since noncellulosic impurities are mainly located at the surface (cuticle and primary wall) of cotton fiber covering cellulose (secondary wall), the above results indicate that for 1 min plasma-treated fabric cotton fibers, the surface is still covered with noncellulosic impurities. As the plasma treatment time increases, more and more impurities are removed from cotton fiber and thus more hydroxyl groups of the cellulose are now present at the surface, which is reflected in the spectra of 2 and 3 min plasma-treated fabric.[144]

In the spectra of 4 and 5 min plasma-treated fabric, two clear bands are visible after HCl vapor treatment at 1641 and 1749 cm^{-1} with intensity of band at 1749 cm^{-1} higher than that of 1641 cm^{-1}. This indicates that plasma treatment is creating carbonyl functional groups (COOH, COO^{-}), apart from eliminating noncellulosic impurities. Formation of such functional groups is well documented in the literature.[123,131–134,137] Such functional groups can be produced due to the oxidation of cellulose (advanced oxidation based on ozone and atomic oxygen) by active species of plasma, especially primary alcohol at C_6 (Fig. 6.11) carbon of glucose in cellulose can be oxidized to carboxylic acid.[146] Since primary alcohol groups can only be oxidized to carboxylic acid, and no other impurities except cellulose have primary alcohol groups, the formation of carboxylate indicates the presence of cellulose on the surface. Thus, it can be inferred that in the case of plasma treatment time exceeding 3 min, most of the impurities are removed from the surface, and polar functional groups such as COO^{-} are created by subsequent cellulose oxidation which can further contribute in the wettability of cotton. Since during plasma treatment, removal of impurities and formation of carbonyl functional groups are simultaneously occurring, we never observed nil intensity around 1749 cm^{-1}. However, the present results are satisfactory to trace removal of noncellulosic impurities by air DBD treatment. Thus, it is clear from ATR-FTIR results that the plasma treatment removes noncellulosic impurities from the surface of cotton fiber, and then polar functional groups are generated by subsequent oxidation of cellulose.[144]

In this study, optical emission spectroscopy is used for the identification of various active species in the plasma discharges, which indicates that ozone is produced in the air DBD.[144] Degradative effects of ozone on cuticle layer are well established in the literature, which includes wax erosion and changes in chemical composition, thickness, and structure of cuticle layer.[147] Also, the treatment with ozone gas is reported to remove wax and the starch sizing from the gray cotton samples and increase the water absorbency of the gray and desized cotton samples. Ozone treatment is also reported to oxidize cellulose surface-generating carboxylate anions.[148–150] Thus, removal of noncellulosic impurities from gray cotton by air DBD treatment reported in this study is inferred due to etching by energetic species in combination with ozone-based advanced oxidation of surface noncellulosic impurities.

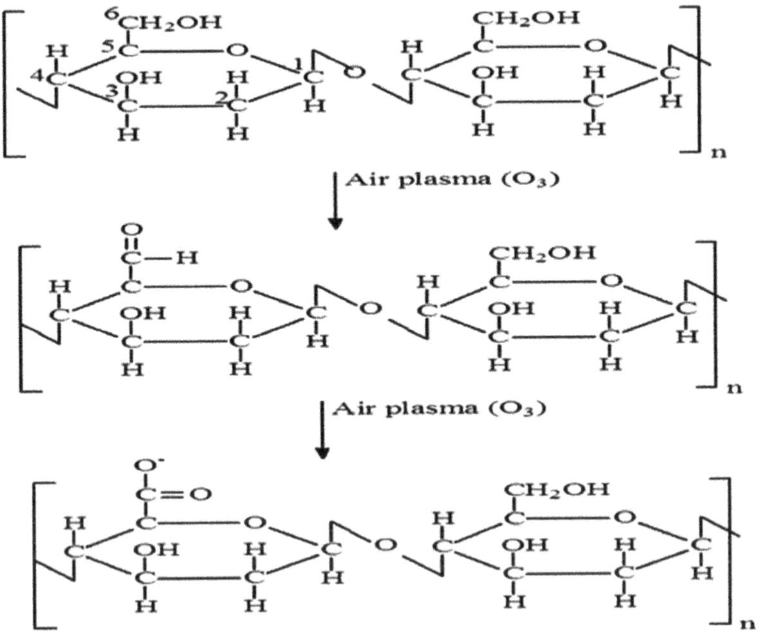

FIGURE 6.11   Reaction Mechanism for Formation of Carboxylate by Oxidation of Primary Alcohol Groups at C_6 Carbon of Glucose in Cellulose.

6.5.2   Plasma Treatment for Removal of Size From Cotton Textile

The enzymatic desizing is now industrially used as cleaner treatment for removal of starch; however, for desizing of PVA, plasma treatment may be a suitable alternative. Various studies have been carried out for plasma-aided desizing of PVA. The pretreatment with atmospheric pressure He/O_2 APPJ is reported to improve $NaHCO_3$ desizing of blended sizes of starch phosphate and PVA on cotton fabrics and its combination with ultrasound has shown higher effectiveness in desizing and provides an alternative approach that decreases the water, energy, and chemicals consumption.[151,152] The treatment with argon/oxygen DBD is reported for desizing of PVA from cotton fabric.[153] Air/He and air/O_2/He atmospheric plasma treatments are applied to desize PVA on cotton, of which Air/O_2/He plasma is more effective than air/He plasma on PVA desizing.[154] Plasma treatment is also studied for desizing of gray cotton fabric sized with starch and PVA, and results showed that the plasma has positive effects on size removal, hydrophility, and the pilling values of the PVA and also starch-sized fabrics.[155] Near-atmospheric pressure oxygen DC plasma is reported for desizing of starch from gray cotton fabric.[156] Atmospheric pressure plasma treatment of gray cotton denim fabric is reported for desizing and improvement in color fading in subsequent enzymatic color-fading process.[157]

6.5.3 Plasma Treatment to Improve Dyeing/Printing Cotton Textile

The plasma treatment at atmospheric pressure and low pressure is extensively explored to improve dye uptake of cotton textile. The low-pressure air and oxygen plasma is reported to generate carboxyl groups on the cotton surface that results in the improvement in wettability and dye uptake. The direct dye (chloramine Fast Red K) uptake of the treated samples increases almost linearly with the increase in fiber carboxyl group content caused by plasma-based advanced oxidation of cotton surface by species such as atomic oxygen radicals.[158] The plasma pretreatment coupled with the ink-jet printing technique could improve the final properties of printed cotton fabric.[159] The activation of cotton fabric surface by atmospheric pressure argon plasma and subsequent grafting with two different amine compounds, ethylenediamine and triethylenetetramine, is reported to improve dyeing properties of cotton with acid dyes.[160] Similarly, air corona plasma treatment of cotton fabric is reported to improve dyeing with reactive dyes.[161] The plasma treatment with low-pressure radio-frequency (RF) oxygen plasma is reported for grafting of acrylic acid on the cotton fabric. Improvement in dyeing with Berberine, a natural cationic dye, is observed. The samples dyed after acrylic acid grafting showed the highest antibacterial activity.[162] Treatment of cotton fabric with air plasma and dichlorodifluoromethane (DCFM) plasma is reported to improve dyeing behavior of cotton fabrics with reactive and natural dyes. The DCFM plasma treatment also improves hydrophobicity of cotton fabric.[163] The plasma treatment with low-pressure oxygen is also reported for coating of printing medium containing (i) sodium alginate alone and (ii) sodium alginate/chitosan mixture. It is revealed that with 2 min plasma treatment, significant improvement in ink-jet printing property can be obtained.[164] Similar results are also obtained with atmospheric pressure plasma treatment.[165] The treatment of cotton fabric with APPJ is recently reported for improvement in pigment application for coloration of cotton textile.[166]

6.5.4 Plasma Treatment for Flame-Retardant Cotton Textile

Pretreatment with APPJ is reported to enhance flame-retardant property by improving flame-retardant finishing and may compensate for the reduction in tensile and tearing strength caused by flame-retardant agents.[167] The simultaneous grafting and polymerization of fire-retardant monomers on cotton fabric induced by argon plasma have been investigated with four acrylate monomers containing phosphorus, diethyl (acryloyloxyethyl) phosphate, diethyl-2-(methacryloyloxyethyl) phosphate, diethyl (acryloyloxymethyl) phosphonate, and dimethyl (acryloyloxymethyl) phosphonate for potential flame retardancy for cotton fabrics.[168] Pretreatment by cold plasma has been found to be effective for improvement in thermal stability and/or flame retardancy of cotton fabric using suspensions of nanoparticles (namely, hydrotalcite and nanometric silica).[169] Two different properties to cotton fabrics by means of the cold plasma technique are achieved involving Ar plasma-induced graft polymerization (PIGP) of flame-retardant monomers (acrylate phosphate and phosphonates derivatives) combined to a water-repellent treatment – CF_4 plasma treatment or Ar PIGP of 1,1,2,2, tetrahydroperfluorodecylacrylate (AC8).[170]

6.5.5 Plasma Treatment for Improvement in Hydrophobicity of Cotton Textile

Self-cleaning and allied applications require hydrophobic surface properties for textiles. Processed cotton being hydrophilic in nature, plasma polymerization provides an innovative way to provide hydrophibicity to cotton. Treatment of cotton textile with polymerizing gas hexafluoroethane (C_2F_6) plasma is reported to improve hydrophobicity of cotton.[171] Also plasma treatment with hexafluoropropene (C_3F_6) results in incorporation of fluorine atoms on the surface by forming $-CF$, $-CF_2$, $-CF_3$ groups on the surface. This resulted in remarkable improvement in hydrophobicity; after plasma treatments of only 1 min, contact angles can reach 120° or higher, while wet-out time can be as high as 60 min. Also tensile strength improved after the plasma treatment.[172] The effect of RF inductively coupled SF_6 plasma on the surface characteristics of cotton fabric is studied for improvement in mechanical strength and hydrophobicity. Overall good and durable hydrophobic properties on cotton can be achieved only for the longer treatments used but the mechanical strength of the fabrics is compromised to a significant extent.[173] Water repellency can also be achieved with plasma polymerization of hexamethyldisiloxane and hexamethyldisilane monomers on to the cotton surface by low–pressure, low-temperature RF (RF 13.56 MHz) plasma. For the cotton fabric, water repellency property was enhanced with a heat treatment after hexamethyldisiloxane (HMDSO) plasma polymerization. This may result from the orientation of Si groups. Flame-retardant properties of cotton fabrics were improved by plasma treatments.[174] Improvement in hydrophobicity is demonstrated by plasma treatment with 3:1 HMDSO/toluene glow plasma discharge at low pressure for the hydrophobic and super-hydrophobic coating on cotton clothes.[175] The plasma equipment is also developed for atmospheric pressure deposition of hydrophobic siloxane coatings on cotton using hexamethyldisiloxane as a siloxane precursor.[176]

6.5.6 Plasma Treatment for Antimicrobial Cotton Textile

The use of cotton as surgical garments has evolved as a standard practice, and their primary purpose is to protect surgical zone from microbial invasion and fluid-borne pathogens. Cotton is widely used for dressing in medical treatment. Cotton fibers due to their hydrophilic nature support the growth of micro-organisms such as bacteria and fungi. The plasma treatment of cotton provides an innovative mean for imparting antimicrobial properties. The plasma treatment of cotton fabric with water vapor plasma is reported to enhance loading and adhesion of nanosilver to cotton fabric, which contributed to antimicrobial effectiveness to *Pseudomonas aeruginosa* and *Escherichia coli.*[177] Similarly, nitrogen and oxygen plasma treatment is reported to increase absorption of silver particle. The results showed that nitrogen plasma-treated samples can absorb more silver particles than oxygen-treated samples and thus provide antibacterial activity to cotton.[178] The atmospheric pressure plasma treatment is reported for improved loading of chemical antimicrobial agents and silver particles.[179,180] The plasma treatment with APPJ is reported to increase the surface roughness of cotton textile and improve the loading of zinc oxides on the surface. The plasma-treated specimens had better overall tensile properties even after antimicrobial treatment.[181] The antimicrobial activity was imparted to the RF oxygen plasma-treated cotton fabric using methanolic extract of neem (*Azadirachta indica*) leaves containing Azadirachtin. The antimicrobial activity of these samples has been analyzed and compared with the activity of the cotton fabric treated with neem extract alone. The investigation revealed that the surface modification due to RF oxygen plasma was found to increase the hydrophilicity due to carboxyl functional groups and hence the antimicrobial activity of the cotton fabric when treated with Azadirachtin.[182] The plasma treatment followed by enzymatic treatment is reported to have synergetic effect to provide durable antimicrobial activity using methanolic extract of neem leaves. Thus, plasma treatments of cotton fulfill the need for eco-friendly antimicrobial textiles.[183]

6.6 CONCLUSION

The main purpose of this review is to offer an overview to readers about wide variety of applications of emerging plasma technologies, aiming for reducing environmental impacts of conventional processing of cotton textile. This review offers a background of plasma chemistry as an innovative tool for processing of cotton textile. It can be concluded that a large number of applications can be developed for processing of cotton textiles including imparting novel properties such as super-hydrophobicity, antimicrobial, and antiflammable properties. The NTP indeed has great potential to become a cost-effective as well as environmental friendly textile treatment technology. However, the plasma technology has few challenges to completely replace conventional processing of cotton textile. In many studies, helium is used as carrier gas to produce atmospheric pressure plasma; however, its cost and requirement in large quantity pose a problem in commercialization of plasma technology. To overcome the skepticism from industry, this would require more research in this field. In order to understand plasma chemistry that takes place at the surface and how plasma species interacts with the surface of cotton textile, in-depth study is essential. A wide range of plasma sources are used in different studies; however, there are very few scientific reports available in using plasma technology at industrial scale reactors. Apart from that, a comparative study of different types of plasmas for cotton textile processing should be carried out to come to the conclusion to select appropriate plasma source for textile processing.

ACKNOWLEDGMENTS

The authors are greatly thankful to Department of Science and Technology, Government of India, for the financial support. The authors are thankful to the all staff members of Institute for Plasma Research, Gandhinagar, and Manipal University, Jaipur, for their co-operation and support.

KEYWORDS

- **Cotton**
- **natural fiber**
- **textile**
- **nonthermal plasma**
- **surface modification**
- **plasma chemistry**
- **environmental friendly**
- **surface properties**

REFERENCES

1. Mollah, M. Y. A.; Schennach, R.; Patscheider, J.; Promreuk, S.; Cocke, D. L. Plasma chemistry as a tool for green chemistry, environmental analysis and waste management. *J. Hazard. Mater.* 2000, 79, 301–320.
2. Roth, J. R. *Industrial Plasma Engineering: Volume 1: Principles*; Industrial Plasma Engineering; Taylor & Francis: London, 2002.
3. Denes, F. S.; Manolache, S. Macromolecular plasma-chemistry: an emerging field of polymer science. *Prog. Polym. Sci.* 2004, 29, 815–885.
4. Kim, H.-H. Nonthermal plasma processing for air-pollution control: a historical review, current issues, and future prospects. *Plasma Process. Polym.* 2004, 1, 91–110.
5. National Research Council. *Plasma Processing of Materials: Scientific Opportunities and Technological Challenges*; The National Academies Press: Washington, DC, 1991.
6. National Research Council. *Plasma Science: From Fundamental Research to Technological Applications*; The National Academies Press: Washington, DC, 1995.
7. Bogaerts, A.; Neyts, E.; Gijbels, R.; van der Mullen, J. Gas discharge plasmas and their applications. *Spectrochimica Acta B.* 2002, 57, 609–658.
8. Attri, P.; Arora, B.; Choi, E. H. Utility of plasma: a new road from physics to chemistry. *RSC Adv.* 2013, 3, 12540–12567.

9. Samukawa, S.; Hori, M.; Rauf, S.; Tachibana, K.; Bruggeman, P.; Kroesen, G.; Whitehead, J. C.; Murphy, A. B.; Gutsol, A. F.; Starikovskaia, S.; Kortshagen, U.; Boeuf, J. P.; Sommerer, T. J.; Kushner, M. J.; Czarnetzki, U.; Mason, N. The 2012 plasma roadmap. *J. Phys. D Appl. Phys.* 2012, 45, 253001.

10. Taylor, P.; Pirzada, S. Thermal plasma processing of materials: a review. *Adv. Perform. Mater.* 1994, 1, 35–50.

11. Boulos, M. I. Thermal plasma processing. *IEEE Trans. Plasma Sci.* . 1991, 19, 1078–1089.

12. Yoshida, T. The future of thermal plasma processing for coating. *Pure Appl. Chem.* 1994, 66, 1222–1230.

13. Bessmertnyi, V. S. Plasma treatment of glasses (a review). *Glass Ceram.* 2001, 58, 121–124.

14. Taga, Y. Review of plasma thin-film technology in automobile industry. *Surf. Coat. Technol.* 1999, 112, 339–346.

15. Anders, A. Plasma and ion sources in large area coating: a review. *Surf. Coat. Technol.* 2005, 200, 1893–1906.

16. Merche, D.; Vandencasteele, N.; Reniers, F. Atmospheric plasmas for thin film deposition: a critical review. *Thin Solid Films.* 2012, 520, 4219–4236.

17. Randhawa, H. Review of plasma-assisted deposition processes. *Thin Solid Films.* 1991, 196, 329–349.

18. Mitchel, S.; Bell Alexis, T. A review of recent advances in plasma polymerization. *Plasma Polym.* 2009, 108, 1–33.

19. Lundin, D.; Pedersen, H. High power pulsed plasma enhanced chemical vapor deposition: a brief overview of general concepts and early results. *Phys. Proced.* 2013, 46, 3–11.

20. Shah, A.; Moulin, E.; Ballif, C. Technological status of plasma-deposited thin-film silicon photovoltaics. *Solar Energy Mater. Solar Cells.* 2013, 119, 311–316.

21. Chu, P. K.; Chan, C. Applications of plasma immersion ion implantation in microelectronics – a brief review. *Surf. Coat. Technol.* 2001, 136, 151–156.

22. Anders, A. From plasma immersion ion implantation to deposition: a historical perspective on principles and trends. *Surf. Coat. Technol.* 2002, 156, 3–12.

23. Pelletier, J.; Anders, A. Plasma-based ion implantation and deposition: a review of physics, technology, and applications. *IEEE Trans. Plasma Sci.* 2005, 33, 1944–1959.

24. Chu, P. K. Progress in direct-current plasma immersion ion implantation and recent applications of plasma immersion ion implantation and deposition. *Surf. Coat. Technol.* 2013, 229, 2–11.

25. Okada, K. Plasma-enhanced chemical vapor deposition of nanocrystalline diamond. *Sci. Tech. Adv. Mat.* 2007, 8, 624.

26. Ishigaki, T.; Li, J.-G. Synthesis of functional nanocrystallites through reactive thermal plasma processing. *Sci. Tech. Adv. Mat.* 2007, 8, 617.

27. Bryjak, M.; Gancarz, I.; Smolinska, K. Plasma nanostructuring of porous polymer membranes. *Adv. Colloid Interface Sci.* 2010, 161, 2–9.

28. Ostrikov, K. Control of energy and matter at nanoscales: challenges and opportunities for plasma nanoscience in a sustainability age. *J. Phys. D Appl. Phys.* 2011, 44, 174003.

29. Keshri, A. K.; Agarwal, A. Plasma processing of nanomaterials for functional applications – a review. *Nanosci. Nanotechnol Lett.* 2012, 4, 228–250.

30. Wang, L.; Zhang, J.; Jiang, W. Recent development in reactive synthesis of nanostructured bulk materials by spark plasma sintering. *Int J. Refract. Metals Hard Mater.* 2013, 39, 103–112.

31. Kylián, O.; Choukourov, A.; Biederman, H. Nanostructured plasma polymers. *Thin Solid Films.* 2013, 548, 1–17.

32. He, B.; Yang, Y.; Yuen, M. F.; Chen, X. F.; Lee, C. S.; Zhang, W. J. Vertical nanostructure arrays by plasma etching for applications in biology, energy, and electronics. *Nano Today.* 2013, 8, 265–289.

33. Kali, R.; Mukhopadhyay, A. Spark plasma sintered/synthesized dense and nanostructured materials for solid-state Li-ion batteries: overview and perspective. *J. Power Sources.* 2014, 247, 920–931.

34. Wang, J.-J.; Choi, K.-S.; Feng, L.-H.; Jukes, T. N.; Whalley, R. D. Recent developments in DBD plasma flow control. *Prog. Aerosp. Sci.* 2013, 62, 52–78.

35. Kizling, M. B.; Järås, S. G. A review of the use of plasma techniques in catalyst preparation and catalytic reactions. *Appl. Catal. A.* 1996, 147, 1–21.

36. Liu, C.; Vissokov, G. P.; Jang, B. W.-L. Catalyst preparation using plasma technologies. *Catal. Today.* 2002, 72, 173–184.

37. Hammer, T.; Kappes, T.; Baldauf, M. Plasma catalytic hybrid processes: gas discharge initiation and plasma activation of catalytic processes. *Catal. Today.* 2004, 89, 5–14.

38. Istadi; Amin, N. A. S. Co-generation of synthesis gas and C2+ hydrocarbons from methane and carbon dioxide in a hybrid catalytic-plasma reactor: a review. *Fuel.* 2006, 85, 577–592.

39. Petitpas, G.; Rollier, J.-D.; Darmon, A.; Gonzalez-Aguilar, J.; Metkemeijer, R.; Fulcheri, L. A comparative study of non-thermal plasma assisted reforming technologies. *Int. J. Hydrogen Energy.* 2007, 32, 2848–2867.

40. Tao, X.; Bai, M.; Li, X.; Long, H.; Shang, S.; Yin, Y.; Dai, X. CH_4–CO_2 reforming by plasma – challenges and opportunities. *Prog. Energy Combust. Sci.* 2011, 37, 113–124.

41. Nozaki, T.; Okazaki, K. Non-thermal plasma catalysis of methane: principles, energy efficiency, and applications. *Catal. Today.* 2013, 211, 29–38.

42. Hessel, V.; Anastasopoulou, A.; Wang, Q.; Kolb, G.; Lang, J. Energy, catalyst and reactor considerations for (near)-industrial plasma processing and learning for nitrogen-fixation reactions. *Catal. Today.* 2013, 211, 9–28.

43. Hessel, V.; Cravotto, G.; Fitzpatrick, P.; Patil, B. S.; Lang, J.; Bonrath, W. Industrial applications of plasma, microwave and ultrasound techniques: nitrogen-fixation and hydrogenation reactions. *Chem. Eng. Process.* 2013, 71, 19–30.

44. Chen, H. L.; Lee, H. M.; Chen, S. H.; Chao, Y.; Chang, M. B. Review of plasma catalysis on hydrocarbon reforming for hydrogen production – interaction, integration, and prospects. *Appl. Catal. B Environ.* 2008, 85, 1–9.

45. Hui, R.; Wang, Z.; Kesler, O.; Rose, L.; Jankovic, J.; Yick, S.; Maric, R.; Ghosh, D. Thermal plasma spraying for SOFCs: applications, potential advantages, and challenges. *J. Power Sources.* 2007, 170, 308–323.

46. Starikovskiy, A.; Aleksandrov, N. Plasma-assisted ignition and combustion. *Prog. Energy Combust. Sci.* 2013, 39, 61–110.

47. Ohl, A.; Schröder, K. Plasma-induced chemical micropatterning for cell culturing applications: a brief review. *Surf. Coat. Technol.* 1999, 11(6–119), 820–830.

48. Moisan, M.; Barbeau, J.; Moreau, S.; Pelletier, J.; Tabrizian, M.; Yahia, L. Low-temperature sterilization using gas plasmas: a review of the experiments and an analysis of the inactivation mechanisms. *Int. J. Pharm.* 2001, 226, 1–21.

49. Siow, K. S.; Britcher, L.; Kumar, S.; Griesser, H. J. Plasma methods for the generation of chemically reactive surfaces for biomolecule immobilization and cell colonization – a review. *Plasma Process. Polym.* 2006, 3, 392–418.

50. Moreau, M.; Orange, N.; Feuilloley, M. G. J. Non-thermal plasma technologies: new tools for biodecontamination. *Biotechnol. Adv.* 2008, 26, 610–617.

51. Bazaka, K.; Jacob, M. V.; Crawford, R. J.; Ivanova, E. P. Plasma-assisted surface modification of organic biopolymers to prevent bacterial attachment. *Acta Biomater.* 2011, 7, 2015–2028.

52. Coad, B. R.; Jasieniak, M.; Griesser, S. S.; Griesser, H. J. Controlled covalent surface immobilisation of proteins and peptides using plasma methods. *Surf. Coat. Technol.* 2013, 233, 169–177.

53. Kong, M. G.; Kroesen, G.; Morfill, G.; Nosenko, T.; Shimizu, T.; van Dijk, J.; Zimmermann, J. L. Plasma medicine: an introductory review. *N. J Phys.* 2009, 11, 115012.

54. Schlegel, J.; Köritzer, J.; Boxhammer, V. Plasma in cancer treatment. *Clin. Plasma Med.* 2013, 1, 2–7.

55. von Woedtke, T.; Reuter, S.; Masur, K.; Weltmann, K.-D. Plasmas for medicine. *Phys. Rep.* 2013, 530, 291–320.

56. Isbary, G.; Zimmermann, J. L.; Shimizu, T.; Li, Y.-F.; Morfill, G. E.; Thomas, H. M.; Steffes, B.; Heinlin, J.; Karrer, S.; Stolz, W. Non-thermal plasma – more than five years of clinical experience. *Clin. Plasma Med.* 2013, 1, 19–23.

57. Emmert, S.; Brehmer, F.; Hänßle, H.; Helmke, A.; Mertens, N.; Ahmed, R.; Simon, D.; Wandke, D.; Maus-Friedrichs, W.; Däschlein, G.; et al. Atmospheric pressure plasma in dermatology: ulcus treatment and much more. *Clin. Plasma Med.* 2013, 1, 24–29.

58. Kramer, A.; Lademann, J.; Bender, C.; Sckell, A.; Hartmann, B.; Münch, S.; Hinz, P.; Ekkernkamp, A.; Matthes, R.; Koban, I.; et al. Suitability of tissue tolerable plasmas (TTP) for the management of chronic wounds. *Clin. Plasma Med.* 2013, 1, 11–18.

59. Robert, E.; Vandamme, M.; Brullé, L.; Lerondel, S.; Pape, A. L.; Sarron, V.; Riès, D.; Darny, T.; Dozias, S.; Collet, G.; et al. Perspectives of endoscopic plasma applications. *Clin. Plasma Med.* 2013, 1, 8–16.

60. Martines, E.; Brun, P.; Brun, P.; Cavazzana, R.; Deligianni, V.; Leonardi, A.; Tarricone, E.; Zuin, M. Towards a plasma treatment of corneal infections. *Clin. Plasma Med.* 2013, 1, 17–24.

61. Boulais, E.; Lachaine, R.; Hatef, A.; Meunier, M. Plasmonics for pulsed-laser cell nanosurgery: fundamentals and applications. *J. Photochem. Photobiol. C Photochem. Rev.* 2013, 17, 26–49.

62. Nishida, Y.; Liu, C.-M.; Fan, F.-Y.; Iwasaki, K.; Ou, K.-L. Application of high-electric-field plasma to medical fields. *J. Exp. Clin. Med.* 2013, 5, 5–11.

63. Ladizesky, N. H.; Ward, I. M. A review of plasma treatment and the clinical application of polyethylene fibres to reinforcement of acrylic resins. *J. Mater. Sci. Mater. Med.* 1995, 6, 497–504.

64. Loh, J. H. Plasma surface modification in biomedical applications. *Med. Dev. Technol.* 1999, 10, 24–30.

65. Yang, Y.; Kim, K.-H.; Ong, J. L. A review on calcium phosphate coatings produced using a sputtering process – an alternative to plasma spraying. *Biomaterials.* 2005, 26, 327–337.

66. Yang, J.; Cui, F.; Lee, I. S.; Wang, X. Plasma surface modification of magnesium alloy for biomedical application. *Surf. Coat. Technol.* 2010, 205(Suppl 1), S182–S187.

67. Surmenev, R. A. A review of plasma-assisted methods for calcium phosphate-based coatings fabrication. *Surf. Coat. Technol.* 2012, 206, 2035–2056.

68. Chu, P. K.; Chen, J. Y.; Wang, L. P.; Huang, N. Plasma-surface modification of biomaterials. *Mater. Sci. Eng.* 2002, 36, 143–206.

69. Desmet, T.; Morent, R.; Geyter, N. D.; Leys, C.; Schacht, E.; Dubruel, P. Nonthermal plasma technology as a versatile strategy for polymeric biomaterials surface modification: a review. *Biomacromolecules.* 2009, 10, 2351–2378.

70. Morent, R.; De Geyter, N.; Desmet, T.; Dubruel, P.; Leys, C. Plasma surface modification of biodegradable polymers: a review. *Plasma Process Polym.* 2011, 8, 171–190.

71. Jacobs, T.; Morent, R.; De Geyter, N.; Dubruel, P.; Leys, C. Plasma surface modification of biomedical polymers: influence on cell-material interaction. *Plasma Chem. Plasma Process.* 2012, 32, 1039–1073.

72. Park, G. Y.; Park, S. J.; Choi, M. Y.; Koo, I. G.; Byun, J. H.; Hong, J. W.; Sim, J. Y.; Collins, G. J.; Lee, J. K. Atmospheric-pressure plasma sources for biomedical applications. *Plasma Sources Sci. Technol.* 2012, 21, 043001.

73. Liston, E. M. Plasma treatment for improved bonding: a review. *J. Adhes.* 1989, 30, 199–218.

74. Liston, E. M.; Martinu, L.; Wertheimer, M. R. Plasma surface modification of polymers for improved adhesion: a critical review. *J. Adhes. Sci. Technol.* 1993, 7, 1091–1127.

75. Li, R.; Ye, L.; Mai, Y.-W. Application of plasma technologies in fibre-reinforced polymer composites: a review of recent developments. *Compos. A.* 1997, 28, 73–86.

76. Dilsiz, N. Plasma surface modification of carbon fibers: a review. *J. Adhes. Sci. Technol.* 2000, 14, 975–987.

77. Pankaj, S. K.; Bueno-Ferrer, C.; Misra, N. N.; Milosavljević, V.; O'Donnell, C. P.; Bourke, P.; Keener, K. M.; Cullen, P. J. Applications of cold plasma technology in food packaging. *Trends Food Sci. Technol.* 2014, 35, 5–17.

78. Pekárek, S. Non-thermal plasma ozone generation. *Acta Polytech.* 2003, 43, 47–51.

79. Mizuno, A. Industrial applications of atmospheric non-thermal plasma in environmental remediation. *Plasma Phys. Control. Fusion.* 2007, 49, A1.

80. Huang, H.; Tang, L. Treatment of organic waste using thermal plasma pyrolysis technology. *Energy Convers. Manage.* 2007, 48, 1331–1337.

81. Heberlein, J.; Murphy, A. B. Thermal plasma waste treatment. *J. Phys. D Appl. Phys.* 2008, 41, 053001.

82. Gomez, E.; Rani, D. A.; Cheeseman, C. R.; Deegan, D.; Wise, M.; Boccaccini, A. R. Thermal plasma technology for the treatment of wastes: a critical review. *J. Hazard. Mater.* 2009, 161, 614–626.

83. Yang, L.; Wang, H.; Wang, H.; Wang, D.; Wang, Y. Solid waste plasma disposal plant. *J. Electrostat.* 2011, 69, 411–413.
84. Galeno, G.; Minutillo, M.; Perna, A. From waste to electricity through integrated plasma gasification/ fuel cell (IPGFC) system. *Int. J. Hydrogen Energy.* 2011, 36, 1692–1701.
85. Fabry, F.; Rehmet, C.; Rohani, V.; Fulcheri, L. Waste gasification by thermal plasma: a review. *Waste Biomass Valor.* 2013, 4, 421–439.
86. Tang, L.; Huang, H.; Hao, H.; Zhao, K. Development of plasma pyrolysis/gasification systems for energy efficient and environmentally sound waste disposal. *J. Electrostat.* 2013, 71, 839–847.
87. Cheng, H.-H.; Chen, S.-S.; Wu, Y.-C.; Ho, D.-L. Non-thermal plasma technology for degradation of organic compounds in wastewater control: a critical review. *J. Environ. Eng. Manage.* 2007, 17, 427–433.
88. Zhang, J.; Chen, J.; Li, X. Remove of phenolic compounds in water by low-temperature plasma: a review of current research. *J. Water Res. Prot.* 2009, 2, 99–109.
89. Wang, X.; Zhou, M.; Jin, X. Application of glow discharge plasma for wastewater treatment. *Electrochim. Acta.* 2012, 83, 501–512.
90. Jiang, B.; Zheng, J.; Qiu, S.; Wu, M.; Zhang, Q.; Yan, Z.; Xue, Q. Review on electrical discharge plasma technology for wastewater remediation. *Chem. Eng. J.* 2014, 236, 348–368.
91. Liu, C.; Xu, G.; Wang, T. Non-thermal plasma approaches in CO_2 utilization. *Fuel Process. Technol.* 1999, 58, 119–134.
92. Chang, J.-S. Recent development of plasma pollution control technology: a critical review. *Sci. Tech. Adv. Mat.* 2001, 2, 571.
93. Chae, J.-O. Non-thermal plasma for diesel exhaust treatment. *J. Electrost.* 2003, 57, 251–262.
94. Chen, H. L.; Lee, H. M.; Chen, S. H. Review of packed-bed plasma reactor for ozone generation and air pollution control. *Ind. Eng. Chem. Res.* 2008, 47, 2122–2130.
95. Durme, J. V.; Dewulf, J.; Leys, C.; Langenhove, H. V. Combining non-thermal plasma with heterogeneous catalysis in waste gas treatment: a review. *Appl. Catal. B Environ.* 2008, 78, 324–333.
96. Vandenbroucke, A. M.; Morent, R.; Geyter, N. D.; Leys, C. Non-thermal plasmas for non-catalytic and catalytic VOC abatement. *J. Hazard. Mater.* 2011, 195, 30–54.
97. Vandenbroucke, A. M.; Morent, R.; De Geyter, N.; Leys, C. Decomposition of toluene with plasma-catalysis: a review. *J. Adv. Oxid. Technol.* 2012, 15, 232–241.
98. Preis, S.; Klauson, D.; Gregor, A. Potential of electric discharge plasma methods in abatement of volatile organic compounds originating from the food industry. *J. Environ. Manage.* 2013, 114, 125–138.
99. Mizuno, A. Generation of non-thermal plasma combined with catalysts and their application in environmental technology. *Catal. Today.* 2013, 211, 2–8.
100. Moore, S. B.; Ausley, L. W. Systems thinking and green chemistry in the textile industry: concepts, technologies and benefits. *J. Clean. Prod.* 2004, 12, 585–601.
101. Hasanbeigi, A.; Price, L. A review of energy use and energy efficiency technologies for the textile industry. *Renew. Sustain. Energ. Rev.* 2012, 16, 3648–3665.
102. Verma, A. K.; Dash, R. R.; Bhunia, P. A review on chemical coagulation/flocculation technologies for removal of colour from textile wastewaters. *J. Environ. Manage.* 2012, 93, 154–168.
103. Roy Choudhury, A. K. Green chemistry and the textile industry. *Textile Prog.* 2013, 45, 3–143.
104. Radetic, M.; Jovancic, P.; Puac, N.; Petrovic, Z. L. Environmental impact of plasma application to textiles. *J. Phys.* 2007, 71, 012017.
105. Kale, K. H.; Desai, A. N. Atmospheric pressure plasma treatment of textiles using non-polymerising gases. *Indian J. Fibre Textile Res.* 2011, 36, 289–299.
106. Tomasino, C.; Cuomo, J. J.; Smith, C. B.; Oehrlein, G. Plasma treatments of textiles. *J. Ind. Textiles.* 1995, 25, 115–127.
107. Samanta, K.; Jassal, M.; Agrawal, A. K. Atmospheric pressure glow discharge plasma and its applications in textile. *Indian J. Fibre Textile Res.* 2006, 31, 83–98.
108. Hegemann, D. Plasma polymerization and its applications in textiles. *Indian J. Fibre Textile Res.* 2006, 31, 99–155.

109. Morent, R.; Geyter, N. D.; Verschuren, J.; Clerck, K. D.; Kiekens, P.; Leys, C. Non-thermal plasma treatment of textiles. *Surf. Coat. Technol.* 2008, 202, 3427–3449.

110. Buyle, G. Nanoscale finishing of textiles via plasma treatment. *Mater. Technol.* 2009, 24, 46–51.

111. Parvinzadeh, M.; Ebrahimi, I. Atmospheric air-plasma treatment of polyester fiber to improve the performance of nanoemulsion silicone. *Appl. Surf. Sci.* 2011, 257, 4062–4068.

112. Chaudhry, M. R. *Cotton Production and Processing.* In *Industrial Applications of Natural Fibres*; Müssig, J., Ed., Stevens, C. V., Series Ed.; John Wiley & Sons, Ltd: Hoboken, 2010; pp 219–234.

113. Khadi, B. M.; Santhy, V.; Yadav, M. S. *Cotton: An Introduction.* In *Cotton; Biotechnology in Agriculture and Forestry*; Springer: Berlin, HD, 2010; Vol. 65, pp 1–14.

114. Commissioner, Office of the Textile. *Compendium of International Textile Statistics – 2009-10*; Ministry of Textile: India, 2009.

115. Losonczi, L. *Bioscouring of Cotton Fabric*; Budapest Unvirsity of Technology and Economics: Budapest, 2004.

116. Agrawal, P. B. *The Performance of Cutinase and Pectinase in Cotton Scouring*; University of Twente: Enschede, 2005.

117. Varadarajan, P. V.; Iyer, V.; Saxana, S. Wax on cotton fibre: its nature and distribution – a review. *J. Indian Soc. Cotton Improv.* 1990, 15, 123–127.

118. Vaughn, K. C.; Turley, R. B. The primary walls of cotton fibers contain an ensheathing pectin layer. *Protoplasma.* 1999, 209, 226–237.

119. Ridley, B. L.; O'Neill, M. A.; Mohnen, D. Pectins: structure, biosynthesis, and oligogalacturonide-related signaling. *Phytochemistry.* 2001, 57, 929–967.

120. Shukla, S.; Jain, D.; Verma, K.; Verma, S. Pectin-based colon-specific drug delivery. *CYS.* 2011, 2, 83–89.

121. Warwicker, J. O.; Cotton, Silk and Man-Made Fibres Research Association. *A Review of the Literature on the Effect of Caustic Soda and Other Swelling Agents on the Fine Structure of Cotton*; Pamphlet; Cotton, Silk and Man-Made Fibres Research Association: Manchester, 1966.

122. Karmakar, S. R.; Ed. Preface. Chemical Technology in the Pre-treatment Processes of Textiles. In *Textile Science and Technology*; Elsevier: Amsterdam, 1999; Vol. 12, pp v–vi.

123. Tian, L.; Nie, H.; Chatterton, N. P.; Branford-White, C. J.; Qiu, Y.; Zhu, L. Helium/oxygen atmospheric pressure plasma jet treatment for hydrophilicity improvement of grey cotton knitted fabric. *Appl. Surf. Sci.* 2011, 257, 7113–7118.

124. Tzanov, T.; Calafell, M.; Guebitz, G. M.; Cavaco-Paulo, A. Bio-preparation of cotton fabrics. *Enzyme Microb. Technol.* 2001, 29, 357–362.

125. Agrawal, P. B.; Nierstrasz, V. A.; Bouwhuis, G. H.; Warmoeskerken, M. M. C. G. Cutinase and pectinase in cotton bioscouring: an innovative and fast bioscouring process. *Biocatal. Biotransform.* 2008, 26, 412–421.

126. Thorsen, W. J. Modification of the cuticle and primary wall of cotton by corona treatment. *Textile Res. J.* 1974, 44, 422–428.

127. Goto, T.; Wakita, T.; Nakanishi, T.; Ohta, Y. Application of low temperature plasma treatment to the scouring of gray cotton fabric. *Sen'i Gakkaishi.* 1992, 48, 133–137.

128. Pandiyaraj, K. N.; Selvarajan, V. Non-thermal plasma treatment for hydrophilicity improvement of grey cotton fabrics. *J. Mater. Process. Technol.* 2008, 199, 130–139.

129. Inbakumar, S.; Morent, R.; Geyter, N.; Desmet, T.; Anukaliani, A.; Dubruel, P.; Leys, C. Chemical and physical analysis of cotton fabrics plasma-treated with a low pressure DC glow discharge. *Cellulose.* 2010, 17, 417–426.

130. Wang, L.; Xiang, Z.-Q.; Bai, Y.-L.; Long, J.-J. A plasma aided process for grey cotton fabric pretreatment. *J. Clean. Prod.* 2013, 54, 323–331.

131. Prabaharan, M.; Carneiro, N. Effect of low-temperature plasma on cotton fabric and its application to bleaching and dyeing. *Indian J. Fibre Textile Res.* 2005, 30, 68–74.

132. Karahan, H. A.; Özdoğan, E. Improvements of surface functionality of cotton fibers by atmospheric plasma treatment. *Fibers Polymers.* 2008, 9, 21–26.

133. Sun, S.; Sun, J.; Yao, L.; Qiu, Y. Wettability and sizing property improvement of raw cotton yarns treated with He/O$_2$ atmospheric pressure plasma jet. *Appl. Surf. Sci.* 2011, 257, 2377–2382.

134. Sun, S.; Qiu, Y. Influence of moisture on wettability and sizing properties of raw cotton yarns treated with He/O$_2$ atmospheric pressure plasma jet. *Surf. Coat. Technol.* 2012, 206, 2281–2286.

135. Bhat, N. V.; Bharati, R. N.; Gore, A. V.; Patil, A. J. Effect of atmospheric pressure air plasma treatment on desizing and wettability of cotton fabrics. *Indian J. Fibre Textile Res.* 2011, 36, 42–46.

136. Sun, D.; Stylios, G. K. Effect of low temperature plasma treatment on the scouring and dyeing of natural fabrics. *Textile Res. J.* 2004, 74, 751–756.

137. Wang, Q.; Fan, X.-R.; Cui, L.; Wang, P.; Wu, J.; Chen, J. Plasma-aided cotton bioscouring: dielectric barrier discharge versus low-pressure oxygen plasma. *Plasma Chem. Plasma Process.* 2009, 29, 399–409.

138. Ma, P.; Huang, J.; Cao, G.; Xu, W. Influence of temperature on corona discharge treatment of cotton fibers. *Fibers Polymers.* 2010, 11, 941–945.

139. Sun, S.; Yu, H.; Williams, T.; Hicks, R. F.; Qiu, Y. Eco-friendly sizing technology of cotton yarns with He/O$_2$ atmospheric pressure plasma treatment and green sizing recipes. *Textile Res. J.* 2013, 83, 2177–2190.

140. Kan, C.-W.; Lam, C.-F.; Chan, C.-K.; Ng, S.-P. Using atmospheric pressure plasma treatment for treating grey cotton fabric. *Carbohydr. Polym.* 2014, 102, 167–173.

141. Wang, Q.; Fan, X.; Gao, W.; Chen, J. Characterization of bioscoured cotton fabrics using FT-IR ATR spectroscopy and microscopy techniques. *Carbohydr. Res.* 2006, 341, 2170–2175.

142. Chung, C.; Lee, M.; Choe, E. K. Characterization of cotton fabric scouring by FT-IR ATR spectroscopy. *Carbohydr. Polym.* 2004, 58, 417–420.

143. Hou, W.-C.; Chang, W.-H.; Jiang, C.-M. Qualitative distinction of carboxyl group distributions in pectins with ruthenium red. *Bot. Bull. Acad. Sinica.* 1999, 40, 115–119.

144. Dave, H.; Ledwani, L.; Chandwani, N.; Chauhan, N.; Nema, S. K. The removal of impurities from gray cotton fabric by atmospheric pressure plasma treatment and its characterization using ATR-FT-IR spectroscopy. *J. Textile Inst.* 2014, 105, 586–596.

145. Raja, K. S.; Vasanthi, N. S.; Saravanan, D.; Ramachandran, T. Use of bacterial lipase for scouring of cotton fabrics. *Indian J. Fibre Textile Res.* 2012, 37, 299–302.

146. Nithya, E.; Radhai, R.; Rajendran, R.; Shalini, S.; Rajendran, V.; Jayakumar, S. Synergetic effect of DC air plasma and cellulase enzyme treatment on the hydrophilicity of cotton fabric. *Carbohydr. Polym.* 2011, 83, 1652–1658.

147. Schreuder, M. D. J.; Van Hove, L. W. A.; Brewer, C. A. Ozone exposure affects leaf wettability and tree water balance. *New Phytol.* 2001, 152, 443–454.

148. Prabaharan, M.; Nayar, R. C.; Kumar, N. S.; Rao, J. V. A study on the advanced oxidation of a cotton fabric by ozone. *Color. Technol.* 2000, 116, 83–86.

149. Eren, H. A.; Ozturk, D. The evaluation of ozonation as an environmentally friendly alternative for cotton preparation. *Textile Res. J.* 2011, 81, 512–519.

150. Gashti, M. P.; Pournaserani, A.; Ehsani, H.; Gashti, M. P. Surface oxidation of cellulose by ozone-gas in a vacuum cylinder to improve the functionality of fluoromonomer. *Vacuum.* 2013, 91, 7–13.

151. Li, X.; Qiu, Y. The effect of plasma pre-treatment on NaHCO$_3$ desizing of blended sizes on cotton fabrics. *Appl. Surf. Sci.* 2012, 258, 4939–4944.

152. Li, X.; Qiu, Y. The application of He/O$_2$ atmospheric pressure plasma jet and ultrasound in desizing of blended size on cotton fabrics. *Appl. Surf. Sci.* 2012, 258, 7787–7793.

153. Peng, S.; Gao, Z.; Sun, J.; Yao, L.; Qiu, Y. Influence of argon/oxygen atmospheric dielectric barrier discharge treatment on desizing and scouring of poly (vinyl alcohol) on cotton fabrics. *Appl. Surf. Sci.* 2009, 255, 9458–9462.

154. Cai, Z.; Qiu, Y.; Zhang, C.; Hwang, Y.-J.; Mccord, M. Effect of atmospheric plasma treatment on desizing of PVA on cotton. *Textile Res. J.* 2003, 73, 670–674.

155. Oktav Bulut, M.; Devirenoğlu, C.; Oksuz, L.; Bozdogan, F.; Teke, E. Combination of grey cotton fabric desizing and gassing treatments with a plasma aided process. *J. Textile Inst.* 2014, 0, 1–14.

156. Prasath, A.; Sivaram, S. S.; Vijay Anand, V. D.; Dhandapani, S. Desizing of starch containing cotton fabrics using near atmospheric pressure, cold dc plasma treatment. *J. Inst. Eng. India Ser E.* 2013, 94, 1–5.

157. Kan, C. W.; Yuen, C. W. M. Effect of atmospheric pressure plasma treatment on the desizing and subsequent colour fading process of cotton denim fabric. *Coloration Technol.* 2012, 128, 356–363.

158. Malek, R. M. A.; Holme, I. Effect of plasma treatment on some properties of cotton. *Iranian Polymer J.* 2003, 12, 271–280.

159. Yuen, C. W. M.; Kan, C. W. Influence of low temperature plasma treatment on the properties of ink-jet printed cotton fabric. *Fibers Polymers.* 2007, 8, 168–173.

160. Karahan, H. A.; Özdoğan, E.; Demir, A.; Ayhan, H.; Seventekin, N. Effects of atmospheric plasma treatment on the dyeability of cotton fabrics by acid dyes. *Coloration Technol.* 2008, 124, 106–110.

161. Patiño, A.; Canal, C.; Rodríguez, C.; Caballero, G.; Navarro, A.; Canal, J. Surface and bulk cotton fibre modifications: plasma and cationization. Influence on dyeing with reactive dye. *Cellulose.* 2011, 18, 1073–1083.

162. Haji, A. Eco-friendly dyeing and antibacterial treatment of cotton. *Cellul. Chem. Technol.* 2013, 47, 303–308.

163. Bhat, N.; Netravali, A.; Gore, A.; Sathianarayanan, M.; Arolkar, G.; Deshmukh, R. Surface modification of cotton fabrics using plasma technology. *Textile Res. J.* 2011, 81, 1014–1026.

164. Kan, C. W.; Yuen, C. W. M.; Tsoi, W. Y.; Chan, C. K. Ink-jet printing for plasma-treated cotton fabric with biomaterial. *ASEAN J. Chem. Eng.* 2011, 11, 1–7.

165. Kan, C. W.; Yuen, C. W. M.; Tsoi, W. Y. Using atmospheric pressure plasma for enhancing the deposition of printing paste on cotton fabric for digital ink-jet printing. *Cellulose.* 2011, 18, 827–839.

166. Man, W. S.; Kan, C. W.; Ng, S. P. The use of atmospheric pressure plasma treatment on enhancing the pigment application to cotton fabric. *Vacuum.* 2014, 99, 7–11.

167. Lam, Y. L.; Kan, C. W.; Yuen, C. W. M. Effect of zinc oxide on flame retardant finishing of plasma pre-treated cotton fabric. *Cellulose.* 2011, 18, 151–165.

168. Tsafack, M. J.; Levalois-Grützmacher, J. Flame retardancy of cotton textiles by plasma-induced graft-polymerization (PIGP). *Surf. Coat. Technol.* 2006, 201, 2599–2610.

169. Alongi, J.; Tata, J.; Frache, A. Hydrotalcite and nanometric silica as finishing additives to enhance the thermal stability and flame retardancy of cotton. *Cellulose.* 2011, 18, 179–190.

170. Tsafack, M. J.; Levalois-Grützmacher, J. Towards multifunctional surfaces using the plasma-induced graft-polymerization (PIGP) process: flame and waterproof cotton textiles. *Surf. Coat. Technol.* 2007, 201, 5789–5795.

171. Sun, D.; Stylios, G. K. Fabric surface properties affected by low temperature plasma treatment. *J. Mater. Process. Technol.* 2006, 173, 172–177.

172. Li, S.; Jinjin, D. Improvement of hydrophobic properties of silk and cotton by hexafluoropropene plasma treatment. *Appl. Surf. Sci.* 2007, 253, 5051–5055.

173. Kamlangkla, K.; Paosawatyanyong, B.; Pavarajarn, V.; Hodak, J. H.; Hodak, S. K. Mechanical strength and hydrophobicity of cotton fabric after plasma treatment. *Appl. Surf. Sci.* 2010, 256, 5888–5897.

174. Kilic, B.; Aksit, A. C.; Mutlu, M. Surface modification and characterization of cotton and polyamide fabrics by plasma polymerization of hexamethyldisilane and hexamethyldisiloxane. *Int. J. Cloth. Sci. Technol.* 2009, 21, 137–145.

175. Cho, S. C.; Hong, Y. C.; Cho, S. G.; Ji, Y. Y.; Han, C. S.; Uhm, H. S. Surface modification of polyimide films, filter papers, and cotton clothes by HMDSO/toluene plasma at low pressure and its wettability. *Curr. Appl. Phys.* 2009, 9, 1223–1226.

176. Nättinen, K.; Nikkola, J.; Minkkinen, H.; Heikkilä, P.; Lavonen, J.; Tuominen, M. Reel-to-reel inline atmospheric plasma deposition of hydrophobic coatings. *J. Coat. Technol. Res.* 2011, 8, 237–245.

177. Gorjanc, M.; Bukosek, V.; Gorensek, M.; Vesel, A. The Influence of water vapor plasma treatment on specific properties of bleached and mercerized cotton fabric. *Textile Res. J.* 2009, 80, 557–567.

178. Shahidi, S.; Rashidi, A.; Ghoranneviss, M.; Anvari, A.; Rahimi, M. K.; Bameni Moghaddam, M.; Wiener, J. Investigation of metal absorption and antibacterial activity on cotton fabric modified by low temperature plasma. *Cellulose.* 2010, 17, 627–634.
179. Kan, C. W.; Lam, Y. L.; Yuen, C. W. M.; Luximon, A.; Lau, K. W.; Chen, K. S. Chemical analysis of plasma-assisted antimicrobial treatment on cotton. *J. Phys.* 2013, 441, 012002.
180. Arik, B.; Demir, A.; Özdoğan, E.; Gülümser, T. Effects of novel antibacterial chemicals on low temperature plasma functionalized cotton surface. *Tekstil Ve Konfeksiyon.* 2011, 4, 356–363.
181. Kan, C.-W.; Lam, Y.-L. Low stress mechanical properties of plasma-treated cotton fabric subjected to zinc oxide-anti-microbial treatment. *Materials.* 2013, 6, 314–333.
182. Vaideki, K.; Jayakumar, S.; Thilagavathi, G.; Rajendran, R. A study on the antimicrobial efficacy of RF oxygen plasma and neem extract treated cotton fabrics. *Appl. Surf. Sci.* 2007, 253, 7323–7329.
183. Nithya, E.; Radhai, R.; Rajendran, R.; Jayakumar, S.; Vaideki, K. Enhancement of the antimicrobial property of cotton fabric using plasma and enzyme pre-treatments. *Carbohydr. Polym.* 2012, 88, 986–991.

CHAPTER 7

Ligand-Free Palladium Nanoparticles Catalyzed Hiyama Cross-Coupling of Aryl and Heteroaryl Halides in Ionic Liquids

Chanchal Premi, Ananya Srivastava, and Nidhi Jain*

Department of Chemistry, Indian Institute of Technology, New Delhi 110016, India
*Email: njain@chemistry.iitd.ac.in

CONTENTS

ABSTRACT

We exhibit for the first time to utilize the 1-butyl-3-methylimidazolium fluoride [bmim]F as an activator of the organosilanes with simple handling, storage, and work-up in contrast to traditional fluorine sources such as tetrabutylammonium fluoride (TBAF), which is requisite for Hiyama coupling. Palladium nanoparticles 2–5 nm in size stabilized in nitrile-functionalized 3-(3-cyanopropyl)-1-methyl-1H-imidazol-3-ium hexafluorophosphate {[CN-bmim]PF$_6$} work as an in situ catalyst for carbon–carbon bond forming reactions of aromatic and heterocyclic halides with aryl- and vinyltrimethoxysilanes. A range of biphenyl derivatives, substituted styrenes, and aromatic heterocycles were obtained in 76–98% yield. Elevated yields of the cross-coupled products were obtained at a low catalyst loading of only 4 mol%, and the catalyst could be reused and recycled up to four times with only a slight loss in catalytic activity.

GRAPHICAL REPRESENTATION

7.1 INTRODUCTION

Transition metal-catalyzed cross-coupling reactions have emerged as a convincing tool for the building of carbon–carbon (C–C) and carbon–heteroatom bonds, particularly for pharmaceuticals and agrochemicals.[1] Particularly, palladium catalysts such as $Pd(PPh_3)_4$ and Pd–carbene complexes, phosphane–palladacycles, and many Pd salts in the presence of an excess amount of PPh_3 are used in C–C reactions.[2] Phosphane and nonphosphane ligands stabilize Pd^0 and influence its catalytic activity in cross-coupling reactions with zinc-, boron-, tin-, and silicon-containing organometallics.[3] Of these materials, organosilanes as transmetalation reagents present various advantages above organoboranes and organostannanes in terms of their low cost, chemical stability, ready availability, and low toxicity.[4] Also, silicon waste produced from a reaction can be transformed into harmless SiO_2 and all of these points have made Hiyama coupling attractive from green and user-friendly points of view.[5] Hiyama coupling has been achieved by the combined use of a homogeneous Pd catalyst and a phosphane or an N-heterocyclic carbene ligand derived from an imidazolium salt.[6] Several of these ligands turn out to be air and moisture susceptible, and thus unstable and expensive.

Recently, a few ligand-free Hiyama coupling reactions performed with homogeneous Pd catalysts such as $Pd(OAc)_2$, $PdCl_2$, and palladacycles or heterogeneous Pd nanoparticles (NPs) have been reported.[7] Thus, ligand-free system is the simplest and cheapest substitute for cross-coupling reactions. In a review, Reetz and de Vries[8] reported that the use of homeopathic doses of palladium (0.01–0.1 mol%) could effectively promote a ligand-free Heck reaction with aryl iodides and bromides. The only restriction is that at low Pd concentrations, the reaction proceeds slowly, whereas at higher concentrations, Pd black is rapidly formed. However, in either case, coupling does not occur with aryl chlorides. The existence and task of Pd NPs have been established in the Jeffery system, wherein tetraalkylammonium halides stabilize nanosized Pd colloids (1–5 nm) generated from $Pd(OAc)_2$ (5 mol%), thus preventing undesirable agglomeration by forming a monomolecular layer around the metal core and promoting phosphane-free Pd catalysis.[9] Ever since their commencement, catalysis by Pd NPs has gained substantial interest due to their high surface-to-volume ratio, which facilitates high turnover frequencies at low catalyst loadings.[10] NPs, however, are thermodynamically and kinetically unstable; hence, some kind of stabilization by way of dendrimers, polymers, or surfactants needs to be provided to prevent agglomeration of the particles.[11] Dupont et al.[12] demonstrated the preparation of transition metal NPs from ionic liquids (ILs) without the use of additional stabilizers. It has been proposed that ILs form preorganized structures as a result of cooperative hydrogen bonding between the cation and the anion, and this induces structural directionality and provides electrosteric stabilization to the NPs.[13] Although Pd-NP-catalyzed Heck, Suzuki, and Sonogashira couplings in imidazolium-based ILs have been reported, there have been no reports on IL-stabilized Pd NPs for Hiyama coupling.[14]

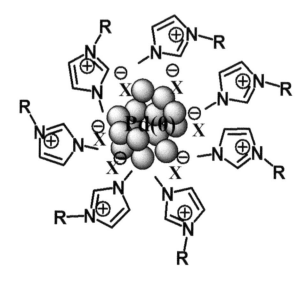

FIGURE 7.1   Ionic Liquid Stabilizing Palladium Nanoparticles.

Herein, we report the catalytic activity of Pd NPs starting from a Pd^{II} precursor in 3-(3-cyanopropyl)-1-methyl-1*H*-imidazol-3-ium hexafluorophosphate {[CN-bmim] PF$_6$} for the Hiyama coupling of trimethoxyaryl and vinylsilanes with aryl and heterocyclic halides in the presence of 1-butyl-3-methylimidazolium fluoride {[bmim] F} as the activator. The procedure accommodates various functional groups to yield a diverse range of biphenyl derivatives, substituted styrenes, and aromatic heterocycles. The method offers distinct advantages over conventional methods of coupling, as it occurs with a reusable nanocatalytic system and a highly convenient-to-handle silane activator, it is versatile in terms of substrate scope, the reaction is cleaner, and the products are obtained in high yields. To the best of our knowledge, this is the first example demonstrating the use of IL-stabilized Pd NPs as a catalyst for Hiyama cross-coupling reactions under phosphane-free conditions.

7.2   RESULTS AND DISCUSSION

The ILs [CN-bmim]PF$_6$ (**1**), [bmim]F (**2**), and 1-butyl-3-methylimidazolium acetate {[bmim]OAc} (**3**) were synthesized as reported in the literature (Fig. 7.1).[15] For IL1, mixture of 4-bromobutyronitrile and 1-methylimidazole was refluxed at 70°C for 24 h, which leads to the formation of viscous, pale yellow 3-(3-cyanopropyl)-1-methyl-1*H*-imidazol-3-ium bromide {[CN-bmim]Br}, which upon anion exchange with potassium hexafluorophosphate yielded **1** as a colorless liquid. Similarly, ILs **2** and **3** were synthesized by anion exchange of 1-butyl-3-methylimidazolium bromide {[bmim]Br} with AgF and NaOAc, respectively. The presence of fluorine in **2** was established by ^{19}F NMR spectroscopy, which showed a singlet at $\delta = -121.8$ ppm. Purity of ILs **1–3** was established by NMR spectroscopy (99% pure) (Supporting Information).

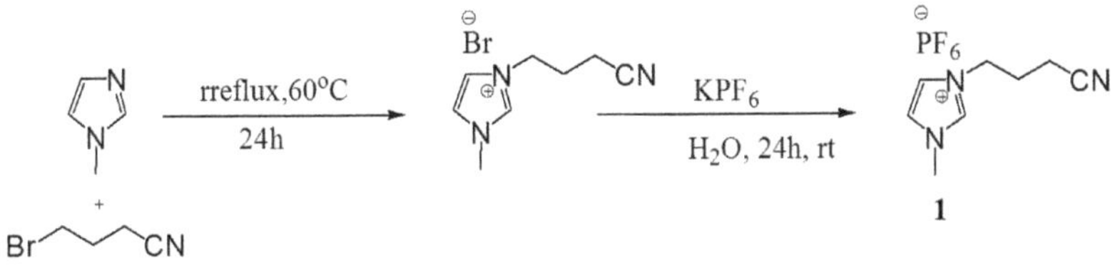

Scheme 1

SCHEME 7.1 Synthesis of 3-(3-cyanopropyl)-1-methyl-1H-imidazol-3-ium bromide {[CN-bmim] Br}and 3-(3-cyanopropyl)-1-methyl-1H-imidazol-3-ium hexafluorophosphate {[CN-bmim]PF$_6$}.

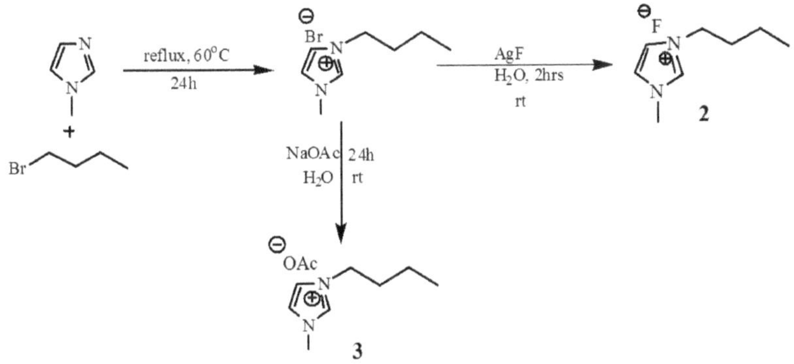

SCHEME 7.2 Synthesis of 1-butyl-3-methylimidazolium bromide{[bmim]Br}, 1-butyl-3-methylimidazolium acetate {[bmim]OAc}, and 1-butyl-3-methylimidazolium fluoride{[bmim]F}.

The Pd NP catalyst was prepared by a literature method.[16] Pd(OAc)$_2$ was added to **1** and diluted with acetonitrile (10 mL), and the contents were heated and vigorously stirred at 80°C for 2 h. The color of the solution changed from orange to yellow and finally to colorless; at this point, acetonitrile was evaporated under reduced pressure. The reaction mixture was heated further at 140°C for 6 h, which resulted in a dark brown–black suspension of Pd^0 in **1** (Scheme 7.1). In the literature, basic ILs with tertiary aliphatic amines as pendant groups were shown to act as reducing agents in the redox process leading to the formation of Pd^0 NPs from Pd(OAc)$_2$.[17] In another report, thermolytic decomposition of Pd(OAc)$_2$ at 100°C in polar solvents such as propylene carbonate has been reported to yield Pd NPs that are 8–10 nm in size.[18] To establish the role of ILs in reduction and stabilization of the NPs, we screened other ILs such as [bmim]F (**2**) and [bmim]OAc (**3**) under same reaction conditions. We found that no black-colored Pd^0 solution was obtained in ILs **2** and **3** at temperatures of 100 and 140°C (Table 7.1, entries 1 and 2). However, in IL **1**, Pd NPs 25–50 nm in size (Table 7.1, entry 4) were isolated at 140°C. This indicated that conversion of Pd^{II} into Pd^0 was not just a thermal decomposition of $Pd(OAc)_2$ but that it was the nitrile group on the imidazolium side chain that affected the reduction.

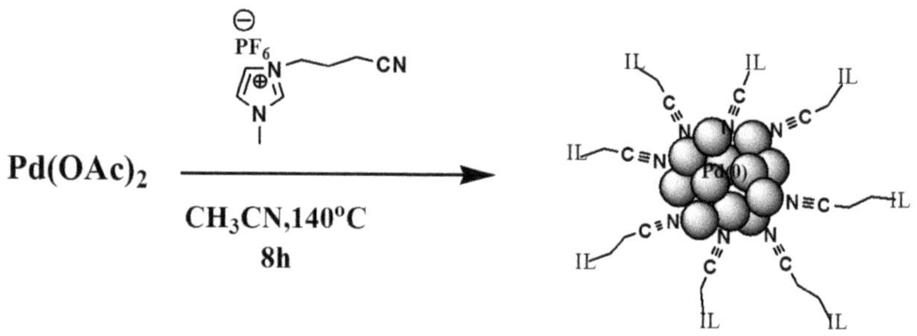

FIGURE 7.2   Synthesis of Palladium Nanoparticles.

The size and shape of the Pd NPs were determined by different characterization techniques such as transmission electron microscopy (TEM), scanning electron microscopy (SEM), and dynamic light scattering (DLS), as shown in Figure 7.2. TEM sample was prepared by drop casting the $Pd[CN\text{-}bmim]PF_6$ solution on copper TEM grid in hexane (1 mL); it showed uniformly dispersed NPs with an average diameter of around 2–5 nm. Similar results were obtained from SEM and DLS analyses, which showed particles to be in the range from 5 to 10 nm. The formation of Pd^0 was confirmed by powder X-ray diffraction (PXRD, Fig. 7.3a). The XRD pattern of the Pd NPs showed three planes at (111), (200), and (220) with lattice constant $a = 3.871$ Å. This could be readily indexed to crystalline Pd in terms of both the peak position and the relative intensity and matched JCPDS file 87-0638.[1] The EDX spectrum showed the presence of Pd, C, N, P, F, and Fe, which suggests the existence of IL 1-stabilized Pd NPs (Fig. 7.4).

The formation of the NPs was followed by recording the UV/vis spectrum at each step of reaction, monitoring changes every 1 h. It is well known that metal NPs absorb photons in the UV/vis region as a result of coherent oscillation of the electrons in the conduction band induced by the interacting electromagnetic fields. Initially, the solution of $Pd(OAc)_2$ in $[bmim]PF_6$ and acetonitrile (brownish orange color) showed an absorption maximum at 240 nm. After 2 h, the color of the solution changed to yellow, and the appearance of a new peak at 260 nm was observed. This peak eventually disappeared when the color of the solution changed to dark brown, and a typical UV/vis pattern indicative of the formation of Pd^0 NPs was observed (Supporting Information).

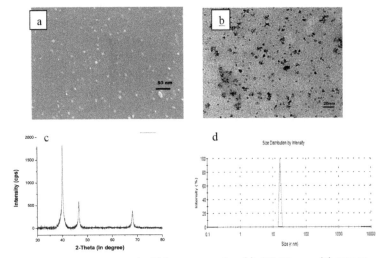

FIGURE 7.3 Characterization of Pd(0) Nanoparticles: (a) SEM Image, (b) TEM Image, (c) XRD Image, and (d) DLS Showing Intensity Distributed Particle Size.

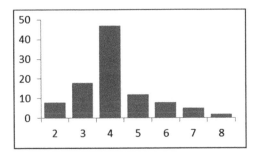

FIGURE 7.4 Histogram Showing Distribution of Pd NPs in Contrast to Size.

TABLE 7.1 Generation of Pd NPs in Ionic Liquids.[a]

Entry	Palladium Precursor	Ionic Liquid	T [°C]	Pd NPs [nm]
1	Pd(OAc)$_2$	2	100, 140	–
2	Pd(OAc)$_2$	3	100, 140	–
3	Pd(OAc)$_2$	[CN-bmim]F	100	–
4	Pd(OAc)$_2$	[CN-bmim]F	140	25–50
5	Pd(OAc)$_2$	1	140	2–5
6	PdCl$_2$	[CN-bmim]F	140	80–100
7	PdCl$_2$	1	140	20–25

[a]Palladium precursor (4 mg), ionic liquid (2.0 mL).

The preformed Pd NPs generated in **1** is now used as catalyst in Hiyama coupling by diluting it with acetonitrile followed by the addition of the arylhalide, trimethoxyphenylsilane, and [bmim]F as activators. The cross-coupling of iodobenzene with trimethoxyphenylsilane and Pd NPs in **1** as catalyst was investigated by using different activators such as TBAF, NaOAc, [bmim]OAc, and [bmim]F. Although the reaction failed to occur with NaOAc, in the presence of TBAF, biphenyl was obtained in 98% yield (Table 7.2, entry 2), and **2** and **3** (2.0 equiv.) afforded the product in 98 and 72% yield, respectively. This suggests that fluorine-based **2** is a better desilyate than oxygen-based **3** (Table 7.2, entries 7 and 8). Optimization of the reaction conditions for the Hiyama coupling of iodo-, bromo-, and chlorobenzene with trimethoxyphenylsilane was carried out by varying the activator, catalyst ratio, time, and temperature (Table 7.2). Optimization of the reaction conditions showed that trimethoxyphenylsilane (**2** equiv.) and Pd(OAc)$_2$ (4 mol%) in **1** acetonitrile solution were required for the aryl iodide (1 equiv.) to yield 98% of the coupled product at 70°C over 8 h.

A low catalyst loading of 4 mol% of the Pd^{II} precursor was sufficient to catalyze the reaction by virtue of NP formation and stabilization by the ionic framework of **1**. However, upon decreasing the Pd concentration from 4 mol% to 3 and 2 mol%, the yield of biphenyl dropped to 82 and 50%, respectively (Table 7.2, entries 4 and 3). With bromo- and chlorobenzene, the yield of the product was 98 and 82%, respectively, but the reactions required slightly elevated temperatures (Table 7.2, entries 12 and 13).

TABLE 7.2 Optimization of the Catalytic Conditions for the Hiyama Coupling of Aryl Halides

Entry	X	Pd(OAc)$_2$ (mol%)	Activator (equiv.)	T (°C)	Yield[b] (%)
1	I	4.0	NaOAc (2.0)	70	–
2	I	4.0	TBAF (2.0)	70	98
3	I	2.0	**2** (1.5)	70	50
4	I	3.0	**2** (1.5)	70	82
5	I	4.0	**2** (1.5)	70	90
6	I	7.0	**2** (1.5)	70	90
7	I	4.0	**2** (2.0)	70	98,76[c]
8	I	4.0	**3** (2.0)	70	72
9	I	4.0	**2** (3.0)	70	98
10	Br	4.0	**2** (2.0)	70	82[d]
11	Br	4.0	**2** (2.0)	80	98
12	Cl	4.0	**2** (2.0)	100	–
13	Cl	4.0	**2** (2.0)	120	82

[a] Reaction conditions: aryl halide (1.0 mmol), arylsilane (2.0 mmol), preformed Pd NPs from Pd(OAc)$_2$ and **1** (2.0 mL), CH$_3$CN (2.0 mL), activator, 8 h.
[b] Isolated yield.
[c] CH$_2$Cl$_2$ was used as a cosolvent instead of CH$_3$CN.
[d] Isolated yield after a reaction time of 24 h.

The scope of the Hiyama coupling was explored as the best reaction conditions were set by screening different aryl halides and a few heterocyclic halides. Table 7.3 summarizes the reactions performed under our optimized conditions in yields ranging from 76 to 98%. The yield was slightly higher for substrates bearing electron-donating substituents such as $-CH_3$ and $-OCH_3$ in the *para* position (Table 7.3, entries 4 and 5) than for substrates bearing electron-withdrawing substituents such as $-NO_2$, $-CF_3$, $-CN$, and $-COCH_3$ (Table 7.3, entries 6–9); the yield was particularly low with pentafluorophenyl iodide (Table 7.3, entry 10). Coupling occurred efficiently with heterocyclic halides as well to give the aromatic heterocycles in 79–85% yield (Table 7.3, entries 14–16).

TABLE 7.3

Entry	Aryl halide	Product	(%) Isolated yield
1	⬡–I	⬡–Ph	98 (91[c],86[d])
2	⬡–Br	⬡–Ph	98[e]
3	⬡–Cl	⬡–Ph	80[g]
4	H_3C–⬡–I	H_3C–⬡–Ph	97
5	H_3CO–⬡–I	H_3CO–⬡–Ph	96
6	O_2N–⬡–I	O_2N–⬡–Ph	89[f]
7	F_3C–⬡–I	F_3C–⬡–Ph	82[f]
8	NC–⬡–I	NC–⬡–Ph	85[f]
9	(pentafluorophenyl iodide)	(pentafluorophenyl Ph)	76[f]

Entry	Aryl halide	Product	(%) Isolated yield
10	NC—⟨⟩—Br	NC—⟨⟩—Ph	88[f]
11	Br / OH (naphthalene)	Ph / OH (naphthalene)	86[f]
12	(thiophene)—I	(thiophene)—Ph	79[f]
13	(pyridine) Br	(pyridine) Ph	85[f]
14	(quinoline) Br	(quinoline) Ph	81[f]
15	NC—⟨⟩—Cl	NC—⟨⟩—Ph	76

[a] Reaction conditions: aryl halide (1.0 mmol), trimethoxyphenylsilane (2.0 mmol), preformed Pd NPs from Pd(OAc)$_2$ (4.0 mol%), [bmim]F (2.0 equiv.), 1 (2.0 mL), CH$_3$CN (2.0 mL), stirred at 70°C for 8 h.
[b] Isolated yield.
[c] Yield after first recycle.
[d] Yield after second recycle.
[e] Yield after third recycle.
[f] Yield after fourth recycle.
[g] Reaction performed at 80°C.
[h] Reaction with Pd-(OAc)$_2$ (4 mol%) added directly instead of preformed Pd NPs.
[i] Reaction with Pd(OAc)$_2$ (4 mol%) and PPh$_3$ (20 mol%) added directly instead of preformed Pd NPs.
[j] Reaction performed at 120°C.

The protocol was further extended to the Hiyama coupling of vinylsiloxanes to yield substituted styrenes. The vinylation reaction of alkenyl stannanes and potassium trifluorovinylborates has been mainly used to yield styrenes through Heck coupling.[19]

$$R\text{—Ar—I} + \diagup\text{Si(OMe)}_3 \xrightarrow[\substack{\text{bmimCN[PF}_6], 60\text{-}70°C \\ 15\ \text{min-30 min}}]{\substack{\text{Pd (OAc)}_2 (0.04\text{moleq.}) \\ \text{bmim[F](2moleq.)}}} R\text{—Ar—}\diagup$$

Alkenylsilanes has found limited utility in styrene synthesis because of their lesser reactivity and coupling has been achieved essentially only with aryl iodides. With aryl bromides and chlorides, only a few procedures involving the use of vinyltrialkoxysilanes have been reported. One method had used Pd(OAc)$_2$, an imidazolium salt (3 mol%), and TBAF at 80°C.[20] The second route employs the use of π-allyl palladium chloride (2.5 mol%), N-dicyclohexylphosphanyl-N-methylpiperazine, and TBAF in DMF at 110°C.[21] Recently, a fluoride-free coupling is reported in aqueous sodium hydroxide by using tetrabutylammonium bromide as an additive under microwave and thermal conditions at 120C.[22] We have demonstrated the cross-coupling of tri-methoxyvinylsilane with substituted aryl halides to yield substituted styrene. The reactions gave high yields of the products in the range from 81 to 98%. The reaction time was particularly short (15–30 min) with aryl iodides, whereas with bromide and chloride derivatives, reaction time of 12 and 16 h, respectively, was required to achieve greater conversions (Table 7.4).

Progress of reaction was observed by thin layer chromatography after completion of reaction, the products were extracted with diethyl ether, and reaction mixture was washed with water to remove **2** leaving behind the catalyst immobilized in **1**. To the recovered Pd catalyst in **1**, fresh iodobenzene trimethoxysilane and **2** were added, and the contents were heated at 70°C for 8 h to give biphenyl in 91% yield (Table 7.3, entry 1). The reusability of the catalyst was checked up to four recycles, and biphenyl was obtained in 86, 80, and 70% yield in the second, third, and fourth cycles, respectively (Table 7.3, entry 1).

TABLE 7.4 Pd-NP-catalyzed Hiyama Coupling of Aryl Iodides with Trimethoxyvinylsilane.[a]

$$R\!-\!\!\bigcirc\!\!-I \ + \ \diagup\!\!\diagdown Si(OMe)_3 \ \xrightarrow[60-70^{\circ}C, \ 15-30 \ min]{\begin{array}{c} Pd\,(OAc)_2(4 \ mol\%.) \\ \mathbf{2} \ (2 \ eq.), \ \mathbf{1}, \ CH_3CN \end{array}} \ \bigcirc\!\!\!=\!\!R$$

Entry	Aryl Halide	Product	Isolated yield [%][b]
1	H₃C–⟨⟩–I	H₃C–⟨⟩–⟍	98
2	H₃CO–⟨⟩–I	H₃CO–⟨⟩–⟍	98
3	O₂N–⟨⟩–I	O₂N–⟨⟩–⟍	97
4	H₃COC–⟨⟩–Br	H₃COC–⟨⟩–⟍	86
5	NC–⟨⟩–Cl	NC–⟨⟩–⟍	75

[a] Reaction condition: aryl halide (1 mol eq.), vinyl silane (2 mol eq.), Pd(OAc)₂ (0.05 mol eq.), bmim[F] (2 mmol), bmimCN[PF₆] (2 mL), CH₃CN (2.0 mL) stirred at 60°C for 15–30 min.
[b] Isolated yield.
[c] Reaction carried out for 12 h.
[d] Reaction carried out for 16 h.

Following the established mechanism, oxidative addition of the aryl halide to Pd⁰ leads to the formation of aryl–Pd complex **4**. The agglomeration of these ligand-free Pd⁰ species toward the catalytically inactive bulk material is avoided because IL **1** stabilizes the nanomeric structures responsible for the catalytic activity. Complex **4** undergoes transmetalation wherein the nucleophile is transferred to palladium in the presence of [bmim]F to produce intermediate complex **6**. Reductive elimination of **6** gives the coupled product and regenerates palladium(0) for the next catalytic cycle (Fig. 7.5).

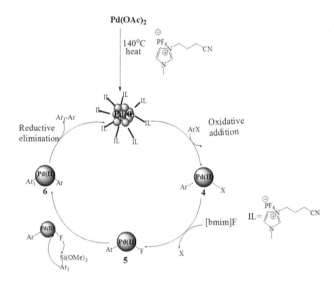

FIGURE 7.5 Proposed Mechanism for Hiyama Coupling.

7.3 EXPERIMENTAL SECTION

7.3.1 Materials and Methods

Chemicals were either purchased or purified by standard techniques without special instructions. The products were purified by column chromatography on silica gel (230–400 mesh, SRL). ^{1}H NMR and ^{13}C NMR spectra were measured with a Bruker DPX-300 MHz spectrometer (^{1}H 300 MHz, ^{13}C 75 MHz) by using CDCl$_3$ and D$_2$O as the solvents and tetramethylsilane as the internal standard at room temperature. Chemical shifts are given in δ relative to TMS. The ESI-MS for **1–3** were performed with a MICROTOF-II mass instrument. High-resolution transmission electron microscope experiments were conducted with a JEM 2100 F at an accelerating voltage of 200 kV. The scanning electron microscopy (SEM) image was captured with a Zeiss Evo series SEM model EVO 50. The size of the NPs was determined by Malvern DLS.

UV/vis spectroscopy was recorded with a Lambda Bio 20, Perkin-Elmer. PXRD was taken with a powder X-ray diffractometer (Bruker). DSC analysis was carried out with a Q-100 DSC instrument (TA instruments, USA) in the temperature range from −50 to 500°C at a heating rate of 10°C min^{-1} under a N$_2$ atmosphere. Calibration was performed with indium as a standard, for which the melting point was determined as 156.6°C well within the acceptable limit. GC–MS was recorded with a Perkin–Elmer Clarus 600C with a gas flow of 1.0 mL min^{-1} and a heating rate of 20°C min^{-1} with an initial temperature of 50°C and a final temperature of 310°C.

7.3.2 Synthesis of 3-(3-Cyanopropyl)-1-methyl-1*H*-imidazol-3-ium hexafluorophosphate {[CN-bmim]PF$_6$} (1)

To 1-methyl-1*H*-imidazole(2 mL, 1 equiv.), 4-bromobutyronitrile (2.5 mL, 1.2 equiv.) was added, and the reaction mixture was heated at reflux for 24 h at 70°C. After completion of the reaction as monitored by TLC, the mixture was washed with diethyl ether and ethyl acetate to remove trace amounts of unreacted 4-bromobutyronitrile, and the pure product was obtained in 92% yield. Anion exchange was carried out by following a literature method. To a stirred solution of 3-(3-cyanopropyl)-1-methyl-1*H*-imidazol-3-ium bromide {[CN-bmim]Br; 2 mL, 1 equiv.} in water (5 mL) was added potassium hexafluorophosphate (1.6 g, 1 equiv.), and the contents were stirred for 24 h at room temperature. The reaction mixture was filtered through Celite to remove potassium bromide and water was evaporated at 55°C under reduced pressure. IL **1** was dissolved in dichloromethane, dried with molecular sieves (4 Å) to remove any trace amounts of water, and finally concentrated under reduced pressure to obtain pure **1** in 80% yield.

7.3.3 Synthesis of Pd NPs

A solution of palladium acetate (4 mg, 0.04 mmol) and **1** (2 mL) in acetonitrile (5 mL) was stirred at 80°C for 1 h. After 1 h, the color of the solution changed from reddish brown to yellow and finally turned colorless after 2 h. Acetonitrile was evaporated under reduced pressure, and the contents were vigorously stirred for another 6 h at 140°C at which point the solution turned brownish-black. Pd NPs 2–5 nm in size were formed, and this was confirmed by UV, DLS, TEM, SEM, and PXRD. The catalyst immobilized in **1** was used as such for Hiyama coupling.

7.3.4 Characterization of Pd NPs

The formation of Pd NPs was confirmed by UV, PXRD, and EDAX analysis. The size of the NPs was determined by DLS, TEM, and SEM. To prepare the sample for SEM/EDAX, the Pd NPs were centrifuged at 6000 rpm and washed with water and then with absolute ethanol to remove most of **1**. The precipitate was then redispersed in dry acetone and sonicated for about 1 h. The sample was prepared on a glass slide with the help of spin coating to get a uniform distribution of particles. The sample was then coated in a sputter coater (EMITECH K 550×) with a gold layer in vacuo. Samples for TEM were prepared by placing a drop of a colloidal dispersion of $Pd[CNbmim]PF_6$ in hexane on the carbon-coated copper grid, followed by evaporation of the solvent at room temperature. DLS was recorded in acetonitrile at a sample concentration of $0.1 \, mg \, mL^{-1}$.

7.3.5 General Procedure for Hiyama Coupling of Aryl and Heterocyclic Halides

Preformed Pd NPs were used as in situ catalysts for the coupling reaction. The aryl halide (1 mmol), trimethoxyphenylsilane (2 mmol), **2** (2 mmol), and acetonitrile (2 mL) were added to the catalytic solution of Pd NPs in **1**, and the contents were stirred under an argon atmosphere at 70C for 8 h. Upon completion of the reaction as monitored by TLC, the product was extracted with diethyl ether (5 5 mL). The organic layer was washed with brine, dried with $MgSO_4$, and concentrated in vacuo. Purification was done by silica gel column chromatography (ethyl acetate/hexane) to afford the C–C coupled products. All products obtained herein are known compounds, and their structures were confirmed by ^1H NMR and ^{13}C NMR spectroscopy and mass spectrometry (see the Supporting Information for full details).

7.3.6 Recycling Experiment

Upon completion of the reaction, the product was extracted with diethyl ether to leave behind Pd NPs immobilized in **1**. The reaction mixture was repeatedly washed with water to remove $Si(OMe)_3F$ and residual **2**. As **1** is water immiscible, this limited the amount of Pd NPs leached. Finally, acetonitrile (2 mL) was added, and the contents were sonicated for 1 h and reused for the next reaction by addition of a fresh batch.

7.4 CONCLUSION

In summary, successful protocol for the ligand-free Hiyama coupling of aryl and heterocyclic halides with aryl- and vinylsilanes catalyzed by Pd NPs in nitrile-functionalized IL **1** was developed. Nitrile-functionalized ILs proved to be the best media for the generation and stabilization of the Pd NPs, and their usefulness is apparent upon catalyst recycling. Further, [bmim]F was employed for the first time as a silane activator with the advantages of being easy to handle, store, and remove during work-up. The system allows easy isolation of the products with the recovery and reuse of the catalyst up to four times, which make the overall process highly economical and cost effective.

ACKNOWLEDGMENTS

This work has been financially supported by the Department of Science and Technology, India, under the project RP02555. C. P. thanks Council of Scientific and Industrial Research, India, for graduate fellowship.

KEYWORDS

- **Palladium Nanoparticles**
- **b.Hiyama Cross-Coupling,**
- **c. Catalytic Actvity**

REFERENCES

1. (a) Beletskaya, I. P.; Cheprakov, A. V., The heck reaction as a sharpening stone of palladium catalysis, *Chem. Rev.* 2000, 100, 3009–3066; (b) Wu, X.-F.; Anbarasan, P.; Neumann, H.; Beller, M., From noble metal to nobel prize: palladium-catalyzed coupling reactions as key methods in organic synthesis, *Angew. Chem. Int. Ed.* 2010, 49, 9047–9050; (c) Leadbeater, N. E., Cross coupling: when is free really free, *Nat. Chem.* 2010, 2, 1007–1009.

2. (a) Rosner, T.; Bars, J. L.; Pfaltz, A.; Blackmond, D. G., Kinetic studies of Heck coupling reactions using palladacycle catalysis: experimental and kinetic modelling of the role of dimer species, *J. Am. Chem. Soc.* 2001, 123, 1848–1855; (b) Dupont, J.; Pfeffer, M.; Spencer,Palladacycles-An old organometallic family revisited: new simple and efficient catalyst presursors for homogenous catalysis, *J. Eur. J. Inorg. Chem.* 2001, 8, 1917–1927; (c) Bedford, R. B., Palladacyclic catalysis in C-C and C-heteroatom bond-forming reactions, *Chem. Commun.* 2003, 1787–1796.

3. (a) Morales-Morales, D.; Redón, R.; Yung, C.; Jensen, C. M., High yield olefination of a wide scope of aryl chlorides catalysed by the phosphinito palladium PCP pincer complex:[PdCl {C6H3(OPPri2) 2-2,6}], *Chem. Commun.* 2000, 1619–1620; (b) Selvakumar, K.; Zapf, A.; Beller, M., New palladium carbine catalysts for the Heck reaction of aryl chlorides in ionic liquids, *Org. Lett.* 2002, 4(18), 3031–3033; (c) Martın, R.; Buchwald, S. L., An improved protocol for the Pd-catalyzed α-arylation of aldehydes with aryl halides, *Org. Lett.* 2008, 10(20), 4561–4564; (d) Datta, G. K.; Schenck, H.; Hallberg, A.; Larhed, M., Selective terminal Heck arylation of vinyl ethers with aryl chlorides: a combined experimental- computational approach including synthesis of betaxolol, *J. Org. Chem.* 2006, 71, 3896–3903.

4. (a) Mowery, M. E.; DeShong, P., Improvement in cross coupling reactions of hypervalent siloxane derivatives, *Org. Lett.* 1999, 1, 2137; (b) Horn, K. A., Regio-and stereochemical aspects of the palladium-catalyzed reactions of silanes, *Chem. Rev.* 1995, 95, 1317–1350; (c) Gouda, K.; Hagiwara, E.; Hatanaka, Y.; Hiyama, T.,Cross-coupling reactions of aryl chlorides with organochlorosilanes: Highly effective methods for arylation of alkenylation of aryl chlorides, *J. Org. Chem.* 1996, 61, 7232–7233.

5. (a) H. E. Elsayed-Ali, E. Waldbusser, U.S. Patent 200400767552004; (b) E. R. Boller, Process for coating ferrous metals, U.S. Patent 35089131970, 1960.

6. (a) Denmark, S. E.; Regens, C. S., Palladium-catalyzed cross-coupling reactions of organosilanols and their salts: Practical alternatives to boron-and tin-based methods, *Acc. Chem. Res.* 2008, 41(11), 1486–1499; (b) Denmark, S. E.; Werner, N. S., On the stereochemical course of palladium-catalyzed cross-coupling of allylic silanolate salts with aromatic bromides, *J. Am. Chem. Soc.* 2010, 132, 3612–3620; (c) Denmark, S. E.; Smith, R. C.; Chang, W.-T.; Muhuhi, J. M., Cross-coupling reactions of aromatic and heteroaromatic silanolates with aromatic and heteroaromatic halides, *J. Am. Chem. Soc.*

2009, 131, 3104–3118; (d) Nakao, Y.; Imanaka, H.; Sahoo, A. K.; Yada, A.; Hiyama, T., Alkenly-and aryl [2-(hydroxymethyl) phenyl] dimethylsilanes: An entry to tetraorganosilicon reagents for the silicon-based cross-coupling reaction, *J. Am. Chem. Soc.* 2005, 127, 6952–6953; (e) Hatanaka, Y.; Hiyama, T., Cross-coupling of organosilanes with organic halides mediated by a palladium catalyst and tris(diethylamino) sulfonium difluorotrimethylsilicate, *J. Org. Chem.* 1988, 53, 918–920; (f) Lee, J.-Y.; Fu, G. C., Room-temperature Hiyama cross-couplings of arylsilanes with alkyl bromides and iodides, *J. Am. Chem. Soc.* 2003, 125, 5616–5617; (g) Raders, S. M.; Kingston, J. V.; Verkade, J. G., Advantages use of t Bu2P-N P (I BuNCH2CH2) 3N in the Hiyama coupling of aryl bromides and chlorides, *J. Org. Chem.* 2010, 75, 1744–1747; (h) Penafiel, I.; Pastor, I. M.; Yus, M., NHC-Ligand effectiveness in the Fluorine-free Hiyama reaction of aryl halides, *Eur. J. Org. Chem.* 2013, 1479–1484.

7. (a) Yanase, T.; Monguchi, Y.; Sajiki, H., Ligand-free Hiyama cross-coumpling reaction catalysed by palladium on carbon, *RSC Adv.* 2012, 2, 590–594; (b) Gordillo, A.; Jesus, E.; Mardomingo, C. L. *Chem. Commun.* 2007, 39, 4056–4058; (c) Alacid, E.; Najera, C., Aqueous sodium hydroxide promoted cross-coupling reactions of alkenyltrialkoxysilanes under ligand-free conditions, *J. Org. Chem.* 2008, 73, 2315–2322; (d) Srimani, D.; Sawoo, S.; Sarkar, A., Convenient synthesis of palladium nanoparticles and catalysis of Hiyama coupling reaction in water, *Org. Lett.* 2007, 9(18), 3639–3642; (e) Dey, R.; Chattopadhyay, K.; Ranu, B. C., Palladium (0) nanoparticle catalysed cross-coupling of allyl acetates and aryl and vinyl siloxanes, *J. Org. Chem.* 2008, 73, 9461–9464; (f) Alonso, D. A.; Najera, C., Oxime-derived palladacycles as source of palladium nanoparticles, *Chem. Soc. Rev.* 2010, 39, 2891–2902.

8. Reetz, M. T.; de Vries, Ligand-free Heck reactions using low Pd-loading, J. G. *Chem. Commun.* 2004, 1559–1563.

9. (a) Jeffery, T.; David, M., [Pd/base/QX] catalyst systems for directing Heck-type reactions, *Tetrahedron Lett.* 1998, 39, 5751–5754; (b) Jeffery, T., On the efficiency of tetraalkylammonium salts in Heck type reactions, *Tetrahedron.* 1996, 52, 10113–10130.

10. (a) Shenhar, R.; Rotello, V. M., Nanoparticles: scaffolds and building blocks, *Acc. Chem. Res.* 2003, 36, 549–561; (b) Reetz, M. T.; Westermann, Phosphanfreie palladium-katalysierte kupplungen: die entscheidende von Pd- Nanoteilchen, E. *Angew. Chem.* 2000, 112, 170; (c) Reetz, M. T.; Westermann, Phosphane-free palladium-catalyzed coupling reactions: the decisive role of Pd nanoparticles, E. *Angew. Chem. Int. Ed.* 2000, 39, 165–168; (d) Astruc, D.; Lu, F.; Aranzaes, Nanopartikel als regenerierbare katalysatoren: an der nahtstelle zwischen homogener und heterogener katalyse, J. R. *Angew. Chem.* 2005, 117, 8062; (e) Astruc, D.; Lu, F.; Aranzaes,, Nanoparticles as recyclable catalysts: the frontier between homogeneous and heterogeneous catalysis, J. R. *Angew. Chem. Int. Ed.* 2005, 44, 7852–7872; (f) Luska, K. L.; Moores, A., Improved stability and catalytic activity of palladium nanoparticles catalysts using phosphine-functionalized imidazolium ionic liquids, *Adv. Synth. Catal.* 2011, 353, 3167–3177.

11. (a) Crooks, R. M.; Zhao, M.; Sun, L.; Chechi, V.; Yeung, L. K., Dendrimer-encapsulated metal nanoparticles: synthesis, characterization, and applications to catalysis, *Acc. Chem. Res.* 2001, 34, 181–190; (b) Roucoux, A.; Schulz, J.; Patin, H., Reduced transition metal colloids: a novel family of reusable catalysts?, *Chem. Rev.* 2002, 102(10), 3757–3778.

12. Dupont, J.; Scholten, On the structural and surface properties of transition-metal nanoparticles in ionic liquids, J. D. *Chem. Soc. Rev.* 2010, 39, 1780–1804.

13. Dupont, J., From molten salts to ionic liquids: a "nano" journey, *Acc. Chem. Res.* 2011, 44, 1223–1230.

14. (a) Raluy, E.; Favier, I.; Vinasco, A. M. L.; Pradel, C.; Martin, E.; Madec, D.; Teuma, E.; Go´mez, M., A smart palladium catalyst in ionic liquid for tandem processes, *Phys. Chem. Chem. Phys.* 2011, 13, 13579–13584; (b) Cassol, C. C.; Umpierre, A. P.; Machado, G.; Wolke, S. I.; Dupont, The role of Pd nanoparticles in ionic liquid in the Heck reaction, J. J. *Am. Chem. Soc.* 2005, 12, 3298–3299; (c) Calo, V.; Nacci, A.; Monopoli, A.; Ieva, E.; Cioffi, N., Copper-bronze catalysed heck reaction in ionic liquids, *Org. Lett.* 2005, 7, 617–620; (d) Mo, J.; Xu, L.; Xiao, J., Ionic liquid-promoted, highly regioselective Heck arylation of electron-rich olefins by aryl halides, *J. Am. Chem. Soc.* 2005, 127, 751–760; (e) Xiao, J.-C.; Twamley, B.; Shreeve, An ionic liquid-coordinated palladium complex: A highly efficient and recyclable catalyst for the Heck reaction, J. M. *Org. Lett.* 2004, 6(21), 3845–3847;

(f) Zhao, D.; Fei, Z.; Geldbach, T. J.; Scopelliti, R.; Dyson, P. J., Nitrile-functionalized pyridinium ionic liquids: synthesis, characterization, and their application in carbob-carbon coupling reactions, *J. Am. Chem. Soc.* 2004, 126, 15876–15882; (g) Fei, Z.; Zhao, D.; Pieraccini, D.; Ang, W. H.; Geldbach, T. J.; Scopelliti, R.; Chiappe, C.; Dyson, P. J., Development of nitrile-functionalized ionic liquids for CC coupling reactions: Implication of carbine and nanoparticle catalysts, *Organometallics.* 2007, 26, 1588–1598; (h) Cal, V.; Nacci, A.; Monopoli, A.; Cotugno, P., Heck reactions with palladium nanoparticles in ionic liquids: Coupling of aryl chlorides with deactivated olefins, *Angew. Chem. Int. Ed.* 2009, 48, 6101–6103.

15. (a) Swatloski, R. P.; Holbrey, J. D.; Rogers, R. D., Ionic liquids are not always green: hydrolysis of 1-butyl-3-methylimidazolium hexafluorophosphate, *Green Chem.* 2003, 5, 361–363; (b) Maiti, A.; Pagoria, P. F.; Gash, A. E.; Han, T. Y.; Orme, C. A.; Gee, R. H.; Fried, L. E., Solvent screening for a hard-to-dissolve molecular crystal, *Phys.Chem.Chem.Phys.* 2008, 10, 5050–5056.

16. (a) Venkatesan, R.; Prechtl, M. H. G.; Scholten, J. D.; Pezzi, R. P.; Machado, G.; Dupont, J., Palladium nanoparticle catalysts in ionic liquids: synthesis, characterization and selective partial hydrogenation of alkynes to Z-alkenes, *J. Mater. Chem.* 2011, 21, 3030–3036; (b) Zhao, D. B.; Fei, Z. F.; Geldbach, T. J.; Scoppelliti, R.; Dyson, P. J., Nitrile-functionalized pyridinium ionic liquids: synthesis, characterization and their applications in carbon-carbon coupling reactions, *J. Am. Chem. Soc.* 2004, 126, 15876–15882.

17. Ye, C.; Xiao, J.-C.; Twamley, B.; LaLonde, A. D.; Norton, M. G.; Shreeve, J. M., Basic ionic liquids: facile solvents for carbon-carbon bond formation reactions and ready access to palladium nanoparticles, *Eur. J. Org. Chem.* 2007, 5095–5100.

18. (a) Reetz, M. T.; Westermann, E.; Lomer, R.; Lohmer, G., A highly active phosphine-free catalyst system for Heck reractions of aryl bromides, *Tetrahedron Lett.* 1998, 39, 8449; (b) de Vries, A. H. M.; Mulders, J. M. C. A.; Mommers, J. H. M.; Henderckx, H. J. W.; de Vries, J. G., Homeopathic ligand-free palladium as a catalyst in the Heck reaction. A comparison with a palladacycle, *Org. Lett.* 2003, 5, 3285; (c) Reetz, M. T.; Maase, M., Redox-controlled size-selective fabrication of nanostructured transition metal colloids, *Adv. Mater.* 1999, 11, 773.

19. Darses, S.; Genêt, J. P., Potassium trifluoro (organo) borates: new perspectives in organic chemistry, *Eur. J. Org. Chem.* 2003, 4313–4327.

20. Lee, H. M.; Nolan, S. P., Efficient cross-coupling reactions of aryl chlorides and bromides with phenyl-or vinyltrimethoxysilane mediated by a palladium/imidazolium chloride system, *Org. Lett.* 2000, 2, 2053–2055.

21. Clarke, M. L., First microwave-accelerated Hiyama coupling of aryl-and vinylsiloxane derivatives: clean cross-coupling of aryl chlorides within minutes, *Adv. Synth. Catal.* 2005, 347, 303–307.

22. Alacid, E.; Nájera, C., The first fluoride-free Hiyama reaction of vinylsiloxanes propmoted by sodium hydroxide in water, *Adv. Synth. Catal.* 2006, 348, 2085–2091.

23. Navaladian, S.; Viswanathan, B.; Varadarajan, T. K.; Viswanath, R. P., A rapid synthesis of oriented palladium nanoparticles by UV irradiation, *Nanoscale Res. Lett.* 2009, 4, 181–186.

CHAPTER 8

Acyclic and Macrocyclic Schiff Base-Based Chelating Ligands for Uranyl Ion (Uo$_2^{2+}$) Complexation

Summan Swami and Rahul Shrivastava*

Department of Chemistry, Manipal University, Jaipur, India
*Email: rahul.shrivastava@jaipur.manipal.edu

CONTENTS

ABSTRACT

The hexavalent uranyl ion $\{UO_2^{2+}, U(VI)\}$ is an exceptionally stable oxidation state among four different oxidation states $[U(III), U(IV), U(V), \text{and } U(VI)]$ of uranium in aqueous systems. Studies using cell and animals reveal that chronic exposure of uranium ion causes the possibility of genetic, reproductive, and neurological disorders on human population due to its radioactivity and chemotoxicity. Schiff base-based chelators are well known for the formation of stable nontoxic complex with uranyl ion; therefore, a great deal of attention has been focused on the Schiff base derivatives and their complexes with uranyl ion. This chapter summarizes the interaction and complexation studies of the uranyl ion with various Schiff base-based acyclic and cyclic chelators.

8.1 INTRODUCTION

Uranium is the most commonly occurring radionuclide in nature. Uranium occurs in low concentration in rocs, water, and soil, generally varying between 0.5 and 5 ppm. Other sources of uranium contamination include mining (ore extraction), milling (physical and chemical extraction of uranium from the ore), manufacturing, and other anthropogenic activities, such as the use of phosphate fertilizers and combustion from coal and other fuels.[1,2] There are four different oxidation states of uranium in aqueous systems: U(III) (highly unstable), U(IV), U(V) (unstable), and U(VI). The hexavalent uranyl ion $\{UO_2^{2+}, U(VI)\}$ was proved to be most stable form in aqueous solution and therefore, the predominant form of uranium in contaminated ground water and soils. Uranium is also the most commonly used nuclear fuel in fission reactors for commercial purposes. With the continuous increasing demand of energy sources, the actinides have become important industrial elements due to their ability to generate tremendous energy through fission reactions. Major concerns of these industries are the biological hazard associated with nuclear fuels and their wastes. Uranium can enter the human body via inhalation, ingestion, and wounds,[3] and in the body, they are chelated by complexing agents such as proteins or carbonates. After chelation, toxic species are distributed and retained in target organs, which poses a threat to human populations due to its radioactivity and chemotoxicity.[4–6] Studies using cells and animals suggest the possibility of genetic, reproductive, and neurological effects from chronic exposure to uranium wastes.[7] It is generally accepted that radiation and chemical toxicity of uranium can increase the risk of cancers, such as bone cancer and lung cancer, and that uranium can accumulate in kidneys for a longer period and cause renal dysfunction and structural damage. Because of the threat of uranium radioactivity and chemotoxicity to human populations, it is very important to ensure that uranium contamination is under control. The U.S. Environmental Protection Agency has established a maximum contaminant level of 30 µg/L for uranium in drinking water.[9] In the U.S., the total volume of all radionuclide wastes is 5 million m^3 and the volumes of contaminated soil and water have been reported to be 30–80 million m^3 and 1800–4700 million m^3 respectively. There has been immense interest in the scientific community since the last decade in the control of uranium contamination by developing the nontoxic uranium chelators suitable for human use. These non-toxic chelators can form stable complexes with uranium present in the body, and the body can rapidly excrete the poison from blood and target organs. Furthermore, the uranyl chelates must be soluble and stable in physiological fluids in a pH range of 2–9 to be subsequently eliminated from the body after crossing the renal and hepatic barriers. Thus, uranyl concentrations and radiation doses and subsequently, tumor risks may be reduced.[7] Recently, it has been observed that Schiff base-based chelators can form stable non-toxic complex with uranyl ion and, therefore, a great deal of attention has been focused on the Schiff base derivatives and their complexes.

Schiff bases are the important nitrogen analog of an aldehyde/ketone in which the carbonyl group is replaced by azomethine group. It was first synthesized by Hugo Schiff in 1864 in which an aldehyde condensed with primary amine leads to the formation of a Schiff base as a product.[8]

$$\underset{R_2}{\overset{R_1}{>}}C{=}O + H_2N{-}R_3 \longrightarrow \underset{R_2}{\overset{R_1}{>}}C{=}N{\diagup}^{R_3} + H_2O$$

SCHEME 8.1 Synthesis of Schiff Base.[8]

Schiff base-based ligands derived from salicylaldehyde derivatives are an interesting area of research due to their straightforward synthetic methods, and the presence of nitrogen and oxygen donor atoms in the backbones of these ligands makes them suitable ligands to accommodate different metal centers involving various coordination modes, thereby allowing successful synthesis of homo- and heterometallic complexes with varied stereochemistry.[9-11] Schiff base ligands have wide applications in the biological field such as antitumor, antibacterial, antifungal, antidepressants, antimicrobial, antiphlogistic, nematocide, and other medicinal agents.[12,13] Along with these applications, Schiff base metal complexes have also exhibited potential application in catalysis, asymmetric synthesis, epoxidation, electrochemistry, and magnetochemistry.[14,15]

Supramolecular chemistry of Schiff base ligands and their reduced homologs is rapidly growing due to a wide range of complexation modes with almost all types of metal ions and their suitable and straightforward synthetic methods. Actually, the phenomenon of molecular recognition, self-organization, self-assembly, and host–guest chemistry through covalent and noncovalent interactions is fundamental to the understanding and development of supramolecular chemistry. In this direction, several forms of acyclic and macrocyclic Schiff bases and their reduced forms are employed to gain more insights and correctly establish the effect of different donor atoms, their relative position, the number and size of the chelating rings formed, the flexibility, and the geometry around the coordinating moiety on the molecular recognition process and selective binding of hazardous and nonhazardous cations, anions, and neutral species. In this chapter, we emphasized mainly on acyclic and macrocyclic Schiff base chelators and their complexation with toxic uranium ion.

8.2 REVIEW OF LITERATURE

In this context, Sessler et al.[16] have reported hexadentate pyrrole derivatives and its uranyl complexes **1**. Solid state evidences of these complexes revealed that uranyl is coordinated to all six nitrogen atoms in a planar fashion. On the basis of these results, Sessler et al. synthesized other analogous systems **2** and **3** with a view to determine whether the planarity observed for complex **1** is primarily due to the rigidity of the phenylene diamine rings or whether it is due to the inherent chelating geometry of the macrocycle-bound uranyl ion. They concluded that the rigidity of the phenylene diamine rings in **1** does confine the overall ligand to a planar conformation.

(1)   (2)   (3)

Porphyrins are important class of Schiff base-based macrocyclic compounds, utilized successfully for uranyl ions extraction by Sessler et al.[17] One of the most widely studied analogs of the Schiff base porphyrin is texaphyrin (**4**), and complexes of alkoxy-substituted porphyrins (**5, 6**) with uranyl are used for uranyl extraction. Similarly, calixarenes-supported Schiff bases have been reported for uranyl ion complexation.[18]

5 a: $R_1 = R_2 = O(CH_2)_4(CH_3)$
5 b: $R_1 = R_2 = O(CH_2)_9(CH_3)$
5 c: $R_1 = R_2 = O(CH_2)_{13}(CH_3)$

(4)   (6): R=t-butyl

Casellato et al.[19] have synthesized Schiff base ligands **7** and **8** by the condensation of equimolar amount of 3,3'-(3-oxapentane-1,5-diyldioxy)-bis(2-hydroxybenzaldehyde) or 3,3'-(3,6-dioxaoctane-1,8-diyldioxy)bis(2-hydroxybenzaldehyde) and 1,5-diamino-3-azamethylpentane. Both the ligands contain two dissimilar adjacent coordination sites N_3O_2 Schiff base and O_3O_2 or O_4O_2 crown like;, thus, they can be suitably used for determining the coordination preference of different metal ions toward the "soft" or "hard" coordination sites.

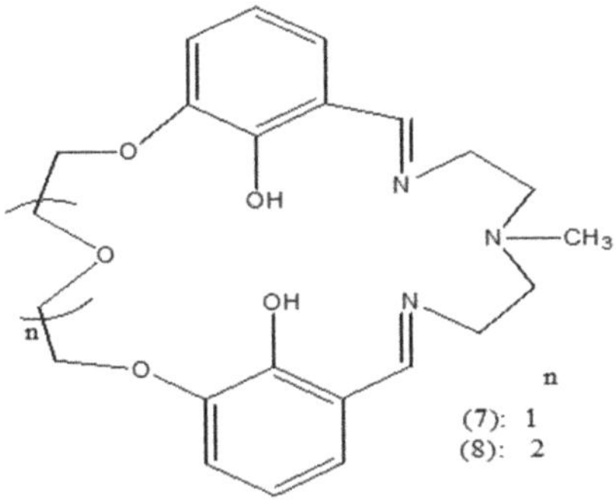

(7): 1
(8): 2

Gatto et al.[20] have reported acylic hydrazine- and semicarbazone-based ligands such as 2,6-diacetylpyridinebis(benzoylhydrazone) **9**, 2,6-diacetylpyridinebis(N4-phenylsemicarbazone) **10**, and 2,6-diacetylpyridine(benzoylhydrazone)(N4-phenyl-semicarbazone) **11** and observed that acetylpyridine benzoylhydrazones and semicarbazones are well appropriate to form stable complexes with uranyl ion.

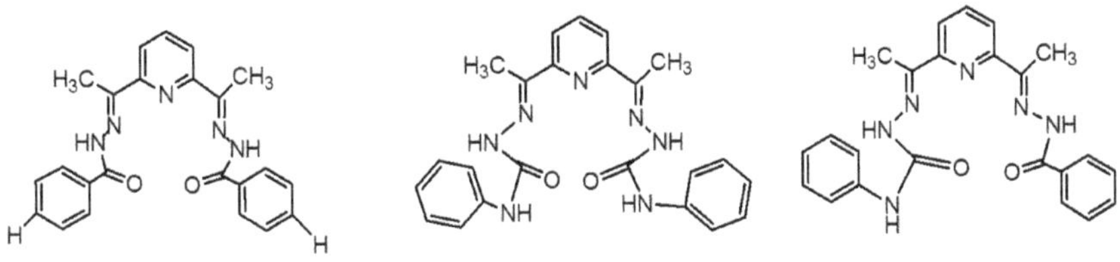

(9) (10) (11)

It is well known that worldwide needs for extraction of toxic ions from wastewater stimulate the global efforts to develop new sensitive sensors for polluting waste such as uranyl, arsenic, and mercuric ions. In this regard, Schiff base chelating agents are some of the most promising choices for the detection of toxic metal ions from wastewater and various matrices. N,N-ethylene bis-(salicilydenimine) (salen) derivatives of Schiff base (**12**) have an ability to form coordination compounds with a wide number of transition metals and toxic uranyl ions due to the presence of suitable tetradentate N,N,O,O coordination site in skeleton of salen derivatives. Bastos et al.[21] have described uranyl ion complex with salen derivatives (**12**) in buffered aqueous solution of 4-(2-hydroxyethyl)-piperazine-1-ethanesulfonic acid.

A new sensitive chelating agent N,N-di(2-hydroxy-4,6-di-tert-butyl-benzyl)imidazolidine **13** has been developed by Albo et al.[22] for the uranyl cation. The reaction of **13** with the uranyl ion involves the opening of the imidazolidine ring to yield a complex of the type UO$_2$L, where L = N,N-di(2-hydroxy-4,6-di-tert-butyl-benzyl)-N-2-hydroxylethane–ethylenediamine. During complex formation in a pH-sensitive system, deep color change was observed.

(12)   (13)

A large number of Schiff base chelating ligands comprising sulfur, nitrogen, and oxygen donor atoms have been synthesized for complexation with transition metal, lanthanide, and actinide ions. These metal complexes have revealed excellent biological activities such as antibacterial, antifungal, and antitumor activities. Chelating behavior of the thiocarbohydrazone ligand **14** toward some transition metal, lanthanide, and actinide ions has been studied by Shebl et al.[23] Their reported ligand as well as some metal complexes showed antibacterial and antifungal activities against selected kinds of bacteria and fungi.

(14)

For selective extraction of uranyl ion in the presence of other lanthanides, tetradentate ligand N,N′-bis(3,5-di-t-butylsalicylidene)-4,5-dimethyl-1,2-phenylenediamine has been synthesized. It has been found that **15** selectively binds to UO$_2^{+2}$ rather than to lanthanides due to its cavity.[12] Similarly, salphenH$_2$ derivatives such as N,N-(propylenedioxy)-benzenebis-(salicylideneimine) **16** and N,N-4,5 (propylenedioxy)-benzenebis-(3,5-di-tert-butylsalicylideneimine) **17** have been developed for a uranyl ion by Kim et al.[24] It has been observed that lipophilic behavior of these ligands enhanced the selectivity for a uranyl ion over the other metal ion. Further, salen- and salphenH$_2$-based hexadentate Schiff base ligands toward complexation with uranyl ion as well as 3d-transition metal ions have also been reported by Salmon et al.[25,26] They have found that ligands **18** and **19** form complex with uranyl ion in the outer O_4 coordination sites while 3d metal ions in the inner N_2O_2 coordination sites. The same research group also reported N,N'-bis(3-hydroxysalicylidene)-2-methyl-1,2-propanediamine] **20** and N,N'-bis-(3 hydroxysalicylidene)-2,2-dimethyl-1,3-propanediamine hexadentate Schiff bases to examine the effect of alkyl substitutions on chelation[27] of uranyl ion.

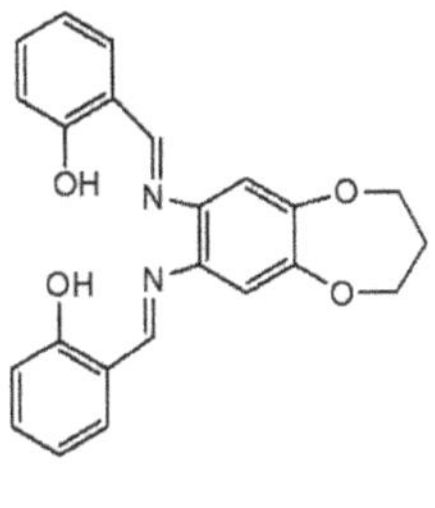

(15) 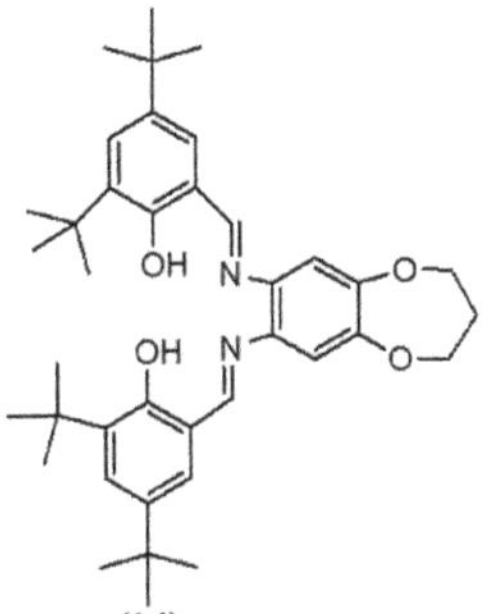   (16) 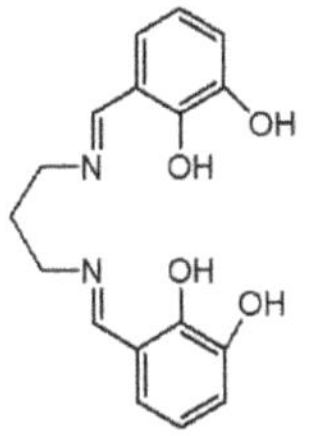   (17)

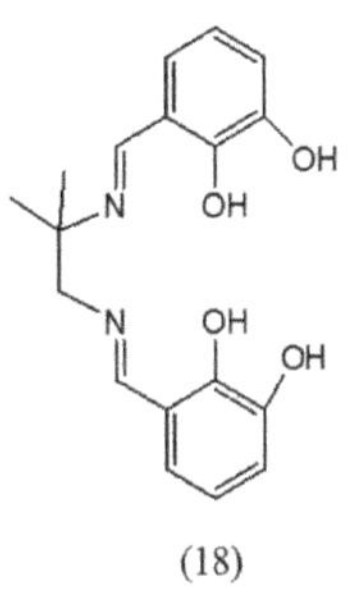

(18) 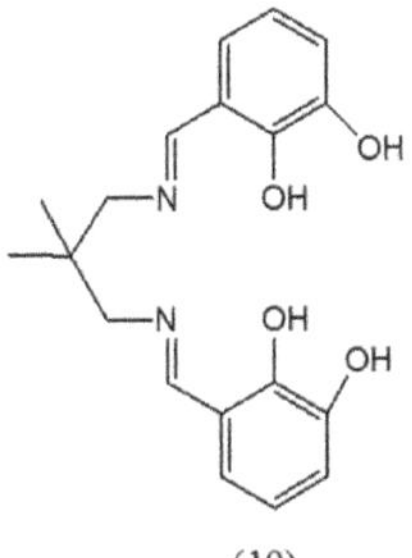   (19) 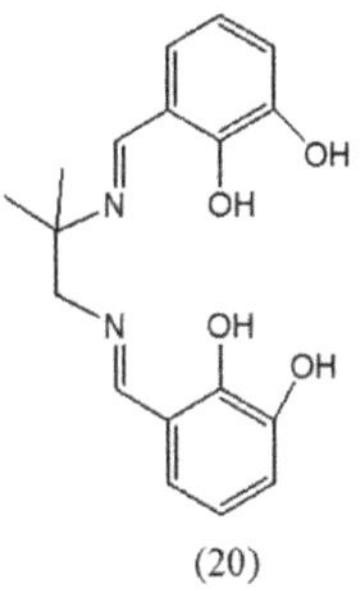   (20)

Apart from these synthetic ligands, vitamin B6 derivatives have the ability to interact with UO_2^{+2}, which allows the chelation of uranium atom and represents a very specific model of absorption of uranium by living bodies.[28] Back et al. have studied that Schiff base N,N'-bis(pyridoxylideneiminato) ethylene **21**, synthesized from the condensation of pyridoxal with ethylenediamine, is an interesting ligand and a close relative of the vitamin B6 for uranyl ion. They enhance chelation ability of ligand **21** by chemical insertion of larger amino groups such as **22** and **23**.[29]

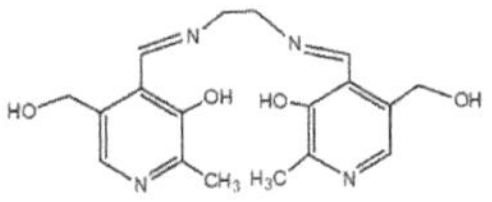

(21) 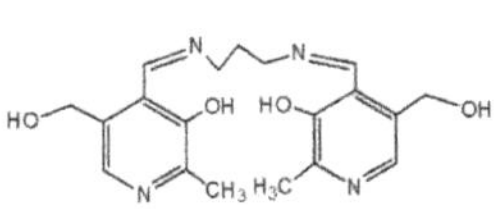   (22) 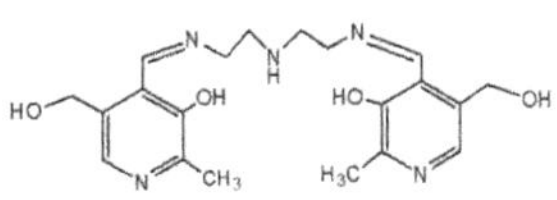   (23)

Sopo et al.[30,31] have reported a series (**24–28**) of aminoalcoholbis(phenolate) $[O,N,O,O]$ donor ligands of a relatively small size, which are potential uranium chelators and have better uranyl ion extraction properties. Further, they have substituted the alcohol part of these aminoalcoholbis(phenolate) molecule with different alkyl groups (**29–33**), and consequently, reduced the number of possible donor atoms from the outer sphere in the coordinating sites. These modifications in the backbone of the ligands make exterior lipophilic after complexation and promote the solubility of the formed complex into organic solvents, which is an important feature in the extraction.

N^1,N^4-diarylidene-S-propyl-thiosemicarbazone and its dichloro derivative (**34, 35**) and Schiff base hydrazone containing the quinoline moiety, **34**, have been reported as potential chelating ligands for uranium atom through an ONNO donor set.[32,33] Additionally, these ligands show better antimicrobial activity as well as toxic metal ion complexation.

Pyrene appendage salen- like Schiff base ligand has reasonable coordination ability with toxic metal ions, and Brancatelli et al.[34] reported a salen–pyrene ligand (**36**) that acts as a tetradentate ligand through its nitrogen and oxygen atoms chelating the UO_2^{2+} ions at the equatorial plane. Geometrically, uranium was situated in seven-coordinated pentagonal–bipyramidal environment, when uranium complexation with ligand.[34]

ACKNOWLEDGMENTS

The authors gratefully acknowledge the assistance received from Material Research Center, MNIT, Jaipur, for Sophisticated Analytical Instrument Facility, IIT Delhi for assessment of research article, and Manipal University Jaipur for teaching assistance SS.

KEYWORDS

- **Uranyl ion**
- **calixarene**
- **Schiff base**
- **chelators**
- **macrocyclic compounds**

REFERENCES

1. Merroun, M.; Selenska-Pobell, S. Bacterial interactions with uranium: an environmental perspective. *J. Contam. Hydrol.* 2008, 102, 285.
2. Guessan, N.; Varionis, A. H.; Resch, C.; Long, P.; Lovley, D. Sustained removal of uranium from contaminated groundwater following stimulation of dissimilatory metal reduction. *Env. Sci. Tech.* 2008, 42, 2999-3004.
3. Zhu, G.; Xiang, X.; Chen, X.; Wang, L.; Hu, H.; Weng, S. Renal dysfunction induced by long term exposure to depleted uranium in rates. *Arch. Toxicol.* 2009, 83, 37-46.
4. Leise, A.; Danesi, P.; Burkart, W. Properties, use and health effects of depleted uranium (DU): a general overview. *J. Environ. Radioact.* 2003, 64, 93.
5. Landa, E.; Gray, US Geological Survey research on the environmental fate of uranium mining and milling wastes. *J. Environ. Geol.* 1995, 26, 19.
6. Suzuki, Y.; Suko, T. Geo-micro biological factors that control uranium mobility in the environment: Update on recent advances in the bioremediation of uranium contaminated sites. *J. Mineral. Petrol. Sci.* 2006, 101, 299.
7. Leydier, A.; Lecercle, D.; Rostaing, S. P.; Reguillon, A. F.; Taran, F.; Lemaire, M. Sequestering agent for uranyl chelation: a new family of CAMS ligands. *Tetrahedron.* 2008, 64, 6662.
8. Schiff, H. *Ann. Suppl.* Eine neue Reihe organischer Diamine. 1864, 3, 343.
9. Zhang, J. J.; Zhou, H. J.; Lachgar, A. Directed assembly of cluster-based supramolecules into one-dimensional coordination polymers. *Angew. Chem. Int. Ed. Engl.* 2007, 46, 4995.
10. Hosseini, M. G.; Mertens, S. F. L.; Ghorbani, M. 'Assimetrical Schiff base as inhibitors of mild steel corrosion in sulphuric acid media. *Mater. Chem. Phys.* 2003, 78, 800.
11. Choudhury, C. R.; Dey, S. K.; Mondal, N.; Mitra, S.; Mahalli, S. O. G.; Malik, K. M. A. *J. Chem. Crystallogr.* 2001, 31, 57.
12. Chandra, S.; Saneetika, X. EPR, magnetic and spectral studies of copper(II) and nickel(II) complexes of schiff base macrocyclic ligand derived from thiosemicarbazide and glyoxal. *Spectrochin. Acta A.* 2004, 60, 147.
13. Tarasconi, P.; Capacchi, S.; Pelosi, G.; Corina, M.; Albertini, R.; Bonati, A.; Dallaglio, P. P.; Lunghi, P.; Pinelli, S. Synthesis, spectroscopic characterization and biological properties of new natural aldehydes thiosemicarbazones. *Biorg. Med. Chem.* 2000, 8, 157.
14. Abdel Aziz, A.A.; Salem, A.N.M.; Sayed, M.A.; Aboaly, M.M. Synthesis, structural characterization, thermal studies, catalytic efficiency and antimicrobial activity of some M(II) complexes with ONO tridentate Schiff base N-salicylidene-o-aminophenol (saphH$_2$). *J. Mol. Struct.* 2012, 1010, 130-138.
15. Colman, J.; Hegedu, L. S. Principal and application of organo-transition metal chemistry; University Scince Book: California, 1980.
16. Sessler, J. L.; Mody, T. D.; Dulay, M. T.; Espinozaa, R.; Lyncha, V. The template synthesis and X-ray characterization of pyrrole-derived hexadentate uranyl(VI) Schiff base macrocyclic complexes. *Inorg. Chim. Acta.* 1996, 246, 23–30.

17. Sessler, J. L.; Melfi, P. J.; Tomat, E.; Callaway, W.; Huggins, M. T.; Gordon, P. L.; Keogh, D. W.; Date, R. W.; Bruce, D. W.; Donnio, B. Schiff base oligopyrrolic macrocycles as ligands for lanthanides and actinides. *J. Alloys Compd.* 2006, 418, 171–177.

18. Ali, A.; Joseph, R.; Mahieu, B.; Rao, C. P. Synthesis and characterization of a (1+1) cyclic Schiff base of lower rim 1,3-diderivative of p-tetr-butylcalix[4]arene and its complexes of VO^{2+}, UO_2^{2+}, Fe^{3+}, Ni^{2+}, Cu^{2+} and Zn^{2+}. *Polyhedron.* 2010, 29, 1035–1040.

19. Casellato, U.; Tamburini, S.; Tomasin, P.; Vigato, P. A. Uranyl(VI) complexes with [1+1] asymmetric compartmental ligands containing a Schiff base and a crown ether like chabmer. *Inorg. Chim. Acta.* 2002, 341, 118–126.

20. Gatto, C. C.; Lang, E. S.; Jagst, A.; Abram, U. Dioxouranium(VI) complexes with 2,6-acetylpyridine-benzoylhydrazones and semicarbazones. *Inorg. Chim. Acta.* 2004, 357, 4405–4412.

21. Bastos, M. B. R.; Moreira, J. C.; Farias, P. A. M. Adsorptive stripping voltammetric behavior of $UO_2(II)$ complex with the Schiff base N, N'-ethylenebis(salicylidenimine) in aqueous medium. *Anal. Chim. Acta.* 2000, 408, 83–88.

22. Albo, Y.; Saphier, M.; Maimon, E.; Zilbermann, I.; Meyerstein, D. A new chelate ligand designed for the uranyl ion. *Coord. Chem. Rev.* 2009, 253, 2049–2055.

23. Shebl, M.; Khalil, S. M. E.; Al-Gohani, F. S. Preparation, spectral characterization and antimicrobial activity of binary and ternary Fe(III), Co(II), Ni(II), Cu(II), Zn(II), Ce(III) and $UO_2(VI)$ complexes of a thiocarbohydrazone ligand. *J. Mol. Struct.* 2010, 980, 78–87.

24. Kim, D. W.; Park, K. W.; Yang, M. H.; Kim, T. H.; Mahajan, R. K.; Kim, J. S. Selective uranyl ion detection by polymeric ion selective electrodes based on salphenH$_2$ derivatives. *Talanta.* 2007, 74, 223–228.

25. Salmon, L.; Thue´ry, P.; Ephritikhine, M. Crystal structure of hetero(bi and tetra-) metallic complexes of compartmental Schiff base uniting uranyl and transition metal (Ni^{2+}, Cu^{2+}) ions. *Polyhedron.* 2003, 22, 2683–2688.

26. Salmon, L.; Thue'ry, P.; Ephritikhine, M. Crystal structure of the first octanuclear uranium(IV) complex with compartmental Schiff base ligands. *Polyhedron.* 2004, 23, 623–627.

27. Salmon, L.; Thue'ry, P.; Ephritikhine, M. Polynuclear uranium(IV) compounds with (μ_3- oxo)U$_3$ or (μ_4-oxo)U$_4$ cores and compartmental Schiff base ligands *Polyhedron.* 2006, 25, 1537–1542.

28. Back, D. F.; de Oliveira, G. M.; Lang, E. S. Chelation of by vitamin B6 complex derivatives: Synthesis and characterization of $[UO_2(\beta$-pyracinide)$_2(H_2O)]$ and $UO_2(Pyr_2en)DMSO]Cl\{Pyr_2$ en=N,N'-et hylenebis(pyridoxylideneiminato)}. A useful modeling of assimilation of uranium by living beings. *J Inorg. Biochem.* 2006, 100, 1698–1704.

29. Back, D. F.; de Oliveira, G. M.; Vargas, J. P.; Lang, E. S.; Tabarelli, G. Chelation of and Th(IV) by N, N'-bis (pyridoxylideneiminato) R (R= n-propyl, diethylamine), new dianionic Schiff bases derived from vitamin B6: Synthesis and structural features of $[Th (pyr_2 pen)_2](pen= 1, 3$-propylendiamine),$[UO_2 (pyr_2 pen)(CH_3 OH)]$ and $[UO_2 (pyr_2 dien)] \cdot 2H_2 O$ (dien= diethylenetriamine). Searching further modelings for heavy metals damage inhibition in living beings. *J. Inorg. Biochem.* 2008, 102, 666–672.

30. Sopo, H.; Vaisanen, A.; Sillanpaa, R. Uranyl ion complexes with long chain aminoalcoholbis(phenolate) [O,N,O,O prime] donor ligands. *Polyhedron.* 2007, 26, 184–196.

31. Sopo, H.; Goljahanpoor, K.; Sillanpaa, R. (phenolate) [O,N,O] donor ligands for uranyl(VI) ion co-ordination: Syntheses, structures, and extraction studies .*Polyhedron.* 2007, 26, 3397–3408.

32. Ozdemir, N.; S'ahin, M.; Bal-Demirci, T.; Ulkuseven, B. The asymmetric ONNO complexes of dioxouranium(VI) with N1,N4-diarylidene-S-propyl-thiosemicarbazones derived from 3,5-dichlo-rosalicylaldehyde; Synthesis, spectroscopic and structural studies. *Polyhedron.* 2011, 30, 515–521.

33. El-Behery, M.; El-Twigry, H. Synthesis, magnetic, spectral, and antimicrobial studies of Cu(II), Ni(II) Co(II), Fe(III), and $UO_2(II)$ complexes of a new Schiff base hydrazone erived from 7-chloro-4-hy-drazinoquinoline.*Spectrochim Acta A Mol Biomol Spectrosc.* 2007, 66, 28–36.

34. Brancatelli, G.; Pappalardo, A.; Sfrazzetto, G. T.; Notti, A.; Geremia, S. Mono-and dinuclear uranyl(VI) complexes with chiral schiff base ligand. *Inorg. Chim. Acta.* 2013, 396, 25–29.

CHAPTER 9

Study of Influence of Operational Parameters on Eliminating Azo Dyes from Textile Effluent by Advanced Oxidation Technology

Preeti Mehta* and Rajeev Mehta

Department of Chemistry, Sangam University, Bhilwara, Rajasthan, India
*Email: preetimehta461@gmail.com

CONTENTS

ABSTRACT

Effluents from dyeing and printing done in textile industries contain complex mixtures of dyes that are highly colored with high biological oxygen demand, chemical oxygen demand, and unbearable odor, which is ethically not acceptable. It is a threat to mankind and has added a new dimension to the problem of water pollution. Heterogeneous photocatalysis is an advanced oxidation process that can be successfully used to oxidize many organic pollutants present in aqueous systems. In this research work, the effect of various parameters on the photocatalytic degradation of commercially available textile azo dye (Reactive Orange 16) in aqueous heterogeneous suspension has been studied. Parameters such as pH, amount of photocatalyst, dye concentration, and nature of photocatalyst as main operational parameters were selected, and their influence on degradation efficiency has been investigated. The progress of reaction was observed spectrophotometrically. The kinetic analysis of photocatalytic degradation reveals that the degradation follows pseudo first-order kinetics according to the Langmuir–Hinshelwood model. The trace quantities of transition metal ions (Fe^{2+}, Cu^{2+}, Mn^{2+}, Zn^{2+}) increase the photocatalytic degradation efficiency to some extent. This may be due to the introduction of new trapping sites by the incorporation of transition metal ions on semiconductor surface. A probable mechanistic pathway has been proposed.

9.1 INTRODUCTION

In the present scenario, environmental pollution is a big problem faced by everyone all over the globe, and water pollution is a major problem. Dyeing, desizing, and scouring are the major sources of water pollution in textile effluent.[1] Being released into the environment, the colored effluents damage the esthetic quality of water and reduce the light penetration, photosynthesis, reoxygenation capacity of water, thereby disturbing the natural activity of aquatic life.[2] Synthetic textile dyes and their intermediate products are toxic or mutagenic, allergic, and carcinogenic to aquatic life and humans.[3,4] Reactive dyes are extensively used for dyeing process in textiles and about 20–40% of these dyes are lost in the effluent. They exhibit a wide variability in chemical structure, primarily based on substituted aromatic and heterocyclic groups. Since reactive dyes are highly soluble in water, their removal from wastewater is difficult by conventional coagulation and the activated sludge process.[5,6] The environmental concern of these potentially carcinogenic pollutants in contaminated water has drawn the attention of many research workers.

In recent years, advanced oxidation processes (AOPs) have emerged as contemporary oxidative technique for degradation of detrimental organic compounds. The utilization of AOPs for the treatment of dyes are based on generation of hydroxyl radicals ($\cdot OH$) that oxidize organic pollutant.[7,8] Common AOPs involve Fenton, Fenton-like process, ozonation, H_2O_2/UV, heterogeneous catalysis, electron beam irradiation, sonolysis, wet-air oxidation, and various combinations of these methods. Among the AOPs, heterogeneous photocatalytic oxidation processes have shown great potential as low cost, environmental friendly, and sustainable treatment technology in the wastewater treatment.

The aim of this work is to assess the photocatalytic treatment of water soluble monoazo group containing Reactive Orange 16 reactive dye. It is used for dying the silk, nylon, and wool. It is mutagenic and genotoxic in nature.

[Molecular formula = $C_{20}H_{17}N_3O_{11}S_3Na_2$ Molecular weight = 617.53, C.I No. = 17757]

9.2   MATERIALS AND METHODS

For the present studies, the commercially available azo dye Reactive Orange 16 and the photocatalyst titanium dioxide (Merck, 99% purity) were used. For the preparation of stock solution, 0.0649 g of dye Reactive Orange 16 (90%) was dissolved in 100 mL of double distilled water so that the concentration of dye solution was 1.0×10^{-3} M. Aqueous solutions of desired concentrations were prepared from the stock solution. The desired pH of the solution was adjusted by the addition of previously standardized sulfuric acid and sodium hydroxide solutions. All laboratory reagents were of analytical grade.

To carry out the photochemical reaction, 100 mL of dye solution of desired concentration (3×10^{-5} M) was taken in 250 mL round bottom flask and appropriate amount of solid TiO_2 catalyst (0.30 g) was added to it. The pH of the reaction mixture was made alkaline (8.5) by adding 0.1 N NaOH and measured with digital pH meter (Systronics, 106). The mixture was then irradiated under light using 2×200 W tungsten lamps to provide energy to excite TiO_2 loading. To ensure thorough mixing of TiO_2 catalyst, oxygen was continuously bubbled with the help of aerator. A water filter was used to cut off thermal radiation. About 3 mL aliquot of the dye solution was withdrawn after a specific time interval and its absorbance was measured using spectrophotometer (Schimadzu, UV-1700 pharmaspec) at $\lambda_{max} = 492$ nm after filtration through a centrifugal machine. The rate of decrease of color with time was continuously monitored. After complete mineralization, the presence of inorganic ions such as sulfate and nitrate was tested by standard procedure. The evolution of CO_2 was tested by passing the evolved gas during the reaction into lime water.

The result of photocatalytic degradation of Reactive Orange 16 is graphically presented in Figure 9.1.

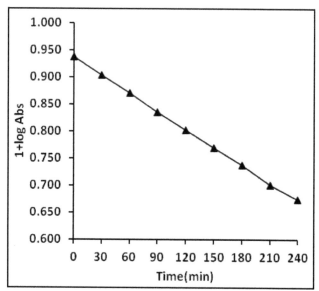

FIGURE 9.1 A Plot Showing a Typical Run of Photochemical Bleaching of Reactive Orange 16 Observed at 492 nm Under Optimum Conditions.

9.3 RESULTS AND DISCUSSION

Control experiments (in the absence of photocatalyst, oxygen, and light) confirm the necessity of photocatalyst, oxygen, and light to follow the photocatalytic path for the photocatalytic degradation of dye. The photocatalytic degradation of Reactive Orange 16 was studied at 492 nm. The optimum conditions for the removal of dye is [Dye] = 3.0×10^{-5} M, pH = 8.5, TiO_2 = 0.30 g. The rate of reaction (k) was determined using the expression:

$$\text{Rate constant } k = 2.303 \times \Delta \left[\log \text{Abs} / \text{time}\right] = 4.24 \times 10^{-5} \text{ s}^{-1}$$

The plot of $1 + \log$ Absorbance was found to be a straight line suggesting that degradation of dye by TiO_2 follows a pseudo first-order rate law. The effect of variation in reaction parameters has been studied such as pH, concentration of the dye, amount of catalyst, nature of photocatalyst, and presence of transition metal ions.

9.3.1 Effect of Variation in pH

The pH of the reaction medium has a significant effect on the surface properties of TiO_2 catalyst. The effect of pH on photocatalytic degradation of Reactive Orange

16 with TiO$_2$ was investigated in the pH range of 6.0–10.0 under visible light source, shown in Figure 9.2.

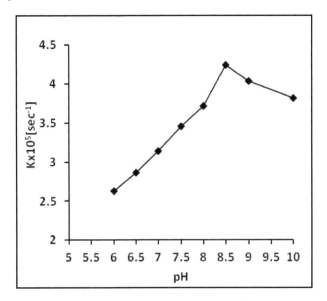

FIGURE 9.2 Effect of pH on the Photocatalytic Bleaching of Reactive Orange 16 by TiO$_2$ [Reactive Orange 16] = 3 × 10^{-5} M, TiO$_2$ = 0.30 g, λ_{max} = 492 nm, irradiation time = 180 min.

It was observed that the rate of photocatalytic bleaching of Reactive Orange 16 increases with an increase in pH up to 8.5. This observation can be explained on the basis that as the pH of solution increases, more OH$^-$ ions are available. These OH$^-$ ions will generate more ˙OH radicals by combining with the positive holes of the semiconductor. These hydroxyl radicals are responsible for photobleaching of dye. After a certain pH value, that is, above pH 8.0, the rate of photobleaching of R.O.16 due to columbic repulsion between the negatively charged surface of photocatalyst and hydroxide anions decreases. This fact could prevent the formation of hydroxyl radicals. This results into a decrease in the rate of photocatalytic bleaching of dye.[9,10]

9.3.2 Effect of Amount of Photocatalyst [TiO$_2$]

The amount of semiconductor powder may also affect the process of dye degradation. Keeping all the factors identical, different amounts of photocatalyst varying from 0.10 to 0.40 g/100 mL were used. It was observed that the rate of dye decolourization increases with increasing catalyst level up to 0.30 g, and beyond this, the rate of reaction becomes almost constant (see Fig. 9.3).

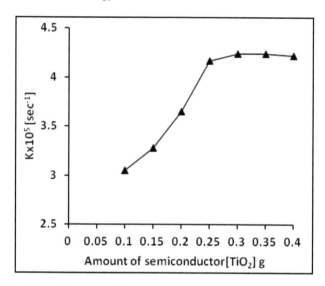

FIGURE 9.3 Effect of Amount of Photocatalyst on the Photocatalytic Degradation of Dye. [Reactive Orange 16] = 3×10^{-5} M, pH = 8.5, λ_{max} = 492 nm, irradiation time = 180 min].

This may be due to the fact that initially the increase in the amount of catalyst increases the number of TiO_2 active sites on the surface that in turn increases the number of $^\bullet OH$ and $O_2^{\bullet-}$ radicals. As a result, the rate of degradation is increased. Above a certain level (saturation point), the number of substrate molecules is not sufficient to fill the active sites of TiO_2 and increase in turbidity of the solution reduces the light transmission through the solution. Hence, further addition of catalyst does not lead to the enhancement of the degradation rate and it remains constant.[11,12]

9.3.3 Effect of Variation in Initial Dye Concentration

The effect of substrate concentration on the degradation of Reactive Orange 16 was studied at different concentrations varying from 1.0×10^{-5} M to 5.0×10^{-5} M at a fixed concentration of TiO_2 = 0.30 g, pH = 8.5. The highest efficiency was observed at lower concentration,[13,14] which decreases with the increase in substrate concentration (see Fig. 9.4).

This can be explained on the basis that the rate of degradation relates to the probability of $^\bullet OH$ radicals formation on the catalyst surface and to the probability of $^\bullet OH$ radicals react with dye molecules. As the initial concentration of dye is increased, the generation of $^\bullet OH$ radicals on the surface of catalyst is reduced as the active sites are covered by dye molecules. Since the relative number of the $^\bullet OH$ radicals attacking the substrate decreases, the photocatalytic efficiency of the reaction also decreases.

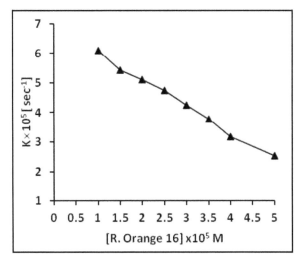

FIGURE 9.4 Effect of Initial Dye Concentrations on Photocatalytic Degradation of Reactive Orange 16: $TiO_2 = 0.30$ g, pH = 8.5, $\lambda_{max} = 492$ nm, irradiation time = 180 min.

9.3.4   Effect of Nature of Semiconductor Photocatalyst

Keeping all the factors identical, semiconductor of different band gaps such as Fe_2O_3, TiO_2, ZnO, and SnO_2 were used to carry out the photocatalytic bleaching of Reactive Orange 16. It was observed that the rate of photobleaching of Reactive Orange 16 decreases with the increase in the band gap of semiconductor (Fig. 9.5). The rate of photobleaching of Reactive Orange 16 is found to be decreasing in the following order:

$$Fe_2O_3 > TiO_2 > ZnO > SnO_2$$

It can be explained on the basis that the semiconductor oxides having $\lambda_{max} > 400$ nm absorb more efficiently in visible region (see Table 9.1).

TABLE 9.1   Effect of Nature of Photocatalyst on the Rate of Photobleaching of Reactive Orange 16

Semiconductor	Band Gap (ev)	λ_{max} (nm)	Region	Rate Constant $K \times 10^5$ (s^{-1})
Fe_2O_3	2.2	564	VIS	4.87
TiO_2	3.2	400	VIS/UV	4.24
ZnO	3.4	387	UV	3.17
SnO_2	3.8	318	UV	0.06

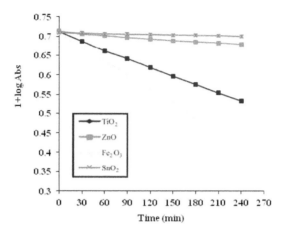

FIGURE 9.5 Effect of Nature of Semiconductor on Photocatalytic Degradation of Reactive Orange 16: [Reactive Orange 16] × 3.0 × 10^{-5} M, pH = 8.5, Amount of Semiconductors = 0.30 g, Irradiation Time = 180 min.

9.3.5 Effect of Various Transition Metal Ions

Effect of various transition metal ions on photocatalytic bleaching of Reactive Orange 16 by TiO$_2$ was studied by taking different transition metal ions such as Cu^{+2}, Fe^{+2}, Mn^{+2}, and Zn^{+2} (see Fig. 9.6). The result shows that the trace quantities of all the added metal ions enhance the rate of photocatalytic bleaching of Reactive Orange 16.

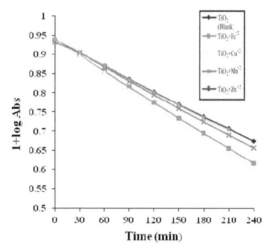

FIGURE 9.6 Effect of Transition Metal Ions on Photocatalytic Degradation of Reactive Orange 16. [Reactive Orange 16] = 3.0 × 10^{-5} M, pH = 8.5, TiO$_2$ = 0.30 g, M^{n+} = 1.0 × 10^{-5} M.

9.3.6 Mechanism

Photocatalysis over a semiconductor oxide such as TiO_2 is initiated by the absorption of photons with energy equal to or greater than the band gap energy of the semiconductor (3.2 eV), producing electron–hole (e^-/h^+) pairs.

$$TiO_2 + h\nu \rightarrow TiO_2^* + h^+(vb) + e^-(cb) \tag{9.1}$$

where cb is conduction band and vb is valence band.

The photo-produced holes and electrons may migrate to the particle surface, where the hole can react with surface bound hydroxyl group (OH^-) and adsorbed water molecules to form hydroxyl radicals $(^\bullet OH)$.

$$h^+ + OH^- \rightarrow {}^\bullet OH \tag{9.2}$$

$$h^+ + H_2O \rightarrow {}^\bullet OH + H^+ \tag{9.3}$$

The presence of oxygen prevents recombination by trapping electrons through the formation of superoxide ions, maintaining electrical neutrality within TiO_2 particles.[15] The final product of the reduction is hydroxyl radicals $(^\bullet OH)$ and hydroperoxy radicals $(HO_2^\bullet)$.

$$e^- + O_2(ads.) \rightarrow O_2^{\bullet-}(ads.) \tag{9.4}$$

$$O_2^- + H^+ \rightarrow HO_2^\bullet \tag{9.5}$$

$$2O_2^- + 2H^+ \rightarrow 2{}^\bullet OH + O_2^\bullet \tag{9.6}$$

$HO_2^\bullet$, ${}^\bullet OH$, and $O_2^\bullet$ are strong oxidizing species, and they react with dye molecules to oxidize them.

In the second pathway, a dye absorbs radiation of suitable wavelength and is excited to its first singlet state followed by intersystem crossing to triplet state.

$${}^1Dye_0 \xrightarrow{h\nu} {}^1Dye_1 \text{(singlet excited state)} \tag{9.7}$$

$${}^1Dye1 \rightarrow {}^3Dye_1 \text{(triplet excited state)} \tag{9.8}$$

The excited dye may be oxidized to product by highly reactive hydroxyl radical $(^\bullet OH)$. The participation of ${}^\bullet OH$ radical as an active oxidizing species was confirmed using its scavenger, that is, 2-propanol, where the rate of bleaching was drastically reduced. Initially, the ${}^\bullet OH$ radicals attack on the azo linkage of the dye molecule and abstract a hydrogen atom or add itself to double bond. After continuous irradiation, the complete mineralization of dye occurred via end products. The end products are simple molecules or ions and less harmful to environment.

$$^3Dye_1 + {}^{\cdot}OH / HO_2^{\cdot} / O_2^{\cdot-} \rightarrow \frac{\text{End products}}{CO_2 + H^+, NO_3^-, SO_4^{2-}} \qquad (9.9)$$

The end products were detected and their presence in the reaction mixture was ascertained either by chemical test or by ion selective electrode method. Nitrate ions were detected and confirmed by using nitrate ion selective electrode, which is having a solid-state polyvinyl chloride polymer membrane. Sulfate ions were detected and confirmed by gravimetric analysis in which excess of barium chloride solution was used and sulfate ions were precipitated as $BaSO_4$. CO_2 was confirmed by introducing the gas to freshly prepared lime water. The lime water turns milky which indicates its presence.

$$\text{(9.10)}$$

A = Sodium 6-(acetylomino)-4-hydroxy-3-nitronaphthelene-2-sulfonate; B = Sodium 2-[(4-nitrophenyl)sulphonyl] ethyl sulfate.

The effect of addition of transition metal ions ($M^{n+} = Fe^{2+}, Cu^{2+}, Mn^{2+}, Zn^{2+}$) on photodegradation efficiency of TiO_2 has been investigated. The result shows that the trace quantities of all the added metal ions enhance the rate of photocatalytic bleaching of Reactive Orange 16. The increase in the photocatalytic activity may be due to introduction of new trapping sites by incorporation of transition metal ions. As the surface of catalyst particles is negatively charged in alkaline medium, it permits more metal ions to get adsorbed on the TiO_2 particles surface. As a consequence, the surface of the semiconductor will become positively charged. As Reactive Orange 16 is anionic dye, so it will face more electrostatic attraction with metal ions (M^{n+}) adsorbed on the semiconductor surface.

The electron from TiO_2 conduction band is transferred to metal ion to convert it into its lower oxidation state, in turn transfers this electron to oxygen molecule and thus prevents electron–hole recombination. At the same time, the positively charged vacancies (h^+) remaining in the valence band of TiO_2 can extract electron from hydroxyl ions in the solution to produce the hydroxyl radicals (${}^{\cdot}OH$). These hydroxyl radicals oxidize the dye molecule into colorless products.

$$TiO_2 + h\nu \rightarrow TiO_2^* \left(h_{vb}^+ + e_{cb}^- \right) \qquad (9.11)$$

$$M^{n+} + TiO_2^{*}\left(e_{cb}^{-}\right) \rightarrow M^{(n-1)+} \text{(electron trapping)} \tag{9.12}$$

$$M^{(n-1)+} + O_2 \rightarrow M^{n+} + O_2^{\cdot -} \tag{9.13}$$

$$TiO_2^{*}\left(h_{vb}^{+}\right) + OH^{-} \rightarrow TiO_2 + {}^{\cdot}OH \tag{9.14}$$

$${}^{3}Dye_1 + OH \rightarrow \text{Degradation of the dye} \tag{9.15}$$

The concentration of transition metal ions is very small and large concentrations are adverse.

The whole TiO2-semiconductor photocatalytic processes can be summarized as shown in Figure 9.7.

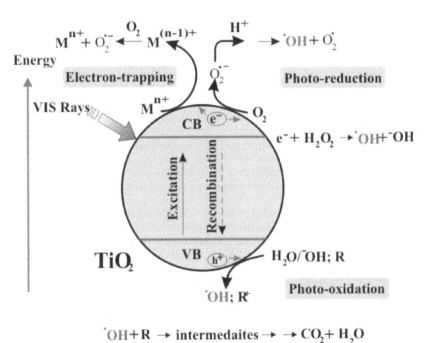

$${}^{\cdot}OH + R \rightarrow \text{intermedaites} \rightarrow \rightarrow CO_2 + H_2O$$
$$R = Dye$$

FIGURE 9.7 TiO2-Semiconductor Photocatalytic Processes. **R** ? Reactive Orange 16.

ACKNOWLEDGMENTS

The authors are thankful to Dr. V. K. Vaidya (Professor, Sangam University, Bhilwara), Dr. R. L. Pitliya (Principal, S.D. College, Bhilwara), and all the faculty members of Department of Chemistry, M.L.V. Govt. College and Sangam University, Bhilwara, for continuous encouragement in accomplishing this work.

KEYWORDS

- Advance oxidation process (AOP)
- photocatalytic degradation
- heterogenous
- Reactive Orange 16
- TiO_2
- transition metal ions

REFERENCES

1. Bouzaida, I.; Ferronato, C.; Chovelon, J. M.; Rammah, M. E.; Herrmann, J. M. Heterogeneous photocatalytic degradation of the anthraquinonic dye, Acid Blue 25 (AB25): a kinetic approach. *J. Photochem. Photobiol. A.* 2004, 168, 23–30.
2. Mariana, N.; Siminiceanu, I.; Yediler, A.; Kettrup, A. Kinetics of decolorization and mineralization of reactive azo dyes in aqueous solution by the UV/H_2O_2. *Dyes Pigm.* 2002, 53, 93–99.
3. Sandhya, S.; Padmavathy, S.; Swaminathan, K.; Subrahmanyam, Y. V.; Kaul, S. N. Microaerophilic-aerobic sequential batch reactor for treatment of azo dyes containing simulated wastewater. *Process Biochem.* 2005, 40, 885–890.
4. Shelke, R. S.; Bharad, J. V.; Madje, B. R.; Ubale, M. B. Studies on the removal of acid dyes from aqueous solutions by Ashoka leaf powder. *Der Chem. Sinica.* 2011, 2(4), 6–11.
5. Nigam, P.; Armour, G.; Banat, I. M.; Singh, D.; Marchant, R. Physical removal of textile dyes from effluents and solid-state fermentation of dye-adsorbed agricultural residues. *Bioresour. Technol.* 2000, 72, 219–226.
6. Van der Zee, F. P.; Lettinga, G.; Field, J. A. Azo dye decolourisation by anaerobic granular sludge. *Chemosphere.* 2001, 44(5), 1169–1117.
7. Muruganandham, M.; Swaminathan, M. Solar driven decolourisation of reactive yellow 14 by advanced oxidation processes in heterogeneous and homogeneous media. *Dyes Pigm.* 2007, 72, 137–143.
8. Surana, M.; Mehta, P.; Pamecha, K.; Kabra, B. V. Heterogeneous photocatalytic treatment of textile dye effluent containing azo dye: direct Crysophenine G. *Der Chem. Sinica.* 2011, 2(2), 177.
9. Neppolian, B.; Choi, H. C.; Sakthivel, S.; Arabindoo, B. Solar/UV-induced photocatalytic degradation of three commercial textile dyes. *Hazard. Mater.* 2002, 80, 303–317.
10. Sauer, T.; Neto, G. C.; Jose, H. J.; Moreira, R. F. P. M. Kinetics of photocatalytic degradation of reactive dyes in a TiO2slurry reactor. *J. Photochem. Photobiol. A. Chem.* 2002, 149, 147–154.
11. Toor, A. P.; Verma, A.; Jotshi, C. K.; Bajpai, P. K.; Singh, V. Photocatalytic degradation of direct yellow 12 dye using UV/TiO_2 in a shallow pond slurry reactor. *Dyes Pigm.* 2006, 68, 53–60.
12. Mrowetz, M.; Pirola, C.; Selli, E. Degradation of organic water pollutants through sonophotocatalysis in the presence of TiO_2. *Ultrason. Sonochem.* 2003, 10, 247–254.

13. Subramani, A. K.; Bryappa, K. S.; Ananda, K. M.; Lokanatha, R.; Ranganathaiah, C.; Yoshimura, M. *Bull. Mater. Sci.* 2000, 30, 37.
14. Abo-Farah, S. A. Comparative study of oxidation of some azo dyes by different advanced oxidation processes: fenton, fenton-like, photo-fenton and photo-fenton-like. *J. Am. Sci.* 2010, 6(11), 128–142.
15. Chatterjee, D. Visible light induced photodegradation of organic pollutants on dye adsorbed TiO2 surface. *Bull. Cata. Soc. Ind.* 2004, 3, 56–58.

CHAPTER 10

Effect of PGRS in Vitro Callus Culture for Production of Secondary Metabolites

Jitendra Mittal[1*], Madan Mohan Sharma[1], Abhijeet Singh[1], and Amla Batra[2]

[1]Department of Biotechnology, Manipal University, Jaipur, Rajasthan, India
[2]Lab. No. 5, Department of Botany, University of Rajasthan, Jaipur, Rajasthan, India
*Email: jitendra.mittal@muj.manipal.edu

CONTENTS

ABSTRACT

The present study was conducted to explore the hidden potential of natural products synthesized in the medicinal plant *Tinospora cordifolia*. This plant is prioritized by National Medicinal Plant Board, New Delhi. Leaf and internodal segments were inoculated on Murashige and Skoog medium fortified with indole-3-butyric acid (1.0 mg/L) produced callus after 4 weeks. The calli were brown due to phenolic substance secreted by the explant. This problem was overcome by using adjuvant activated charcoal (3.0%). Further, we are trying to isolate berberine and tinosporin alkaloids from the callus and enhance the amount of these alkaloids through callus culture.

10.1 INTRODUCTION

Plant cells are chemical factories of natural compounds used as pharmaceuticals, agrochemicals, ingredients of flavors and fragrances, food additives, and pesticides.[1] The natural environment for medicinal plants is vanishing at a faster rate together with environmental pollution. Hence, it is difficult to acquire natural compounds from the phytodiversity. This has encouraged industries as well as scientists to consider the possibilities of investigation into cell cultures as an alternative supply for the production of plant pharmaceuticals.[2] In the search for an alternative to produce desirable medicinal compounds from plants, biotechnological approaches, especially plant tissue culture technology, are found to have potential application over traditional agriculture practices for the industrial production of bioactive plant metabolites.[3]

New medicinally potent compounds have been found valuable by an array of bioassays.[4–6] It has been established that the biosynthetic activity of cultured cells can be improved by regulating abiotic stress (temperature, light, nutrients, carbon, pH, etc.) as well as biotic stress. Plant cells, under in vitro conditions, show physiological and morphological responses to microbial, physical, or chemical factors, which are known as "elicitors." Elicitation is a process of enhanced synthesis of secondary metabolites by the plants to ensure their survival, persistence, and competitiveness. Elicitors have been classified on the basis of their nature and origin. On the basis of nature, elicitors are of two types: biotic (microorganisms) and abiotic (light, temperature, pH, radiations, etc.). While on the basis of origin, elicitors are also of two kinds: exogenous elicitors (originated outside the cell) and endogenous elicitors (originated inside the cell).[7–9] The concentration of some potent natural products is confined in specialized organs of plants. These natural products have been produced through callus culture.[10–13] The possible applications of plant cell cultures for the specific biotransformation of natural compounds have been reported.[14–17] Due to these advances, research in the area of tissue culture technology for the production of bioactive compounds has gained popularity beyond expectations.

In order to obtain high yields of bioactive compounds for commercial production, efforts have been made to utilize a medicinally potent plant *Tinospora cordifolia*

(Willd.) Miersex Hook F. and Thoms belonging to the family Menispermaceae. It is a large, deciduous, climbing shrub found throughout India, especially in the tropical parts, ascending to an altitude of 300 m, and also in certain parts of China.[18] It is also known as heart-leaved Moonseed plant in English and Giloy in Hindi. It is rich with a variety of natural chemical constituents (berberine, tinosporin, cordifolioside).[19,20] These natural products have been used to cure a number of ailments such as viral infections, cancer, diabetes, inflammation, immunomodulatory activity, neurological, psychiatric conditions, vasorelaxant, inflammation, microbial infection, hypertension, and HIV.[21-26]

Owing to the immense medicinal properties, this plant has been overexploited by pharmaceutical companies and folk people for traditional remedies have led to the acute scarcity of this plant to meet the present-day demand. Hence, it requires scientific attention to conserve the gradually depleting *T. cordifolia*. Consequently, regular supply of this medicinal plant for industrial production of natural compounds can be ensured. To fulfill the present-day needs, tissue culture techniques are the alternative method to conserve and enhance the production of secondary products under laboratory conditions. Hence, the present investigation was carried out with the objective of providing new vistas on the cell and tissue culture to elucidate the effect of plant growth regulators on callus cultures for the enhancement of natural products.

10.2 MATERIALS AND METHODS

10.2.1 Establishment of Plant

Different places of Jaipur, such as Kulish Smrity Van, Amanisah Nala region, world Arboretum, and University of Rajasthan Nursery, were surveyed for the selection of mother plant of *T. cordifolia*. Stem cuttings were collected from the Nursery, University of Rajasthan, and established in Manipal University, Jaipur campus garden for regular procurement of explants.

10.2.2 Explant Preparation

Fresh twigs were collected from 1-year-old plant and cut into 1.0–1.5 cm long intermodal segments and leaf segments (1.0 cm^2) for callus development. These explants were washed under running tap water for about 10 min and kept in rankleen (1–2 drops/L) for 2 min followed by washing with distilled water at least thrice.

10.2.3 Nutrient Medium and Hormone

Murashige and Skoog (MS) medium with different concentrations of auxins, such as IBA and NAA (0.5–5.0 mg/L), and cytokinin, namely, BAP and TDZ (0.5–5.0 mg/L), were used for callus production. The MS medium consisted of 95% water, macro- and micronutrients, vitamins, amino acids, and sugars. Besides, MS medium contained sucrose (3.0%) as a carbon source and agar powder (0.8%) as a gelling agent to make this medium semisolid.

10.2.4 Inoculation of Explants

Before inoculation, surface sterilized explants were treated with mercuric chloride (0.1%, w/v) for 1 min under aseptic conditions and washed for 2–3 times with autoclaved distilled water to remove the chances of contamination. These explants cultured on MS medium fortified with various auxins, namely, IBA and NAA (0.5–5.0 mg/L), and cytokinin, such as BAP and TDZ (0.5–5.0 mg/L), for callus production.

10.2.5 Culture Conditions

The cultures were incubated in plant growth chamber under cool, white, fluorescent lights (1500–2000 lx) to provide optimum photoperiod for 16/8 h, 25 ± 2°C temperature and 50 ± 5% relative humidity. Each experiment had 15 replicates and repeated at least three times. Data were documented up to 7 weeks of culture.

10.3 RESULTS AND DISCUSSION

Callus initiation was observed when internodal segments inoculated on basal MS medium supplemented with different concentrations of IBA (0.5, 1, 1.5, 2.5, and 5.0 mg/L). The proliferation efficiency of callus through internodal segments was significantly higher than that of leaf explants within 4–5 weeks. Callus initiations started after the swelling of internodal segments within 1 week of inoculation (Fig. 10.1A). Further, the cells simultaneously absorb nutrients from the medium and callus production was obtained on MS medium augmented with IBA (1.0 mg/L). The produced callus was pale green, spongy, and fragile (Fig. 10.1B and C). Callus turned brown to black after next 3 weeks of subculturing (Fig. 10.1D). It is due to the secretion of phenolic compounds in the medium, which prohibited the further growth of callus. This callus was subcultured on MS medium fortified with IBA (1.0 mg/L) and activated charcoal (1.0–3.0%) to overcome leaching of phenolic compounds (Fig. 10.1E). Subsequently, the callus was subcultured after every 2 weeks on MS medium with IBA (1.0 mg/L) and activated charcoal (3%). Besides, NAA at its varied concentrations tried (0.5–5.0 mg/L) and IBA beyond 1.0 mg/L did not exhibit positive response for callus initiation and its further growth.

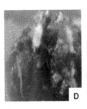

FIGURE 10.1 (A–D) Different Stages of Callus Production Via Internodal Segments Cultured on MS Medium Supplemented with IBA (1.0 mg/L). (A) In Vitro Callus Initiation After 1 week. (B) Further Growth of Callus After 2 weeks. (C) Further Growth of Callus After Subculturing on the Same Medium. (D) Browning of Callus Due to Secretion of Phenolic Substances. (E) Brown Callus Treated with Activated Charcoal (3.0%).

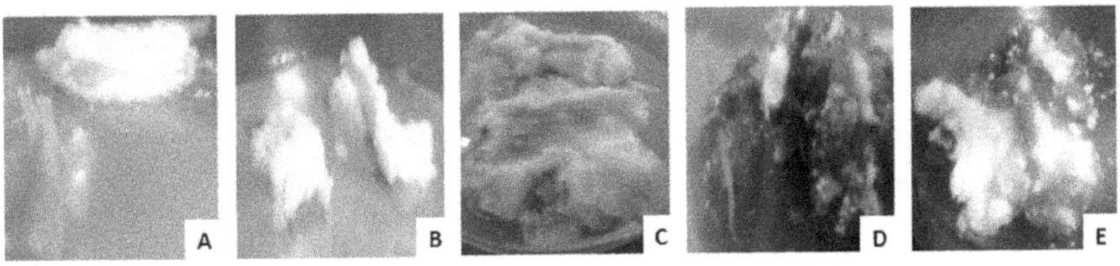

In consonance with our results for callus production, Gururaj et al.[27] also reported callus production on MS medium supplemented with IBA (1.0 mg/L) and 2ip (0.5–1.0 mg/L) hormones via nodal segments explant of *T. cordifolia*. In contrast to our findings, Rao et al.[28] initiated callusing on MS medium fortified with NAA (4.0 mg/L) via leaf explant of *T. cordifolia*.

Till now, we have produced stock callus, which will further be utilized for isolation of secondary metabolites, namely, berberine and tinosporin. The berberine and tinosporin are the alkaloids found in stem and root of this plant species. Berberine is used in diabetes, hypercholesterolemia, viral infection, neurological disorders, etc.,[19] whereas tinosporin is used in inflammation, neurological disorders, etc.[29] The berberine and tinosporin will be isolated by cold extraction method. For this, the dried callus is being processed for further isolation techniques.

KEYWORDS

- **Medicinal plants**
- **callus**
- **natural products**
- **TDZ**
- **IBA**
- **BAP**

ABBREVIATIONS

IAA	indole-3-acetic acid
IBA	indole-3-butyric acid
BAP	6-benzyladenine
TDZ	thidiazuron
2ip	N6-2-iso-pentenyl adenine

REFERENCES

1. Sharma, S.; Rathi, N.; Kamal, B.; Pundir, D.; Kaur, B.; Arya, S. Conservation of biodiversity of highly important medicinal plants of India through tissue culture technology – a review. *Agric. Biol. J. North Am.* 2010, 1(5), 827–833.

2. Mulabagal, V.; Tsay, H. S. Plant cell cultures an alternative and efficient source for the production of biologically important secondary metabolites. *Int. J. Appl. Sci. Eng.* 2004, 2(1), 29–48.

3. Ramanchandra, R. S.; Ravishankar, G. A. Plant cell cultures: chemical factories for secondary metabolites. *Biotechnol. Adv.* 2002, 20, 101–153.

4. Kapur, P.; Wuttke, W.; Jarry, H. Beneficial effects of beta-ecdyson on the joint epiphyseal cartilage tissue and trabecular bone in ovariectomized rats. *Phytomedecine.* 2010, 17, 350–355.

5. Patel, M. B.; Mishra, S. M. Magnoflorine from *Tinospora cordifolia* stem inhibits a-glucosidase and is antiglycemic in rats. *J Funct. Foods.* 2012, 4, 79–86.

6. Wang, T.; Liu, Y. Y.; Wang, X.; Yang, N.; Zhu, H. B.; Zuo, P. P. Protective effects of octacosanol on 6-hydroxydopamine-induced Parkinsonism in rats via regulation of ProNGP and NGF signalling. *Acta. Pharmacol. Sin.* 2010, 31, 765–774.

7. Broeckling, C. D.; Huhman, D. V.; Farag, M. A.; Smith, J. T.; May, G. D.; Mendes, P.; Dixon, R. A.; Sumner, L. W. Metabolic profiling of *Medicago truncatula* cell cultures reveals the effects of biotic and abiotic elicitors on metabolism. *J. Exp. Bot.* 2005, 56(410), 323–336.

8. Qian, Z. G.; Zhao, Z. J.; Xu, Y.; Qian, X.; Zhong, J. J. Novel chemically synthesized salicylate derivative as an effective elicitor for inducing the biosynthesis of plant secondary metabolites. *Biotechnol. Prog.* 2006, 22(1), 331–333.

9. Satdive, R. K.; Fulzele, D. P.; Eapen, S. Enhanced production of azadirachtin by hairy root cultures of *Azadirachta indica* A. Juss by elicitation and media optimization. *J. Biotechnol.* 2007, 128(2), 281–289.

10. Zhou, L.; Liang, Y.; Wang, W.; Tan, H.; Xiao, M.; Fan, C.; Yu, R. Biotransformation of 4-phenylcoumarin by transgenic hairy roots of *Polygonum multiflorum. J. Med. Plant Res.* 2011, 5(17), 4274–4278.

11. Sharma, M. M.; Singh, A.; Verma, R. N.; Ali, D. Z.; Batra, A. Influence of PGRs for the in vitro plant regeneration and flowering in *Portulaca oleracea* (L.): a medicinal and ornamental plant. *Int. J. Bot.* 2011, 7(1), 103–107.

12. Karnawat, M.; Jain, D.; Singh, A.; Malik, C. P. In vitro plant regeneration from different leaf segment of *Verbesina encelioides* and correlation with endogenous level of IAA. *P.T.C. Biotechnol.* 2010, 20(2), 195–201.

13. Sen, A.; Sharma, M. M.; Grover, D.; Batra, A. In vitro regeneration of *Phyllanthus amarus* Schum. and Thonn.: an important medicinal plant. *One Nat.* 2009, 7(1), 110–115.

14. Bourgaud, F.; Gravot, A.; Milesi, S.; Gontier, E. Production of plant secondary metabolites: a historical perspective. *Plant Sci.* 2001, 161, 839–851.

15. Sharma, M. M.; Batra, A. High frequency plantlet regeneration in Indian Ginseng: *Withania somnifera* L. (Dunal). *Physiol. Mol. Biol. Plants.* 2006, 12(4), 289.

16. Sharma, M. M.; Dhingra, M.; Dave, A.; Batra, A. Plant regeneration and stimulation of in vitro flowering in *Eruca sativa* Mill. *Afr. J. Biotechnol.* 2012, 11(31), 7906–7911.

17. Ravishankar, G. A.; Ramachandra, R. S. Biotechnological production of phyto-pharmaceuticals. *J. Biochem. Mol. Biol. Biophys.* 2000, 4, 73–102.
18. Anonymous. *Wealth of India: Raw Materials*; CSIR: New Delhi, 1976; p 10.
19. Patel, M. B.; Mishra, S. Hypoglycemic activity of alkaloidal fraction of *Tinospora cordifolia*. *Phytomedicine*. 2011, 18, 1045–1052.
20. Sangeethaa, M. K.; Priya, C. D. M.; Vasanthia, H. R. Anti-diabetic property of *Tinospora cordifolia* and its active compound is mediated through the expression of Glut-4 in L6 myotubes. *Phytomedicine*. 2013, 20, 246–248.
21. Rout, G. R. Identification of *Tinospora cordifolia* (Willd.) Miers ex Hook F & Thoms using RAPD markers. *Z. Naturforsch C.* 2006, 61, 118–122.
22. Patel, S. S.; Shah, R. S.; Goyal, R. K. Anti-hyperglycemic, anti-hyperlipidemic and antioxidant effects of Dihar, a poly herbal ayurvedic formulation in streptozotocin induced diabetic rats. *Ind. J. Exp. Biol.* 2009, 47, 564–570.
23. Upadhyay, P. R.; Sharma, V.; Anita, K. V. Assesment of the multifaceted immunomodulatory potential of the aqueous extract of *Tinospora cordifolia*. *Res. J. Chem. Sci.* 2011, 1, 71–79.
24. Jayaganthan, P.; Perumal, P.; Balamurugan, T. C.; Verma, R. P.; Singh, L. P.; Pattanaik, A. K.; Kataria, M. Effects of *Tinospora cordifolia* supplementation on semen quality and hormonal profile in rams. *Anim. Reprod. Sci.* 2013, 140, 47–53.
25. Nagarkar, B.; Kulkarni, R.; Bhondave, P.; Kasote, D.; Kulkarni, O.; Harsulkar, A.; Jagtap, S. Comparative hepatoprotective potential of *Tinospora cordifolia*, *Tinospora sinensis* and *Neem-guduchi*. *Br. J. Pharm. Res.* 2013, 3(4), 906–916.
26. Joladarash, D.; Chilkunda, N. D.; Salimath, P. V. Glucose uptake-stimulatory activity of *Tinospora cordifolia* stem extracts in Ehrlich ascites tumor cell model system. *J. Food Sci. Technol.* 2014, 51(1), 178–182.
27. Gururaj, H. B.; Giridhar, P.; Ravishankar, G. A. Micropropagation of *Tinospora cordifolia* (Willd.) Miers ex Hook. F & Thoms – a multipurpose medicinal plant. *Curr. Sci.* 2007, 92(1), 23–26.
28. Rao, B. R.; Kumar, D. V.; Amrutha, R. N.; Jalaja, N.; Vaidyanath, K.; Rao, A. M.; Rao, S.; Polavarapu, R.; Kishor, P. B. K. Effect of growth regulators, carbon source and cell aggregate size on berberine production from cell cultures of *Tinospora cordifolia* Miers. *Curr. Trends Biotechnol. Pharm.* 2008, 2(2), 269–276.
29. Upadhaya, A. K.; Kumar, K.; Kumar, A.; Mishra, H. S. *Tinospora cordifolia* (Willd.) Hook. F. and Thoms. (Guduchi)-alidation of the Ayurvedic pharmacology through experimental and clinical studies. *Int. J. Ayurveda Res.* 2010, 1, 112–121.

CHAPTER 11

Density Functional Theory (DFT): Periodic Advancement and New Challenges

Amrit Sarmah*

Department of Chemistry, Birla Institute of Technology and Science (BITS), Pilani, Rajasthan, India
*Email: amritjorhat2009@gmail.com

CONTENTS

11.1 INTRODUCTION

In the last three decades, density functional theory (DFT)-based electronic structure calculations have provided much needed reliable, cost-effective, and flexible solution for the application of quantum mechanics to solve some of the interesting and challenging problems in chemistry. The significant contribution of DFT has been extended throughout the areas like understanding and designing of catalytic processes in enzymes and zeolites, electron transport, solar energy harvesting and conversion, and drug design in medicine, as well as many other problems of science and technology.[1-9]

11.2 UNDERSTANDING DFT

11.2.1 The Beginning of DFT

DFT is a first principle (quantum mechanical) method that evaluates the electron density directly, without finding the approximate wave function. In this respect, DFT differs from the routine semi-empirical and ab initio calculations. Instead of finding the proper wave function, DFT formalism will concentrate on finding the electron density of a particular system.

Considering the Born–Oppenheimer approximation, the Schrodinger equation for a stationary system of N interacting electrons can be written in the form,

$$\hat{H}\,\varphi(r_1\delta_1, r_2\delta_2, \ldots, r_n\delta_n) = E\,\varphi(r_1\delta_1, r_2\delta_2, \ldots, r_n\delta_n) \tag{11.1}$$

Here E is the energy of the system and $r_1\delta_1, r_2\delta_2, \ldots, r_n\delta_n$ is the electronic wave function, with r_i and δ_i the space and spin coordinates of the electron i. The Hamiltonian of the system is defined as

$$\hat{H} = -\frac{1}{2}\sum_i \nabla_i^2 + \frac{1}{2}\sum_{i\neq j}\frac{1}{|r_i - r_j|} + \sum_i v(r_i) \tag{11.2}$$

We are using the atomic units ($\hbar = m_e = e = 4\pi\varepsilon_0 = 1$) throughout the discussion. The first term on the right hand side of expression (11.2) is the kinetic energy operator $\hat{T}$, the Coulomb electron–electron interaction energy operator $\hat{W}$, represented in the second term, whereas the last term is the potential energy operator $\hat{V}$, of the electrons in the external potential $v(r)$. The Hamiltonian is parametrized by the external potential $v(r)$, thus the energies $E = \psi|\hat{H}|\varphi$ and the electronic wavefunctions ψ that satisfy the Schrodinger equation (11.1) can be considered functionals of this external potential. We will denote $\psi = \psi[v]$ and $E = E[v]$. Now we are in a position to introduce the electron density $\rho(r)$ of the system through the mathematical expression

$$\rho(r) = \Psi|\hat{\rho}(r)|\Psi = N\int \Psi|r_1\sigma_1, r_2\sigma_2, \ldots, r_N\sigma_N|^2 dr_2 \ldots dr_N d\sigma_1 \ldots d\sigma_N \tag{11.3}$$

where the density operator $\rho(r)$ is defined as

$$\hat{\rho}(r) = \sum_i \delta(r - r_i) \tag{11.4}$$

Here, it is possible to write the potential energy due to the external potential $\hat{V}$ using the density operator as,

$$V = \varnothing|\hat{V}|\varnothing = \varnothing\left|\int \hat{\rho}(r)v(r)dr\right|\varnothing = \int \rho(r)v(r)dr \tag{11.5}$$

The one-to-one mapping between the external potential $v(r)$ and the ground-state density $\rho(r)$ is discussed in the next section. From the principle of Legendre transform, these two quantities can now be considered to be conjugate variables in the meaning of the following,

$$\frac{E_0v}{v(r)} \quad \frac{\Psi_0[v]}{v(r)}|\hat{H}_v|\Psi_0[v] \quad \Psi_0[v]|\hat{H}_v|\frac{\Psi_0[v]}{v(r)} \quad \Psi_0[v]\left|\frac{\hat{H}_v}{v\,r}\right|\Psi_0[v] = E_0[v]\frac{1}{v(r)}\Psi_0[v]|\Psi_0[v] \quad \Psi_0[v]^{\hat{}}(r)|\Psi_0[v] \quad [v](r) \tag{11.6}$$

where we have used that the wavefunction $\psi_0[v]$is normalized and the ground-state Eigen function of the Hamiltonian $\hat{H}_v$ with energy $E_0[v]$. We can use the density as basic variable by defining a Legendre transform

$$\mathcal{F}[\rho] = E_0[\rho] - \int \rho(r)v(r)dr = \Psi_0[v]|\hat{V} + \hat{W}|\Psi_0[v] \tag{11.7}$$

where $v(r)$ must be regarded as a functional of $\rho(r)$. The uniqueness of this functional is generated by the one-to-one mapping between the external potential and the ground-state density. The functional $\mathcal{F}[\rho]$ is defined for the so-called υ-representable densities, that is, ground-state densities for a Hamiltonian with external potential v.[4] By using the chain rule of differentiation and the result in equation (11.6), it immediately follows that

$$\frac{\delta\mathcal{F}[\rho]}{\delta\rho(r)} = \frac{\delta E_0[\rho]}{\delta v(r')}\frac{\delta v(r')}{\delta\rho(r)}dr' - \int \rho(r')\frac{\delta v(r')}{\delta\rho(r)} - v(r) = -v(r) \tag{11.8}$$

11.2.2 The Hohenberg–Kohn Theorems

The basic principle of DFT is to replace the many-body electronic wavefunction with the electronic ground-state density as basic quantity.[10] The advantage of ground-state density is a function of only three variables and a relatively simple quantity to deal with both conceptually and practically. On the other hand, a many-body wavefunction is dependent on 3N spatial variables and N spin variables and created a much more complicated situation in terms of computation. The formalism of DFT is in principle exact and is firmly based on the contribution of two theorems derived and executed by Hohenberg and Kohn in 1964.[10] The first theorem explains that the density ρ of a nondegenerate ground state is a fundamental parameter to determine the external potential $v(r)$ (up to an arbitrary constant) to which the many-electron system is subjected. Consequently, the density determines the electronic wavefunction of the system, and thus all the electronic properties of the ground state. The proof of the theorem is rather simple. First, we notice that the external potential $v(r)$ defines a mapping $v \rightarrow \rho$, where $\rho(r)$ is the corresponding nondegenerate ground-state density from the Schrodinger equation. In fact, it is seen that if two potentials $v(r)$ and $v'(r)$ differ by more than a constant, they will not lead to the same wavefunction. From the Schrodinger equation (11.1), we have for the two potentials,

$$(\hat{T} + \hat{V} + \hat{W})|\Psi_0 = E_0|\Psi_0 \tag{11.9}$$

$$(\hat{T} + \hat{V} + \hat{W})|'_0 = E'_0|\Psi'_0 \tag{11.10}$$

If ψ_0 and ψ'_0 were to be the same, then by subtracting (11.10) from (11.9), one would get

$$(\hat{V} + \hat{V}')|\Psi_0 = (E - E_0|\Psi_0 \tag{11.11}$$

where V and $\hat{V}'$ appear to differ only by a constant if ψ_0 does not vanish. However, for "reasonably well behaved" potentials, that is, potentials that do not exhibit in finite barriers, etc., ψ_0 cannot vanish on a set with nonzero measure by the unique continuation theorem.[11] This is in some apparent contradiction with the initial assumption, and it can be concluded that $\psi_0 \neq \psi'_0$. In this instance, it is possible to prove that two potentials $v(r)$ and $v'(r)$ with corresponding Hamiltonians $\hat{H}$ and H', respectively, and nondegenerate ground-state wavefunctions ψ_0 and ψ'_0 give two types of densities $\rho(r)$ and ρ'_r. Implementing the variation theorem, we can have

$$E_0 = \Psi_0|\hat{H}|\Psi_0 < \Psi'_0|\hat{H}|\Psi'_0 = \Psi'_0|\hat{H}' + \hat{V} - \hat{V}'|\Psi'_0 \tag{11.12}$$

and

$$E_0 < E'_0 + \int \rho'(r)[v(r) - v'(r)]dr \tag{11.13}$$

Similarly, interchange of primed and unprimed variables produces

$$E'_0 < E_0 + \int \rho(r)[v'(r) - v(r)]dr \tag{11.14}$$

If $\rho(r)$ and $\rho'(r)$ were to be the same, adding equations (1.13) and (1.14) would result in the inconsistency

$$E_0 + E'_0 < E'_0 + E_0 \tag{11.15}$$

and consequently, this nullifies the possible existence of two different external potentials $v(r)$ and $v'(r)$ corresponding to the same density $\rho(r)$. This defines our mapping $\rho \rightarrow v$, and thus the one-to-one mapping $v \leftrightarrow \rho$ is constructed. The ongoing discussions provide formal proof for the one-to-one mapping between v, ρ, and ψ^0. This concludes that the total energy of a (Coulomb) many-electron system in an external static potential can be expressed in terms of the potential energy due to this external potential and of an energy functional $\mathcal{F}[\rho]$ of the ground-state density,

$$E_0[\rho] = \Psi_0[\rho]|\hat{H}|\Psi_0[\rho] = \int \rho(r)v(r)dr + \mathcal{F}[\rho] \qquad (11.16)$$

with $\mathcal{F}[\rho]$ defined in equation (11.7). It is important to note that this functional is defined independently of the external potential $v(r)$, and thus, it is a universal functional of the density. This indicates a universal nature of the functional (i.e., it can be used for any system provided the explicit form of the functional is known). The second Hohenberg–Kohn theorem states that the exact ground-state density of a system in a particular external potential $v(r)$ minimizes the energy functional

$$E_0 = \min_{\rho}\left\{\mathcal{F}[\rho] + \int \rho(r)v(r)dr\right\} \qquad (11.17)$$

where E_0 is the ground-state energy for the system in an external potential $v(r)$. From the first theorem, it is expected that a trial density $\bar{\rho}(r)$, such that $\bar{\rho}(r) \geq 0$ and $\int \bar{\rho}(r)\,dr = N$, with N a number of electrons, determines its own potential $\bar{v}(r)$, Hamiltonian $\hat{H}$, and wavefunction ψ'. Here, the wavefunction is considered as a trial function for the problem of interest having external potential $v(r)$. Using the variational principle, we have

$$\bar{\Psi}|\hat{H}|\bar{\Psi} = \mathcal{F}[\bar{\rho}] + \int \bar{\rho}(r)v(r)dr \geq \mathcal{F}[\rho] + \int \rho(r)v(r)dr \qquad (11.18)$$

which offers a formal proof for the second Hohenberg–Kohn theorem. Assuming differentiability of $\mathcal{F}[\rho] + \int v(r)\rho(r)dr$, this theorem requires that the ground-state density satisfies the Euler–Lagrange equations

$$0 = \frac{\delta}{\delta\rho(r)}\left\{\mathcal{F}[\rho] + \int \rho(r)v(r)dr - \mu\left(\int \rho(r)dr - N\right)\right\}, \qquad (11.19)$$

where an additional Lagrange multiplier ? is introduced to ensure that the density integrates to the correct number of electrons. It is highly essential to introduce certain degrees of approximations to maintain the exactness of the universal functional, for which no explicit expressions in terms of the density are known.

11.2.3 Kohn–Sham (KS) Theorem

In an important theoretical development, the physical interpretation of DFT was then offered by Kohn and Sham.[12] Basically, the Kohn–Sham (KS) principle introduced an orbital-density description of DFT and that relaxed the constrain of knowing the exact form of $T[\rho]$. This new formalism puts stress on the kinetic energy of a noninteracting system of electrons as a functional of a set of single-particle orbitals, that is, excluding electron–electron interactions, showing the same electron density as the exact electron density. The Hohenberg–Kohn function can be written as

$$F_{HK}[\rho] = T_S[\rho] + J[\rho] + E_{XC}[\rho] \tag{11.20}$$

Here, T_s represents the kinetic energy functional of reference system given by

$$T_S[\rho] = \sum_i \Psi_i \left| -\frac{1}{2}\nabla^2 \right| \Psi_i \tag{11.21}$$

$J[\rho]$ representing the classical columbic interaction energy,

$$J[\rho] = \frac{1}{2} \iint \frac{\rho(r)\rho(r')}{|r-r'|} dr dr' \tag{1.22}$$

and the remaining energy component being assembled in the $E_{XC}[\rho]$ function, that is, the exchange-correlation energy, containing the (small) difference between the true kinetic energy and T_s and the self-interaction correction to equation (11.22). From the above equations, Euler equation can be written as

$$\mu = v_{eff}(r) + \frac{\delta T_s}{\delta \rho} \tag{11.23}$$

where an effective potential has been introduced,

$$v_{eff}(r) = v(r) + \frac{\delta J}{\delta \rho} + \frac{\delta E_{XC}}{\delta \rho} = v(r) + \int \frac{\rho(r)}{|r-r'|} dr + v_{XC}(r) \tag{11.24}$$

containing the exchange-correlation potential $v_{XC}(r)$ defined as

$$v_{XC} = \frac{\delta E_{XC}}{\delta \rho(r)} \tag{11.25}$$

The Kohn–Sham approach provides an alternative solution through the Shrödinger equation. Introduction of more general exchange-correlation potential term in place of exchange-potential term in well-known Hartree–Fock (HF) equations in principle is much simpler because it is only a function of the density. The molecular orbitals are solutions of Kohn–Sham equations (i.e., one electron equations):

$$\left(-\frac{1}{2}\nabla^2 + v_{eff}(r) \right)\Psi_i = E_i \Psi_i \tag{11.26}$$

In principle, DFT is apparently free from any approximations: it is most likely to be exact but still the form of E_{XC} is unknown and for which several strategies for improvement are available.

11.2.4 Time-Dependent DFT

The many-electron wavefunction of a nonrelativistic many-electron system in a time-dependent external potential $v(r,t)$ must satisfy the time-dependent Schrödinger equation,

$$\hat{H}(t)\Psi(r_1\sigma_1,r_1\sigma_1,\ldots,r_N\sigma_N,t) = i\frac{\partial}{\partial t}\Psi(r_1\sigma_1,r_1\sigma_1,\ldots,r_N\sigma_N,t) \qquad (11.27)$$

where the time-dependent Hamiltonian takes the form

$$\hat{H}(t) = -\frac{1}{2}\sum_i \nabla_i^2 + \frac{1}{2}\sum_{i\neq j}\frac{1}{|r_i - r_j|} + \sum_i v(r_i,t) \qquad (11.28)$$

Analogously to the stationary case, we have the kinetic energy operator $\hat{T}$, the Columbic electron–electron interaction energy operator W, and the potential energy operator $\hat{V}(t)$ of the electrons in the time-dependent potential $v(r,t)$. In 1984, Runge and Gross[13] derived the analog of the Hohenberg–Kohn theorem for time-dependent systems by establishing a one-to-one mapping between time-dependent densities and time-dependent potentials for a given initial state.

11.2.5 LDA and GGA Approximation

To compute exchange-correlation (E_{XC}) energy functional, local density approximation (LDA) was initially originated, assuming that the exchange-correlation energy at any point in space is a function of the electron density at that point in space only and can be given by the electron density of a homogeneous electron gas of the same density. It considers an electronic distribution of an infinite number of electrons moving in an infinite volume of a space that is characterized by a uniformly distributed positive charge. This is referred to as the uniform electron gas. It is assumed that in the case of a molecule, the exchange-correlation functional at every point in space is the same as it would be for the uniform electron gas having the same density as at that position. The advantage of this approach is that the exchange functional can be accurately derived for this system:

$$E_x[\rho(r)] = -\frac{9\alpha}{8}\left(\frac{3}{\pi}\right)^{1/3}\int \rho^{4/3}(r)dr \qquad (11.29)$$

where α is $2/3$ for the uniform electron gas. Other models use different values for α: Slater uses a value of 1, and the $X\alpha$ model takes α as $3/4$. The local spin density approximation (LSDA) is an extension of LDA to account for systems including spin polarization (e.g., open-shell systems). Unfortunately, analytical derivation of the correlation functional has not proven possible. However, Vosko, Wilk, and Nusair[14] have fit numerical solutions of the correlation energy of several different uniform electron gases with functionals. These functionals are referred to as VWN.

The basic foundation of the LDA approach depends only on the density at a particular point. Throughout the last two decades, a number of other approaches have been developed that also take into account the gradient of the density. These functionals are popularly known as "nonlocal", "gradient-corrected", or the "generalized gradient approximation" (GGA). They are usually constructed by incorporating additional correction term to the LDA functional. The most popular GGA exchange functional is that developed by Becke[15] (B) as a correction to the LSDA exchange energy. This makes use of an empirical parameter which was obtained by fitting to exactly known exchange energies of the six noble gas atoms He to Rn. One of the widely used GGA correlation functionals is that of Lee, Yang, and Parr[16] (LYP), which contains four empirical parameters fit to the helium atom. The correlation functional of Perdew and Wang[17] (PW91) is also very popular and is a correction to the LSDA energy.

The GGA differs from the LDA in its dependence on the derivative of ρ. Truly nonlocal effects are not included, unlike with the use of the exchange operator in the HF approach. Hybrid methods were developed in order to combine both methodologies. These are based on the adiabatic connection method that relies on the following expression for the exchange-correlation energy:

$$E_{XC} = \int_0^1 \Psi(\lambda)|V_{XC}(\lambda)|\Psi(\lambda)d\lambda$$

(11.30)

where λ describes the extent of interelectronic interaction within the specific range of 0 (none) to 1 (exact). The problem becomes simple by assuming the limit where there the electrons are not interacting, there is no correlation energy, and the exchange energy can be exactly calculated just as in HF calculations to give E_X^{HF}. In this way, equation (11.30) may be written as

$$E_{XC} = E_x^{HF} + z\left(E_{XC}^{DFT} - E_X^{HF}\right)$$

(11.31)

11.3 DIFFERENT CHALLENGES FOR DFT

11.3.1 Improvement of Existing Formalisms

One of the challenging problems with DFT calculations is that it is inevitable to use approximate functionals to calculate the exchange-correlation energy, E_{xc}. In addition, it is not clear how to generate a better functional. There is no progression in levels of theory, which allows the calculation of molecular properties with increasing accuracy, thus allowing extrapolation to the exact answer, as is possible with HF and post-HF methods. Except reasonably well accounted for H-bonding interaction, current functionals are not able to produce impressive outcome (infect poor results) for the other weak interactions such as van der Waals-type interactions and London dispersion.

Early developments of DFT focused on the most basic challenges in chemistry, in particular, the ability to have functionals that could make a reasonable prediction on both the geometries and dissociation energies of molecules. The next major challenge for DFT arose from the need to accurately predict reaction barrier heights in order to determine the kinetics of chemical reactions as well as to describe van der Waals interactions. Whether DFT can predict the small energy differences associated with van der Waals interactions or if additional corrections or nonlocal functionals of the density are needed has been the subject of much debate and current research. The interaction, although a weakest one, is key to the accurate understanding of the biological processes involved in many drug–protein and protein–protein interactions.

11.3.2 Better Correlation Functional

Despite the rapid advancement in theoretical chemistry, to build proper functional remains a very challenging task to date. In principle, a well behaved and standard functional should work for the whole of chemistry, solid-state physics, and biology in different situations and diverse conditions. But in general applications, there are many disappointments in the real-time solutions. These are not breakdowns of the theory itself, but rather are shortcomings of the currently used approximation techniques available for the exchange-correlation functionals. This inherent limitation is clearly observed in case of the two simplest chemical systems (molecules), namely, stretched $H_2^?$ and stretched H_2. Even for these simple systems, the existing functionals are pushed to its limit. The fascinating aspect of DFT, namely, its simplicity will be in trouble if the computations with the proposed functionals become as complicated as full configuration interaction. The simplest functional, the LDA, has its limitation in many areas of chemistry. Although LDA gives good geometries, it massively overbinds molecules. The first step enabling chemists to use DFT satisfactorily is the inclusion of the first derivative of density in the form of the GGA. The next major outbreak came in the early 1990s with the inclusion of a fraction of HF exact exchange in the functional, as described by Becke. This work exhibits the fundamental basis for the development of B3LYP,[16,18] and it is the most widely used of all the contemporary functionals. The concept of hybrid functional (B3LYP) shows promising potential to extend the application of DFT to a wide range of systems along with impressive performance. Development of new functionals that improved upon B3LYP will provide a significant contribution in the progress of DFT-based computation.

11.3.3 Improvement on the Description of Reaction Barriers and Dispersion/van der Waals Interactions

The evaluation of reaction energy barriers for a particular reactive interaction is the basic challenge for LDA/GGA-type functional because the formalism encounters some inherent theoretical limitations that underestimate transition-state barriers by several kilocalories/mole.

There is an urgent need to fix the error before approaching the computation for potential energy surfaces using the functional. However, it is well known that in nature there exists some very important chemical phenomena that, although much smaller in the energetic scale, play a crucial role in large systems. The weak interactions like van der Waals force or London dispersion force are the major obstacles to generate the approximate functionals. It is of fundamental importance for the description of interactions between closed-shell species. The basic understanding of this problem can be seen from simple perturbation theory arguments dating back to London. These predicted that with the increase in the distance between the interacting species ($R \rightarrow \infty$), there should be asymptotic decays as $1/R^6$ for the attractive part of the system energy. The LDA or GGA functional form cannot take care of this behavior of the system due to the local nature of these functionals. We can observe relatively poor performance of most popular functionals on the simple system such as weakly bound dimers. The correct and efficient description of the van der Waals attraction, covalent bonding in chemistry, and transition states all remain firm challenges in the progress of DFT.

11.3.4 Improved DFT Formalism to Take Care of the Effect of Strong Correlation

The challenge of proper definition for strongly correlated systems is a very important frontier for DFT. To fully understand the importance of these systems, they must be looked within the wider realm of electronic structure methods. In general practice, except for Full Configuration Interaction and Valence Bond Theory,[19] most theories struggle to describe a particular system with strong correlation effect. Currently, even for the simplest systems such as infinitely stretched $H_2^{?}$ and infinitely stretched H_2, the limitation of proper theoretical manipulation exposed to all known functionals. Although these systems may seem trivial, they are, in our opinion, some of the great challenges for modern electronic structure theory. The integer nature of electrons is of great importance, and it is key to understand this behavior for DFT. In order to satisfy exact fundamental conditions and not to suffer from systematic errors, the energy functionals must have the correct discontinuous behavior at integer numbers of electrons. The accurate calculation of energy gap along with the correct description of strong correlation is possible with the incorporation of discontinuous behavior of electrons and, from research perspective, should never be ignored in the development of new theories.

11.4 CONCLUSION

This particular study concentrates on the significant contributions of DFT toward chemical science and the probable modifications, adding extra dimension to the accuracy of the formalism that makes it more compact. DFT is the computationally cost-effective solution for higher level computation on relatively large systems. Applications of DFT associated with approximate functionals significantly improve the performance of theoretical computation over a wide realm of chemical science. Currently, DFT becomes the subject of intense interest throughout the globe. The scope of theoretical manipulation for a wide range of properties from energetics and geometries of molecules to reaction barriers and van der Waals interactions is possible with the development of highly advanced theoretical techniques. However, we have also emphasized on some important shortcomings in terms of the contributory effects to get reasonable accuracy in the computations. The mutual interactions of the electrons, that is, electron correlation effect, complicate its theoretical description and manipulation enormously. Time-dependent DFT provides a sophisticated tool to investigate these dynamic properties in atoms, molecules, and clusters. Although some fundamental problems remain and computational techniques still require further upgradation, DFT presents promising growth with impressive outcome throughout the last two decades.

ACKNOWLEDGMENTS

This review is based on the information collected from different sources. The author tried to include maximum number of references discussed throughout the literature. It is purely the author's collective approach to provide some flavors of a highly demanding computational technique to the interested readers. Readers can switch to different books to generate deep understanding on the topic.

FURTHER READINGS

1. Parr, R. G.; Yang, W. *Density – Functional Theory of Atoms and Molecules*; Oxford University Press: New York, 1989.
2. Koch, W.; Holthausen, M. A. *Chemist's Guide to Density Functional Theory*; VCH: Weinheim, 2000.
3. Szabo, A.; Ostlund, N. S. *Modern Quantum Chemistry: Introduction to Advance Electronic Structure Theory*; Dover Publications, Inc: New York, 1982.
4. Jensen, F. *Introduction to Computational Chemistry*; Wiley: 1999.

REFERENCES

1. Kohn, W.; Becke, A. D.; Parr, R. G. Density functional theory of electronic structure. *J. Phys. Chem.* 1996, 100, 12974.
2. Baerends, E. J.; Gritsenko, O. V. A quantum chemical view of density functional theory. *J. Phys. Chem. A.* 1997, 101, 5383.

3. Chermette, H. Density functional theory: a powerful tool for theoretical studies in coordination chemistry. *Coord. Chem. Rev.* 1998, 178, 699.

4. Siegbahn, P. E. M.; Blomberg, M. R. A. Density functional theory of biologically relevant metal centers. *Annu. Rev. Phys. Chem.* 1999, 50, 221.

5. Andrews, L.; Citra, A. Infrared spectra and density functional theory calculations on transition metal nitrosyls. Vibrational frequencies of unsaturated transition metal nitrosyls. *Chem. Rev.* 2002, 102, 885.

6. Sarmah, A.; Saha, S.; Bagaria, P.; Roy, R. K. On the complementarity of comprehensive decomposition analysis of stabilization energy (CDASE) – scheme and supermolecular approach. *Chem. Phys.* 2012, 394, 29.

7. Sarmah, A.; Roy, R. K. Understanding the interaction of aqua-cisplatin with nucleobase guanine over adenine: a density functional reactivity theory based approach. *RSC Adv.* 2013, 3, 2822.

8. Sarmah, A.; Roy, R. K. Understanding the interaction of nucleobases with chiral semi-conducting single-walled carbon nanotubes (SWCNTs): an alternative theoretical approach based on density functional reactivity theory (DFRT). *J. Phys. Chem. C.* 2013, 117, 21539.

9. Saha, S.; Roy, R. K.; Pal, S. CDASE: A reliable scheme to explain the reactivity sequence between Diels Alder pairs. *Phys. Chem. Chem. Phys.* 2010, 12, 9328.

10. Hohenberg, P.; Kohn, W. Inhomogeneous electron gas. *Phys. Rev.* 1964, 136, B864.

11. Lieb, E. H. *Int. J. Quant. Chem.* 1983, 51, 1596.

12. Kohn, W.; Sham, J. Quantum density oscillations in an inhomogeneous electron gas. *Phys. Rev. A.* 1965, 140, 1133.

13. Runge, E.; Gross, E. K. U. Density-functional theory for time-dependent systems. *Phys. Rev. Lett.* 1984, 52, 997.

14. Vosko, S. H.; Wilk, L.; Nusair, M. Accurate spin-dependent electron liquid correlation energies for local spin density calculations: a critical analysis. *Can. J. Phys.* 1980, 58, 1200.

15. Becke, A. D. Density-functional exchange-energy approximation with correct asymptotic behavior. *Phys. Rev. A.* 1988, 33, 8800.

16. Lee, C. T.; Yang, W. T.; Parr, R. G. Development of the Colle-Salvetti correlation-energy formula into a functional of the electron density. *Phys. Rev. B.* 1988, 37, 785.

17. Perdew, J. P.; Parr, R. G.; Levy, M.; Balduz, J. L. Jr. Density-functional theory for fractional particle number: derivative discontinuities of the energy. *Phys. Rev. Lett.* 1982, 49, 1691.

18. Becke, A. D. Density-functional thermochemistry. III. The role of exact exchange. *J. Chem. Phys.* 1993, 98, 5648.

19. Perdew, J. P.; Zunger, A. Self-interaction correction to density-functional approximations for many-electron systems. *Phys. Rev. B.* 1981, 23, 5048.

20. Kohn, W. *Phys. Rev. Lett.* 1983, 51, 1596.

CHAPTER 12

Asbestos, the Carcinogen, and Its Bioremediation

Shabori Bhattacharya[1], Lalita Ledwani[2*], and P. J. John[3]

[1]Manipal University, Jaipur, Centre for Converging Technology, University of Rajasthan, India
[2]Department of Chemistry, Manipal University, Jaipur, India
[3]Department of Zoology, University of Rajasthan, India
*Email: lalitaledwani@gmail.com

CONTENTS

ABSTRACT

Asbestos, a mineral of phyllosilicate nature, known for its unique combination of properties, such as thermal and chemical resistivity, inability to conduct electricity, and low cost of mining and manufacture, has been extensively used since World War II up to the near past, until it was declared a Group I definite carcinogen by International Agency for Research on Cancer, an unit of World Health Organization in the year 1987. This leads to a ban being imposed on asbestos in several parts of the globe. Yet the continued use in some Asian countries, in particular, post the declaration raised severe issues of health hazards. The carcinogenicity of asbestos is attributed to the fiber dimensions, biopersistence, and surface properties that cause mesothelioma, asbestosis, and lung cancer in people exposed to it. Multiple mechanisms operate to induce carcinoma by generating cell transformation, chromosomal aberrations, single- and double-stranded DNA breaks, free radical generation, micronuclei induction in cells, and disturbed immune system. A series of successful experiments on animal models has by far established that asbestos is a complete carcinogen. Efforts to deal with the toxicity of asbestos have led researchers to seek ways of physical, chemical, or biological remediation. This review gives a detailed insight into the carcinogenic effects of asbestos as envisaged by studies on various animal models. It also attempts at summarizing some of the probable detoxification or bioremediation strategies that can be used for asbestos treatment in the near future rendering it safe.

12.1 INTRODUCTION

The Greek word "asbestos", meaning "inextinguishable" or "unquenchable", is said to be first used by Piney. The Webster's Medica dictionary defines asbestos as a mineral that readily separates into long flexible fibers suitable for use as noncombustible, non-conducting chemically resistant material.

12.1.1 Physicochemical Characteristics

Asbestos is a naturally occurring hydrated magnesium phyllosilicate mineral with many of its forms showing a considerable presence of iron. It is broadly divided into two mineralogical groups – the amphibole and the serpentine. This division is based primarily on the chemical composition of asbestos. Amphibole variety includes amosite (brown asbestos), crocidolite (blue asbestos), tremolite, anthophyllite, and actinolite, while serpentine has the single variety – chrysotile (white asbestos).[1] The chemical composition of each of these varieties is shown in Table 12.1.

TABLE 12.1   Chemical Composition of Various Forms of Asbestos.

Sr. no.	Type	Chemical Composition
1.	Chrysotile	$Mg_3Si_2O_5(OH)_4$
2.	Crocidolite	$Na_2(Mg,Fe^{2+})_3Fe_2^{3+}Si_8O_{22}(OH)_2$
3.	Amosite	$[(Mg \cdot Fe^{2+})]7Si_8O_{22}(OH)_2$
4.	Tremolite	$[Ca_2Mg_5Si_8O_{22}(OH)_2]$
5.	Actinolite	$[Ca_2(Mg,Fe^{2+})_5Si_8O_{22}(OH)_2]$
6.	Anthophyllite	$[Mg_7Si_8O_{22}(OH)_2]$

The formulae may be significantly modified in nature due to the occurrence of certain substituting cations such as $Fe^{2+/3+}$, $Al^{3?}$, or Na^+. The crystalline structure of both chrysotile and amphibole asbestos was studied by Fubini et al. They reported chrysotile to be an association of tetrahedral silicate sheet (composition $Si_2O_5)n^{2n-}$ with an octahedral brucite-like sheet of composition $[Mg_3O_2(OH)_4]n^{2n+}$, in which iron may substitute for magnesium, whereas amphiboles had an intrinsically elongated crystal structure with the basic unit formed by a double tetrahedral chain of composition $(Si_4O_{11})n^{6n-}$, sharing oxygen atoms with alternate layers of edge-sharing MO_6 octahedron. "M" is mostly Na^+, Mg^{2+}, Fe^{2+}, Fe^{3+}, or Ca^{2+}.[2] These silicate minerals occur in polyfilamentous bundles with fibers being flexible, thin, and long and easily separable from one another.[3] While chrysotile is curly and pliable, the amphibole varieties are rod-like or needle-shaped and more dustier.[4,5] All forms are fibrillar having width less than 1 μm and length varying from 2 μm to more than 20 μm. The stability of asbestos in natural environment and its biological aggressiveness are related to their fibrous structure and dimensions (Figs 12.1 and 12.2).[6]

FIGURE 12.1   Scanning Electron Microscope Image.

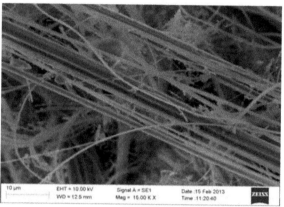

FIGURE 12.2   Scanning Electron Microscope Image of Asbestos Fibers (processed).

A unique combination of characteristics, such as high tensile strength, wear and friction characteristics, thermal, electrical, and acoustic insulation, resistance to heat and chemicals, flexibility to be woven, good adsorption capacity, and low cost,[3,7–9] in asbestos has led to its extensive use in industries since World War II to the recent past. It is used widely in the manufacture of a variety of products, such as AC sheets, AC pipes, brake shoes, brake linings fire proof clothes, and majorly in AC industry.

Following the declaration in the World Health Organization (WHO) in 1987 that asbestos is a definite Group I carcinogen, the use of asbestos has significantly varied around the globe. While some countries have imposed a ban on asbestos mining and use, others have strict regulations to reduce exposure, some have intervened less while still others have not restrained from asbestos use and continue doing so. European Union banned all asbestos products effectively in its member states from January 1, 2005.[10]

12.1.2   Asbestos Toxicity and Diseases

The asbestos fibers, which can be as small as 3–20 μm wide and 0.01 μm thin, can enter air, water, and soil from open pit mining, crushing, milling, screening, transport, wear and tear of asbestos products, disposal of asbestos wastes, weathering of natural deposits, or even just the opening of asbestos-containing bags and pollute the environment.[9,11] They remain suspended in air for long periods and are inhaled by workers or people residing in close vicinity of such sites. Inhalation or ingestion of these fibers eventually leads to asbestos-mediated toxicity in humans. Depending on the fiber dimensions, they are deposited on the airway (nose, throat, trachea)[12] or in the lungs. From lungs, they can be translocated to pleural or various other organs and tissues.[13–16] Continued deposition of asbestos fibers in the lungs or pleural membrane leads to significant health-related hazards.

The International Agency for Research on Cancer, an unit of WHO, reported high incidence of lung, pleural, and peritoneal mesothelioma with occupational exposure to various forms of asbestos.[17] Gastrointestinal and larynx cancers were also reported. The first cohort study to demonstrate lung cancer in asbestos-exposed workers was for a textile industry.[18]

Transmission electron microscopy analysis of asbestos fibers in a textile industry of South Carolina showed that fibers with length ?10 μm and thickness ?0.25 μm were the strongest predictors of lung cancer,[19] but these reports have less evidence. Mesothelioma, as a consequence of asbestos exposure, has been reported from different parts of the world, such as Quebec,[20-22] South Africa,[23] Zimbabwe,[24] Italy,[25] Turkey,[26] Western Australia,[27] and California.[28]

The various types of toxic effects envisaged include:

i)   Inflammation – the asbestos fibers stimulate the release of various inflammatory cell mediators and growth factors that lead to fibrogenesis of lung.[29] An increase in polymorphonuclear inflammatory cells has also been reported.[30]

ii) Mesothelioma – the mesothelial cells lining the pleura, pericardium, and peritoneum are the most affected, leading to diffuse malignant mesothelioma.[31,32]

iii) Asbestosis – deposition of asbestos fibers in lung causes asbestosis and/or lung cancer. The capacity of lung to pump oxygen in requisite amounts to all parts of body is lost.[33]

iv) Bronchogenic carcinoma – tumor in the tracheobronchial epithelium or alveolar epithelial cells has also been reported in people exposed to asbestos fibers, but the severity increases several folds with smoking habit.[34]

The above-mentioned disorders are influenced by the size of asbestos fibers to a large extent. While asbestosis is closely associated with fibers >2 μm in length,[35] mesothelioma is associated with fibers >5 μm long and width <0.1 μm,[36] and lung cancer is associated with fibers having length >10 μm and width <0.15 μm.[19] Progress of cancer is a slow development but continues even after animal is removed from asbestos exposure.[37-39]

12.2   MECHANISMS OF CARCINOGENESIS

The first report of pleural malignancy, due to asbestos, was mentioned by Wedler[40] in the early 1940s, while conducting studies on asbestos workers in Germany. The first published report, although under much controversy then, was in the early 1960s when Wagner et al. published the paper entitled – "Diffuse malignant mesothelioma and asbestos exposure in North – Western Cape Province".[41,42] He produced mesothelial tumors by direct implantation of asbestos fibers in laboratory animals. Asbestos since then has been repeatedly proved by research to be a potent carcinogen for humans inducing mesothelioma and lung carcinomas[43,44] at an early age,[45,46] although manifested after a latency period of 20–25 years. Experiments conducted on various animal models particularly on rats and Syrian hamster cells have successfully established the relationship between asbestos exposure and carcinoma. Wister rats were exposed to 10–15 mg/m^3 of asbestos fibers for 7 h per day for 5 days every week, up to 24 months. With increase in the period of exposure, rats developed lung tumors, thoracic tumors, and mesothelioma.[47] Davis et al.[48] performed similar study, in which they exposed Wister rats to chrysotile, crocidolite, and amosite fibers. Asbestos was reported to induce chromosomal mutations in mammalian cells as shown in hamster embryo cells, rat pleural mesothelial cells, or human lymphocytes.[49–57] It was further shown that asbestos was able to induce transformation in fibroblasts and mesothelial cells in culture.[58–60] Fischer rats showed unexpected mortality when exposed for a period of 10 months to 10 mg fibers/m^3 for 6 h per day for 5 days per week, showing maximum tolerance limits.[61] When Syrian golden hamsters were exposed to amosite for a continued period of 78 weeks, a large number of them developed pleural mesothelioma.[62] Baboons on exposure to amosite and crocidolite for a continued period of 4 years developed mesothelioma.[63,64] Intratracheal administration of asbestos fibers in rat[65,66] and Syrian golden hamster induced the formation of lung carcinoma and mesothelioma.[67,68] Another interesting aspect for asbestos-induced tumors or carcinomas is that cigarette smoke enhances the rate of incidence probably by acting as a co-carcinogen.[44] This type of synergism has also been demonstrated for asbestos with polycyclic aromatic hydrocarbons in some animals.[43,44]

Studies have confirmed that asbestos are biopersistent, and therefore accumulate in lungs, lymph nodes,[69,70] pleura,[71,72] and larynx.[73]

Current research in related area suggests three major hypotheses regarding mechanism of asbestos-induced carcinogenesis:

i)   Oxidative stress hypothesis
ii) Protein adsorption hypothesis
iii) Chromosome tangling hypothesis

12.2.1 Oxidative Stress Hypothesis

This theory implicates iron, a constituent of asbestos structure, as the major player in the induction of carcinogenesis. The phagocytic cells engulf the asbestos fibers but are unable to digest them. This leads to an accumulation of a large concentration of iron species, which catalyzes the production of reactive oxygen species (ROS).[31,32] Free radical generation can take place on asbestos fiber surface either i) by generation of highly reactive hydroxyl ions with H_2O_2 as in phagocytes and macrophages, ii) by reductants within cell that reduce atmospheric O_2 to hydroxyl radicals, or iii) by the homolytic cleavage of carbon–hydrogen bond in biomolecules, thereby generating radicals in target molecules such as peptides and proteins.[74–77] Along with this, the recruitment of inflammatory cells at these sites further causes the generation of reactive nitrogen species (RNS), clastogenic factors, and cytokines that may stimulate and/or damage the neighboring mesothelial cells.[78–81,82,83] As the mesothelial cells try to adapt by oxidative stress, alongside several other pathways such as mitogen-activated protein kinase pathway associated with cell proliferation and apoptosis, DNA repair and control of cell cycle progression in response to DNA damage are also activated.[84,85] All these mechanisms require the presence of iron ions, which was further proved by the fact that chemically synthesized chrysotile, free of iron, was unable to yield free radicals in cell-free extracts,[86] while when supplemented with less than 1 weight% of ferric ions, they became active.[87]

The genetic locus of one of the renal carcinoma target genes, namely, $p16^{INK4A}$, has been observed to be susceptible to oxidative stress by iron. Studies conducted for the expression of the said gene in cancerous mesothelial cell lines found its product, a cyclin-dependent kinase, to be absent in all the cells of asbestos.[88]

Ferrous ions are active in trace amounts[89,90] and conversion of ferric to ferrous ions increases the free radical generation ability.[91] When chelators remove these ions, asbestos fibers can neither generate hydroxyl ions[89,90] nor damage DNA.[74,75] Although ability of amphibole asbestos fibers to generate carboxy radicals is lost, they still have the potential to generate hydroxy radicals, so long as their crystal structure is intact.[92] Thus, various cellular effects observed due to iron-mediated free radical generation include increased fiber uptake by epithelial cells,[93] lipid peroxidation,[94,95] DNA oxidation,[96] tumor necrosis factor release along with cell apoptosis,[84] and inactivation of epidermal growth factor receptor ErbB1.[97]

Co-carcinogenic effects of tobacco smoke along with asbestos also lead to excess ROS and RNS generation, cell injury, apoptosis, and persistent lung inflammation.[82] Excess generation of reactive oxidative species alters the metabolism and detoxification of tobacco smoke carcinogen.[98]

12.2.2 Protein Adsorption Hypothesis

This hypothesis is based on the fact that the surface of asbestos fibers has high affinity for proteins and molecules due to the presence of negative or positive changes on its surface. In some studies, proteins absorbed on asbestos surface were categorized as chromatin/nucleotide/RNA-binding proteins, ribosomal proteins, cytoprotective proteins, cytoskeleton-associated proteins, histones, and hemoglobin.[99] Chrysotile that lack iron could probably concentrate iron by hemolytic activity and thereby induce oxidative DNA damage. Chrysotile fibers with low concentration of iron or even in the absence of iron in their composition have been reported to yield ferruginous bodies by accumulation of iron from disruption of surrounding proteins such as ferritin.[7] These ferruginous bodies with amosite were shown to cause single-strand breaks in DNA.[100] The normal iron homeostasis of body is also disrupted by accumulation of iron in this manner.[101] Ferruginous bodies generated induce various organic changes in cells of lung tissues.[102]

The biopersistence of chrysotile is less as compared to amphibole varieties due to selective leaching being more enhanced in the serpentine variety. This causes loss of magnesium due to phagocytosis by alveolar macrophages[103,104] breaking it into shorter fibers, which are readily cleared from the lungs.

12.2.3 Chromosome Tangling Hypothesis

This theory proposes that asbestos fibers can enter not only cytoplasm but also nucleus and tangle with chromosomes during cell division.[105] The specificity of tangling is a subject of in-depth research. Increased frequency of sister chromatid exchange and chromosomal aberrations were observed in blood lymphocytes of asbestos workers[106–108] and people exposed to it unintentionally, either as residents near asbestos mines and industries or as family members of asbestos workers.[109,110]

12.3 GENOTOXICITY STUDIES IN ANIMAL MODEL

Carcinogenesis is a multistage process consisting of genetic and epigenetic alterations. Asbestos-induced carcinomas have a latency period of 20–30 years before it is manifested, following the first exposure to the fibers.[17] During the latency period, gradual genetic and molecular alterations accumulate in the body system, such as activation of pathways related to resistance of apoptosis, acquired genetic instability, and angiogenesis. Asbestos-related cancer studies have shown that fibers interfere with mitotic apparatus, leading to aneuploidy, polyploidy, or other chromosomal aberrations.[111]

Although asbestos is inactive as a gene mutagen in mammalian cells, it causes neoplastic transformation of treated cells.[49,50] Neoplastic transformation of Syrian hamster embryo cells induced by chrysotile and crocidolite asbestos was dose-dependent.[112] Asbestos induces numerical and structural chromosomal aberrations, as shown by Oshimura et al.[51] in Syrian hamster embryo cells. Diffuse malignant mesothelioma research in mammalian cells revealed complex karyotypes in chromosome-banding patterns.[113,114] Karyotypes showed structural rearrangements and extensive aneuploidy of chromosome, particularly of the short arm of chromosomes 1, 3, and 9 and long arm of chromosome number 6. Most of the diffuse malignant mesothelioma studies showed loss of one copy of chromosome 22.[115] Asbestos was shown to induce anaphase abnormalities, manifested as lagging chromosome, bridges, and sticky chromosome.[31,32] Trisomy of chromosome 11 in asbestos-transformed Syrian hamster cell is yet another reported result.[116] Chromosomal changes in asbestos-treated normal human mesothelial cells alter their growth properties.[57,117] Neoplastic transformation was induced in rat mesothelial cells in culture by asbestos.[59,60]

Asbestos-induced neoplastic transformation is a multistep process involving at least three heritable changes – induction of immortality, activation of transforming oncogenes, and inactivation of a tumor suppressor gene.[118,119] In consistency with the above fact, it was observed that H-ras oncogene was activated in approximately 50% of Syrian hamster tumor cell lines.[120] The ras gene is activated by point mutation and converts immortal preneoplastic cells to the tumorigenic state.[118] The tumor suppresser genes are either lost or inactivated in neoplastic cells, thereby inducing tumor formation. Asbestos-induced immortal hamster cells lose some of the tumor suppressor gene activity with successive passages.[121]

Studies carried out for diffuse malignant mesothelioma have brought to the forefront a number of results related to chromosomal changes. A major event associated with diffuse malignant mesothelioma is deletion or hypermethylation at CDKN2A/ARF locus on chromosome 9p21, which carries three important tumor suppressor genes p15, p16[INK4A], and p14[ARF].[113] Reports state that RASSFIA and GPC3 tumor suppressor genes are silenced, while another tumor suppressor gene NF2 is inactivated in diffuse malignant mesothelioma.[122,123] Point mutations in p53 tumor suppressor gene and loss of heterozygosity were described in lung carcinoma of asbestos-exposed workers.[98]

Several studies conducted in asbestos-treated mesothelial cells have demonstrated base oxidation, DNA breakage (single- and double-stranded breakage), and disturbance in mitotic process,[85,124,125] probably due to ROS/RNS production or phagocytosis of asbestos fibers. Apoptosis is induced in mesothelial cells[126] and alveolar epithelial cells[127] by generation of reactive oxidative species by asbestos. Induction of micronuclei has been reported in primary cultures of human mesothelial cells by Pose et al.[128] Loss of heterozygosity[129] centrosome amplifications and aneuploidy cell formation were seen in asbestos-induced lung carcinoma cells.[130] In vitro genotoxicity studies have demonstrated micronuclei formation and sister chromatid exchange induction in Chinese hamster lung cells. While rats are more susceptible to lung cancer, hamsters show development of malignant pleural mesothelioma showing variation in species-generated effects.[131]

Asbestosis and lung cancer result because of culmination of a chain of events induced by asbestos in alveolar macrophages.[82] The fiber dusts cause secretion of cytokine IL-1β, which triggers release of additional cytokine TNF-alpha, IL-6, and IL-8 along with proliferation and activation of inflammatory response.[82] Mineral fibers have been demonstrated as binding to certain integrin receptors on cells of lung tissue and macrophages.[132,133] Recent investigation identifies NALP3 inflammasome as a key intracellular sensor for sensing initial interactions of asbestos with macrophages in a genetically engineered mice.[134]

ROS/RNS-caused DNA damage studies showed that lesions such as 8-oxodeoxyguanosine or 8-oxoguanine were formed in the DNA of asbestos-exposed various human and animal cell lines.[135–137] A correlation was established between concentration of 8-oxoguanine in DNA of white blood cells to time of asbestos exposure in a asbestos cohort study.[138] Oxidative stress in chrysotile-exposed asbestos workers resulted in increased levels of DNA double-strand breaks.[139] With respect to oxidative damage to DNA, recent research indicates that this may also probably be due to modification or oxidation of cytosine base.[140] The observation that myeloperoxidase activity was found in lungs of rodents exposed to asbestos[141] has been attributed as a probable secondary mechanism responsible for altered epigenetic methylation profiles as seen in human malignant pleural mesothelioma.[142]

Recent research shows that mitochondria are a major cytoplasmic target of asbestos, initiating mitochondria-associated ROS generation that induces nuclear mutagenic events and inflammatory signaling pathways in exposed cells.[143] Appropriate concentration of diallylsulfide, a precursor of glutathione, was seen to protect cells against asbestos-induced toxicity. Glutathione protects cells against oxidative damage.[144] Asbestos was shown to induce TGF-β to stimulate secretion of βIgH3, a cell adhesion protein in human bronchial epithelial cells, which further abrogates the tumorigenic phenotype.[145] Co-exposure of chrysotile with asbestos cement powder led to formation of micronuclei, thiobarbituric acid reactive substance (ROS) and also induced loss of viability in a concentration- and time-dependent manner in V79 cells, a Chinese hamster lung cell line.[146] Chrysotile fibers were reported to induce deletions at red BA/gam loci in gpt delta transgenic mouse primary embryo fibroblasts.[147] An association between asbestos-exposed lung tumors and genomic alterations in 19p13, 2p16, and 9q33.1 chromosomes showing allelic imbalance is reported.[148] Studies show that ferruginous bodies formed by adsorption of ferritin on asbestos fibers exerts peroxidase-like activity and shows cytotoxic activity against mesothelial cells, suggesting an important aspect in asbestos-related diseases.[149]

Effects on the immune system were also mentioned in some research activities. A study of genome-wide gene–asbestos interactions in lung cancer suggest that two pathways of immune function regulation, namely, Fas signaling and antigen processing and presentation pathways might be important in the etiology of asbestos-related lung cancer.[150] Mesothelioma patients show an overexpression of bcl-2 in CD4? peripheral T cells, plasma concentration of interleukin-10, and TGF-β and multiple overrepresentation of T-cell receptor (TCR-Vβ) in peripheral CD3?T cells.[151]

Role of iron in asbestos-induced carcinogenesis was further strengthened when metal chelators, such as deferoxamine and phytic acid, and radical scavengers, such as superoxide dismutase, dimethylthiourea, and glutathione precursor Nacystelyn, could significantly reduce the number of crocidolite- and chrysotile-induced micronuclei in human mesothelial cells by pretreatment and simultaneous treatment, respectively.[128]

12.4 REMEDIATION

Despite the ban having been imposed by several countries on asbestos mining and use post WHO declaring it as a definite Group I carcinogen, many countries across the globe still continue its production, with Russia being one of the major producers. The main sufferers of asbestos-mediated health hazards are the mine and industry workers and people residing around these sites. With the aim of a probable solution to the problem, researchers have been trying physical, and chemical, as well as biological remedies.

12.4.1 Physicochemical Methods

The first measures involve wet separation and proper ventilation of rooms containing sources of asbestos emission.[74,75,152,153] Use of anionic surfactants, such as linear alkyl sulfonate and sodium lauryl sulfate, or nonionic surfactants, such as alcohol alkoxylates, alkyl phenol ethoxylates, and polyoxoethylene esters, to carry out wet removal of asbestos has been done.[154] Asbestos exposure to environment can be further limited by using different binding media in solidification or cementation. Oxalic acid treatment of asbestos coupled with power ultrasound treatment effectively reduced the asbestos fibers to nanosize but the toxicity was retained. Oxalic acid enhanced the reactivity of chrysotile probably by modifying the fiber surface and exposing the poorly coordinated iron ions.[155] Various other chemicals have been used for asbestos digestion as well. But none of these methods could effectively reduce the carcinogenic effect of asbestos. Moreover, the methods are cumbersome, expensive, non-eco-friendly, and unable to carry out on site remediation efficiently. Some reports state that even in strongly acidic environments, the size of asbestos fibers is reduced to nanoscale but toxicity persists.[156]

12.4.2 Bioremediation

This method involves the use of naturally occurring microbes to detoxify or degrade pollutants in the environment. The probable bioremediators for any pollutant are generally found at sites contaminated with it. In a similar search, three fungal species, namely, *Verticellum leptobactrum*, *Paecilomyces lilacinus*, and *Aspergillus fumigatus* were isolated from serpentine rocks in Western Alps.[157] Out of these three, *V. leptobactrum* was a rare species but abundant at these sites. This was therefore investigated as a possible bioweathering agent for asbestos.

As mentioned earlier in the discussion, that iron in asbestos composition is a major factor responsible for its carcinogenicity, removal or extraction of this would prove to be a solution. In vitro experiments have reported that various chelators can extract iron from fibers,[158,159] modify its surface properties,[89,90] and even promote disruption of several sublayers.[160] Another important breakthrough was disruption of magnesia-silicate framework of asbestos by the aposymbiotic lichen-forming fungus *Xanthoparmelia tinctina*.[161] The fungus *Fusarium oxysporum* could effectively extract iron from chrysotile, crocidolite, and amosite by siderophore secretion also.[162] Although no significant changes were observed in protein profiles of *F. oxysporum* following asbestos exposure, increased levels of few proteins suggest that some relevant metabolic pathways might have been induced in an effort to protect cells from oxidative stress.[163] *V.*

leptobactrum was found to be an effective bioremediator with the ability of removing both iron[162] and magnesium.[164] Effect of fungal treatment varies with the type of asbestos being used. Removal of iron by fungi was more effective in crocidolite than in chrysotile. Analysis of chemical composition after fungal treatment of asbestos fibers showed that the fiber surface was modified to a much larger extent than the bulk.[165] Fungi were shown to release two types of chelators for extraction of iron from asbestos: siderophores and organic acids.[166] *P. lilacinus* was another potential fungus identified with the ability of mobilizing iron from crocidolite asbestos.[165] The ability of fibers to generate free radicals after iron mobilization by fungi was studied extensively in various animal models.[89,90,167] A decrease in toxicity was reported after fungal bioremediation of both crocidolite and chrysotile varieties of asbestos.[160,168]

12.5 CONCLUSION

Asbestos, as seen in the above discussion, acts by multiple mechanisms to induce carcinoma. Although fiber size and biopersistence are important factors in this matter, the presence of iron in its structure and also its ability to accumulate iron from surrounding proteins seem to direct the majority of the carcinoma causative factors. These mainly include free radical generation, chromosomal aberrations, cell transformation, and DNA breaks. Chromosomal aberrations induced may lead to suppression of tumor suppression genes or activation of proto-oncogenes.

The answer to asbestos-induced severe health hazard is a strict ban on asbestos mining and use all over the globe followed by decontamination of sites around defunct mines and industries using asbestos, by bioremediation strategies. With some fungal species already identified with this potential, the probability of bacterial and phytoremediation strategies to counter this problem hold ample scope for research.

ACKNOWLEDGMENTS

We owe our sincere thanks to the Director, Centre for Converging Technology, University of Rajasthan, India, for permitting us to use the laboratories for this work. We would also like to acknowledge the help provided by Mine Labour Protection Campaign Organization, India, in providing us with the asbestos samples.

REFERENCES

1. Mossman, B. T.; Kamp, D. W.; Weitzman, S. A. Mechanisms of carcinogenesis and clinical features of asbestos-associated cancers. *Cancer Invest.* 1996, 14, 464–478.
2. Fubini, B.; Otero Areán, C. Chemical aspects of the toxicity of inhaled mineral dusts. *Chem. Soc. Rev.* 1995, 28, 373–381.
3. USGS. *Some Facts about Asbestos (USGS Fact Sheet FS-012–01)*; USGS: Reston, 2001; p 4.
4. Mossman, B. T.; Bignon, J.; Corn, M. Asbestos: scientific developments and implications for public policy. *Science.* 1990, 247, 294–301.

5. Mossman, B. T.; Gee, J. B. L. Asbestos related diseases. *N. Engl. J. Med.* 1989, 320, 1724–1730.

6. Veblen, D. R.; Wylie, A. G. *Mineralogy of Amphiboles and 1:1 Layer Silicates.* In *Health Effects of Mineral Dust Reviews. Mineralogy;* Guthrie, G. D., Mossman, B. T., Eds.; Brookcrafters Inc: Chelsea, MI, 1993; Vol. 28, pp 61–137.

7. Roggli, V. L.; Oury, T. D.; Sporn, T. A. *Asbestos bodies and nonasbestos ferruginous bodies.* In *Pathology of Asbestos-Associated Diseases;* Roggli, V. L., Ed.; Springer: New York, 2004; pp 34–70.

8. Dadson, R. F.; Hammnar, S. P. *Asbestos: Risk Assessment, Epidemiology, and Health Effects;* CRC Press, Taylor & Francis Group: Boca Raton, FL, 2006.

9. NTP. NTP 11th report on carcinogens. *Rep. Carcinog.* 2005, 111–A32.

10. EU. EU Commission Directive 1999/77/EC of 26 July1999. *OJEU.* 1999. [L207/18 – L207/20].

11. ATSDR. *ATSDR, Toxicological Profile for Asbestos (TP-61);* US Dept. of Health & Human Services: Washington DC, 2001.

12. Morgan, A. Acid leaching studies of chrysotile asbestos from mines in the Coalinga region of California and from Quebec and British Columbia. *Ann. Occup. Hyg.* 1997, 41, 249–268.

13. Lippmann, M. Deposition and retention of inhaled fibres: effects on incidence of lung cancer and mesothelioma. *Occup. Environ. Med.* 1994, 51, 793–798.

14. Paoletti, L.; Falchi, M.; Batisti, D.; Zappa, M.; Chellini, E.; Biancalani, M. Characterization of asbestos fibers in pleural tissue from 21 cases of mesothelioma. *Med. Lav.* 1993, 84, 373–378.

15. Boutin, C.; Dumortier, P.; Rey, F.; Viallat, J. R.; De Vuyst, P. Black spots concentrate oncogenic asbestos fibers in the parietal pleura: thoracoscopic and mineralogic study. *Am. J. Respir. Crit. Care Med.* 1996, 153(1), 444–449.

16. Tossavainen, A.; Karjalainen, A.; Karhunen, P. J. Retention of asbestos fibers in the human body. *Environ. Health Perspect.* 1994, 102(5), 253–255.

17. IARC. Some miscellaneous pharmaceutical substances. *IARC Monogr. Eval. Carcinog Risk Chem. Man.* 1977, 13, 1–255.

18. Doll, R. Mortality from lung cancer in asbestos workers. *Br. J. Ind. Med.* 1955, 12, 81–86.

19. Dement, J. M.; Kuempel, E. D.; Zumwalde, R. D.; Smith, R. J.; Stayner, L. T.; Loomis, D. Development of a fibre size-specific job-exposure matrix for airborne asbestos fibres. *Occup. Environ. Med.* 2008, 65, 605–612.

20. Liddell, F. D.; McDonald, A. D.; McDonald, J. C. The1891–1920 birth cohort of Quebec chrysotile miners and millers: development from 1904 and mortality to1992. *Ann. Occup. Hyg.* 1997, 41, 13–36.

21. Hein, M. J.; Stayner, L. T.; Lehman, E.; Dement, J. M. Follow-up study of chrysotile textile workers: cohort mortality and exposure-response. *Occup. Environ. Med.* 2007, 64, 616–625.

22. Bégin, R.; Gauthier, J. J.; Desmeules, M.; Ostiguy, G. Work-related mesothelioma in Québec, 1967–1990. *Am. J. Ind .Med.* 1992, 22, 531–542.

23. Rees, D.; Myers, J. E.; Goodman, K.; Fourie, E.; Blignaut, C.; Chapman, R.; Bachmann, M. O. Case-control study of mesothelioma in South Africa. *Am. J. Ind. Med.* 1999, 35, 213–222.

24. Cullen, M. R.; Baloyi, R. S. Chrysotile asbestos and health in Zimbabwe: I. Analysis of miners and millers compensated for asbestos-related diseases since independence (1980). *Am. J. Ind. Med.* 1991, 19, 161–169.

25. Mirabelli, D.; Calisti, R.; Barone-Adesi, F.; Fornero, E.; Merletti, F.; Magnani, C. Excess of mesotheliomas after exposure to chrysotile in Balangero, Italy. *Occup. Environ. Med.* 2008, 65, 815–819.

26. Baris, I.; Simonato, L.; Artvinli, M.; Pooley, F.; Saracci, R.; Skidmore, J.; Wagner, C. Epidemiological and environmental evidence of the health effects of exposure to erionite fibres: a four-year study in the Cappadocian region of Turkey. *Int. J. Cancer.* 1987, 39, 10–17.

27. Wagner, J. C.; Pooley, F. D. Mineral fibres and mesothelioma. *Thorax.* 1986, 41, 161–166.

28. Pan, X. L.; Day, H. W.; Wang, W.; Beckett, L. A.; Schenker, M. B. Residential proximity to naturally occurring asbestos and mesothelioma risk in California. *Am. J. Respir. Crit. Care Med.* 2005, 172, 1019–1025.

29. Cohen, R. Occupational lung disease: pneumoconiosis. *Occup. Health Nurs.* 1981, 29(4), 10–13.

30. Brody, A. R.; Hill, L. H.; Adkins, B.; O'Connor, R. W. Chrysotile asbestos inhalation in rats: deposition pattern and reaction of alveolar epithelium and pulmonary macrophages. *Am. Rev. Respir. Dis.* 1981, 123, 670–679.

31. Toyokuni, S. Role of iron in carcinogenesis: cancer as a ferrotoxic disease. *Cancer Sci.* 2009, 100, 9–16.

32. Toyokuni, S. Mechanisms of asbestos-induced carcinogenesis. *Nagoya J. Med. Sci.* 2009, 71, 1–10.

33. Kilburn, K. H. Indoor air effects after building renovation and in manufactured homes. *Am. J. Med. Sci.* 2000, 320(4), 249–254.

34. Saracci, R. Asbestos and lung carcinoma: an analysis of the epidemiological evidence on the asbestos-smoking, interation. *Int. J. Cancer.* 1977, 20, 323–331.

35. Dodson, R. F.; Atkinson, M. A.; Levin, J. L. Asbestos fibre length as related to potential pathogenicity: a critical review. *Am. J. Ind. Med.* 2003, 44, 291–297.

36. NIOSH. Asbestos fibres and other elongated mineral particles: state of the science and road map for research, Report. Department of Health and Human Services, Public Health Service, Centers for Disease Control, 2009.

37. Davis, J. M. G. *In Vivo Assays to Evaluate the Pathogenic Effects of Minerals in Rodents.* In *Health Effects of Mineral Dusts. Reviews in Mineralogy*; Guthrie, G. D., Mossman, B. T., Eds.; Mineralogical Society of America: Washington, DC, 1993; Vol. 28, pp 471–487.

38. Reynolds, T. Asbestos-linked cancer rates up less than predicted. *J. Natl. Cancer Instit.* 1992, 84, 560–562.

39. Doll, R.; Peto, J. *Asbestos Effects on Health of Exposure to Asbestos*; N.M. Stationary Office: London, 1985.

40. Wedler, H. W. Lung cancer in asbestos patients. *Dtsch. Arch. Klin. Med.* 1943, 191, 189–209.

41. Wagner, J. C.; Sleggs, C. A.; Marchand, P. Diffuse pleural mesothelioma and asbestos exposure in North Western Cape Province. *Br. J. Ind. Med.* 1960, 17, 260–271.

42. Wagner, J. C. Experimental production of mesothelial tumors of the pleura by implantation of dusts in laboratory animals. *Nature.* 1962, 196, 180.

43. IARC Monograph. *IARC Monographs on the Evaluation of Carcinogenic Risk of Chemicals to Man, Asbestos*; IARC: Lyon, France, 1977; Vol. 14.

44. National Research Council. *Asbestiform Fibers: Nonoccupational Health Risks*; National Academy Press: Washington, DC, 1984.

45. Peto, J.; Henderson, B. E.; Pike, M. C. Trends in mesothelioma incidence in the United States and the forecast epidemic due to asbestos exposure during World War II Quant. *Occup. Cancer.* 1982, 9, 51–60.

46. Peto, J.; Seidman, H.; Selikoff, I. F. Mesothelioma mortality in asbestos" workers: Implications for models of carcinogenesis and risk assessment. *Br. J. Cancer.* 1982, 45, 124–185.

47. Wagner, J. C.; Berry, G.; Skidmore, J. W.; Timbrell, V. The effects of the inhalation of asbestos in rats. *Br. J. Cancer.* 1974, 29, 252–269.

48. Davis, J. M.; Beckett, S. T.; Bolton, R. E.; Collings, P.; Middleton, A. P. Mass and number of fibres in the pathogenesis of asbestos related lung disease in rats. *Br. J. Cancer.* 1978, 37, 673–688.

49. Chamberlain, M.; Thrmy, E. M. Asbestos and glass fibers in bacterial mutation tests. *Mutat. Rts.* 1977, 43, 159–164.

50. Barrett, J. C. *Relationship between mutagenesis and carcinogenesis.* In *Mechanisms of Environmental Carcinogenesis: Role of Genetic and Epigenetic Changes*; Barrett, J. C., Ed.; CRC Press: Boca Raton, FL, 1987; Vol. 1, pp 129–142.

51. Oshimura, M.; Hesterberg, T. W.; Tsutsui, T.; Barrett, J. C. Correlation of asbestos-induced cytogenetic effects with cell transformation of Syrian hamster embryo cells in culture. *Cancer Res.* 1984, 44, 5017–5022.

52. Sincock, A. M.; Seabright, M. Induction of chromosome changes in Chinese hamster cells by exposure to asbestos fibers. *Nature.* 1975, 257, 56–58.

53. Babu, K. A.; Lakkad, B. C.; Nigam, S. K.; Bhatt, D. K.; Karnik, A. S.; Thalmrq, K. N.; Kashyap, S. K.; Chatterjee, S. K. In vitro cytological and cytogenetic effects of an Indian variety of chrysotile asbestos. *Environ. Res.* 1980, 21, 416–422.

54. Valerio, F.; DeFerran, M.; Ottaggio, L.; Repetto, E.; Santi, L. *Cytogenetic effects of Rhodesian chrysotile on human lymphocytes in vitro*. In *Biological Effects of Mineral Fibres*; Wagner, J. C., Ed.; International Agency for Research on Cancer: Lyon, France, 1980; Vol. 1, pp 485–489.

55. Lavappa, K. S.; Fu, M. M.; Epstein, S. S. Cytogenetic studies on chrysotile asbestos. *Environ. Res.* 1975, 10, 165–173.

56. Jaurand, M. C.; Kheuang, L.; Magne, L.; Bignon, J. Chromosomal changes induced by chrysotile fibres or benzo-3,4-pyrene in rat pleural mesothelial cells. *Mutat. Res.* 1986, 169, 141–148.

57. Lechner, J. F.; Tokiwa, T.; LaVeck, M.; Benedict, W. F.; Banks-Schlegel, S.; Yeager, H. Jr.; Banerjee, A.; Harris, C. C. Asbestos-associated chromosomal changes in human mesothelial cells. *Proc. Natl. Acad. Sci. U.S.A.* 1985, 82, 3884–3888.

58. Hesterberg, T. W.; Brody, A. R.; Oshimura, M.; Barrett, J. C. *Asbestos and silica induce morphological transformation of mammalian cells in culture: A possible mechanism*. In *Silica, Silicosis and Cancer*; Goldsmith, D. F., Winn, D. M., Shy, C. M., Eds.; Praeger Press: New York, 1986; pp 177–190.

59. Paterour, M. J.; Bignon, J.; Jaurand, M. C. In vitro transformation of rat pleural esothelial cells by chrysotile and/or benzo-a-pyrene. *Carcinogenesis*. 1985, 6, 523–529.

60. Paterour, M. J.; Renier, A.; Bignon, J.; Jaurand, M. C. Induction of transformation in cultured rat pleural mesothelial cells by chrysotile fibres In: In vitro Effects of Mineral Dusts Third International Workshop, NATO ASl Series; Beck, E.G., Bignon, J. Ed, Springer-Verlag, Berlin, 1985, Vol. G3, pp 203–207.

61. McConnell, E. E.; Kamstrup, O.; Musselman, R.; Hesterberg, T. W.; Chevalier, J.; Miiller, W. C.; Thevenaz, P. Chronic inhalation study of size- separated rock and slag wool insulation fibres in Fischer 344/N rats. *Inhal. Toxicol.* 1994, 6, 571–614.

62. McConnell, E. E.; Axten, C.; Hesterberg, T. W.; Chevalier, J.; Miiller, W. C.; Everitt, J.; Oberdörster, G.; Chase, G. R.; Thevenaz, P.; Kotin, P. Studies on the inhalation toxicology of two fibreglasses and amosite asbestos in the Syrian golden hamster. Part II. Results of chronic exposure. *Inhal. Toxicol.* 1999, 11, 785–835.

63. Goldstein, B.; Coetzee, F. S. Experimental malignant mesothelioma in baboons. *S. Afr. J. Sci.* 1990, 86, 89–93.

64. Webster, I.; Goldstein, B.; Coetzee, F. S.; van Sittert, G. C. H. Malignant mesothelioma induced in baboons by inhalation of amosite asbestos. *Am. J. Ind. Med.* 1993, 24, 659–666.

65. Pott, F.; Ziem, U.; Reiffer, F. J.; Huth, F.; Ernst, H.; Mohr, U. Carcinogenicity studies on fibres, metal compounds, and some other dusts in rats. *Exp. Pathol.* 1987, 32, 129–152.

66. Smith, D. M.; Ortiz, L. W.; Archuleta, R. F.; Johnson, N. F. Long-term health effects in hamsters and rats exposed chronically to man-made vitreous fibres. *Ann. Occup. Hyg.* 1987, 31(4B), 731–754.

67. Pott, F.; Ziem, U.; Mohr, U. Lung carcinomas and mesotheliomas following intratracheal instillation of glass fibres and asbestos. In: Proceedings of the VIth International Pneumoconiosis Conference 20–23 September 1983; Bochum, Germany: International Labour Office, 1984, pp 746–756.

68. Feron, V. J.; Scherrenberg, P. M.; Immel, H. R.; Spit, B. J. Pulmonary response of hamsters to fibrous glass: chronic effects of repeated intratracheal instillation with or without benzo[a]pyrene. *Carcinogenesis*. 1985, 6, 1495–1499.

69. Dodson, R. F.; Williams, MG Jr; Corn, C. J.; Brollo, A.; Bianchi, C. Asbestos content of lung tissue, lymph nodes, and pleural plaques from former shipyard workers. *Ann. Rev. Respir. Dis.* 1990, 142, 843–847.

70. Dodson, R. F.; Atkinson, M. A. Measurements of asbestos burden in tissues. *Ann. N Y Acad. Sci.* 2006, 1076, 281–291.

71. Gibbs, A. R.; Stephens, M.; Griffiths, D. M.; Blight, B. J.; Pooley, F. D. Fibre distribution in the lungs and pleura of subjects with asbestos related diffuse pleural fibrosis. *Br. J. Ind. Med.* 1991, 48, 762–770.

72. Suzuki, Y.; Yuen, S. R. Asbestos tissue burden study on human malignant mesothelioma. *Ind. Health.* 2001, 39, 150–160.

73. Roggli, V. L.; Greenberg, S. D.; McLarty, J. L.; Hurst, G. A.; Spivey, C. G.; Heiger, L. R. Asbestos body content of the larynx in asbestos workers. A study of five cases. *Arch. Otolaryngol.* 1980, 106, 533–535.

74. Hardy, J. A.; Aust, A. E. Iron in asbestos chemistry and carcinogenicity. *Chem. Rev.* 1995, 95, 97–118.

75. Hardy, J. A.; Aust, A. E. The effect of iron binding on the ability of crocidolite asbestos to catalyze DNA single-strand breaks. *Carcinogenesis*. 1995, 16, 319–325.
76. Fubini, B.; Otero, A. C. Chemical aspects of the toxicity of inhaled mineral dusts. *Chem. Soc. Rev.* 1999, 28, 373–381.
77. Kamp, D. W.; Weitzman, S. A. The molecular basis of asbestos induced lung injury. *Thorax*. 1999, 54, 638–652.
78. Manning, C. B.; Vallyathan, V.; Mossman, B. T. Diseases caused by asbestos: mechanisms of injury and disease development. *Int. Immunopharmacol.* 2002, 2, 191–200.
79. Vallyathan, V.; Mega, J. F.; Shi, X. L.; Dalal, N. S. Enhanced generation of free radicals from phagocytes induced by mineral dusts. *Am. J. Respir. Cell Mol. Biol.* 1992, 6, 404–413.
80. Antony, V. B.; Sahn, S. A.; Mossman, B. T.; Gail, D. B.; Kalica, A. Pleural cell biology in health and disease. *Am. Rev. Respir. Dis.* 1992, 145(5), 1236–1239.
81. Kane, A. B. *Mechanisms of Mineral Fibre Carcinogenesis*. In *Mechanisms of Fibre Carcinogenesis*; Kane, A. B., Boffetta, P., Sarracci, R., Wilbourn, J. D., Eds.; IARC Scientific Publications No. 140, International Agency for Research on Cancer: Lyon, France, 1996; pp 11–34.
82. Shukla, A.; Gulumian, M.; Hei, T. K.; Kamp, D.; Rahman, Q.; Mossman, B. T. Multiple roles of oxidants in the pathogenesis of asbestos-induced diseases. *Free Radic. Biol. Med.* 2003, 34, 1117–1129.
83. Bhattacharya, K.; Dopp, E.; Kakkar, P.; Jaffery, F. N.; Schiffmann, D.; Jaurand, M. C.; Rahman, I.; Rahman, Q. Biomarkers in risk assessment of asbestos exposure. *Mutat. Res.* 2005, 579, 6–21.
84. Upadhyay, D.; Kamp, D. W. Asbestos-induced pulmonary toxicity: role of DNA damage and apoptosis. *Exp. Biol. Med. (Maywood)*. 2003, 228, 650–659.
85. Jaurand, M. C. Mechanisms of fiber-induced genotoxicity. *Environ. Health Perspect.* 1997, 105, 1073–1084.
86. Gazzano, E.; Foresti, E.; Lesci, I. G.; Tomatis, M.; Riganti, C.; Fubini, B.; Roveri, N.; Ghigo, D. Different cellular responses evoked by natural and stoichiometric synthetic chrysotile asbestos. *Toxicol. Appl. Pharmacol.* 2005, 206, 356–364.
87. Gazzano, E.; Turci, F.; Foresti, E.; Putzu, M. G.; Aldieri, E.; Silvagno, F.; Lesci, I. G.; Tomatis, M.; Riganti, C.; Romano, C.; Fubini, B.; Roveri, N.; Ghigo, D. Iron-loaded synthetic chrysotile: a new model solid for studying the role of iron in asbestos toxicity. *Chem. Res. Toxicol.* 2007, 20, 380–387.
88. Kratzke, R.; Otterson, G.; Lincoln, C.; Ewing, S.; Oie, H.; Geradts, J.; Kaye, F. Immunohistochemical analysis of the p16INK4 cyclin-dependent kinase inhibitor in malignant mesothelioma. *J. Natl. Cancer Inst.* 1995, 87, 1870–1875.
89. Fubini, B.; Bolis, V.; Cavenago, A.; Volante, M. Physico-chemical properties of crystalline silica dusts and their possible implication in various biological responses. *Scand. J. Work Environ. Health*. 1995, 21, 9–15.
90. Fubini, B.; Mollo, L.; Giamello, E. Free radical generation at the solid/liquid interface in iron containing minerals. *Free Radic. Res.* 1995, 23, 593–614.
91. Gulumian, M.; Bhoolia, D. J.; Du Toit, R. S. Activation of UICC crocidolite: the effect of conversion of some ferric ions to ferrous ions. *Environ. Res.* 1993, 60, 193–206.
92. Tomatis, M.; Prandi, L.; Bodoardo, S.; Fubini, B. Loss of surface reactivity upon heating amphibole asbestos. *Langmuir*. 2002, 18, 4345–4350.
93. Hobson, J.; Wright, J. L.; Churg, A. Active oxygen species mediate asbestos fibre uptake by tracheal epithelial cells. *FASEB J.* 1990, 4, 3135–3139.
94. Ghio, A. J.; Kadiiska, M. B.; Xiang, Q. H.; Mason, R. P. In vivo evidence of free radical formation after asbestos instillation: an ESR spin trapping investigation. *Free Radic. Biol. Med.* 1998, 24, 11–17.
95. Gulumian, M. The ability of mineral dusts and fibres to initiate lipid peroxidation. Part I: parameters which determine this ability. *Redox Rep.* 1999, 4, 141–163.
96. Aust, A. E.; Eveleigh, J. F. Mechanisms of DNA oxidation. *Proc. Soc. Exp. Biol. Med.* 1999, 222, 246–252.
97. Baldys, A.; Aust, A. E. Role of iron in inactivation of epidermal growth factor receptor after asbestos treatment of human lung and pleural target cells. *Am. J. Respir. Cell. Mol. Biol.* 2005, 32, 436–442.
98. Nymark, P.; Wikman, H.; Hienonen-Kempas, T.; Anttila, S. Molecular and genetic changes in asbestos related lung cancer. *Cancer Lett.* 2008, 265, 1–15.

99. Nagai, H.; Ishihara, T.; Lee, W. H.; Ohara, H.; Okazaki, Y.; Okawa, K.; Toyokuni, S. Asbestos surface provides a niche for oxidative modification. *Cancer Sci.* 2011, 102(12), 2118–2125.

100. Lund, L. G.; Williams, M. G.; Dodson, R. F.; Aust, A. E. Iron associated with asbestos bodies is responsible for the formation of single strand breaks in phi X174 RFI DNA. *Occup. Environ. Med.* 1994, 51, 200–204.

101. Ghio, A. J.; Stonehuerner, J.; Richards, J.; Devlin, R. B. Iron homeostasis in the lung following asbestos exposure. *Antioxid. Redox Signal.* 2008, 10, 371–377.

102. Pezerath, H. *The Surface Activity of Mineral Dusts and the Process of Oxidative Stress.* In *Mechanisms in Fibre Carcinogenesis*; Brown, R. C., Hoskins, J. A., Johnson, N. F., Eds.; Plenum Press: New York, 1991; pp 387–395.

103. Gulumian, M. An update on the detoxification processes for silica particles and asbestos fibres: successes and limitations. *J. Toxicol Environ Health B Crit Rev.* 2005, 8, 453–483.

104. Langer, A. M.; Nolan, R. P. Chrysotile: its occurrence and properties as variables controlling biological effects. *Ann. Occup. Hyg.* 1994, 38, 427–451.

105. Wang, N.; Jaurand, M.; Magne, L.; Kheuang, L.; Pinchon, M.; Bignon, J. The interactions between asbestos fibers and metaphase chromosomes of rat pleural mesothelial cells in culture: a scanning and transmission electron microscopic study. *Am. J. Pathol.* 1987, 126, 343–349.

106. Rom, W. N.; Livingston, G. K.; Casey, K. R.; Wood, S. D.; Egger, M. J.; Chiu, G. L.; Jerominski, L. Sister chromatid exchange frequency in asbestos workers. *J. Natl. Cancer Inst.* 1983, 70, 45–48.

107. Fatma, N.; Jain, A. K.; Rahman, Q. Frequency of sister chromatid exchange and chromosomal aberrations in asbestos cement workers. *Br. J. Ind. Med.* 1991, 48, 103–105.

108. Lee, S. H.; Shin, M.; Lee, K. J.; Lee, S. Y.; Lee, J. T.; Lee, Y. H. Frequency of sister chromatid exchange in chrysotile-exposed workers. *Toxicol. Lett.* 1999, 108, 315–319.

109. Donmez, H.; Ozkul, Y.; Ucak, R. Sister chromatid exchange frequency in inhabitants exposed to asbestos in Turkey. *Mutat. Res.* 1996, 361, 129–132.

110. Ramsey, M. J.; Moore, D. H.; Briner, J. F.; Lee, D. A.; Olsen, L.; Senft, J. R.; Tucker, J. D. The ffects of age and lifestyle factors on the accumulation of cytogenetic damage as measured by chromosome painting. *Mutat. Res.* 1995, 338, 95–106.

111. Jaurand, M. C. *Use of In-Vitro Genotoxicity and Cell Transformation Assays to Evaluate the Potential Carcinogenicity of Fibres*; IARC: France, 1996; pp 55–72.

112. Hesterberg, T. W.; Barrett, J. C. Dependence of asbestos- and mineral dust-induced transformation of mammalian cells in culture on fiber dimension. *Cancer Res.* 1984, 44, 2170–2180.

113. Murthy, S.; Testa, J. Asbestos, chromosomal deletions, and tumor suppressor gene alterations in human malignant mesothelioma. *J. Cell Physiol.* 1999, 180, 150–157.

114. Testa, J. R.; Pass, H. I.; Carbone, M. *Molecular biology of mesothelioma.* In *Molecular Biology of Mesothelioma*; Devita, V. T. Jr., Hellman, S., Rosenberg, S. A., Eds.; Lippincott Williams & Wilkins: Philadelphia, 2001; pp 1937–1943.

115. Taguchi, T.; Jhanwar, S.; Siegfried, J.; Keller, S.; Testa, J. Recurrent deletions of specific chromosomal sites in 1p, 3p, 6q, and 9p in human malignant mesothelioma. *Cancer Res.* 1993, 53, 4349–4355.

116. Sanders, C. Pleural mesothelioma in the rat following exposure to 239PuO2. *Health Phys.* 1992, 63, 695–697.

117. Linnnainmaa, K.; Gerwin, B.; Pelin, K.; Jantunen, K.; La Veck, M.; Lechner, J. F.; Harris, C. C. Asbestos-induced mesothelioma and chromosomal abnormalities in human mesothelial cells in vitro. In: Proceedings of the 3rd Joint U.S Finnish NIOSH Science Symposium, Washington, 1986, pp 119–122.

118. Barrett, J. C.; Oshimura, M.; Koi, M. *Role of oncogenes and tumor suppressor genes in a multistep model d carcinogenesis.* In *Critical Molecular Determinants of Carcinogenesis*; Becker, F. F., Slaga, T. J., Eds.; University of Texas Press: Auatin, TX, 1987; Vol. 39, pp 45–56.

119. Barrett, J. C.; Fletcher, W. F. *Cellular and molecular mechanisms of multistep carcinogenesis in cell culture models.* In *Mechaniams of Environmental Carcinogenesis: Multistep Models of Carcinogenesis*; Barrett, J. C., Ed.; CRC Press: Boca Raton, FL, 1987; Vol. 2, pp 78–116.

120. Gilmer, T. M.; Annab, L.; Barrett, J. C. Characterization of activated protooncogenes in chemically transformed Syrian hamster embryo cells. *Mol. Carcinog.* 1988, 1(3), 180–188.
121. Koi, M.; Barrett, J. C. Loss of tumor suppression function during chemically induced neoplastic progression of Syrian hamster embryo cells. *Proc. Natl. Acad. Sci. U.S.A.* 1986, 83, 5992–5996.
122. Apostolou, S.; Balsara, B. R.; Testa, J. R. *Cytogenetics of malignant mesothelioma.* In *Malignant Mesothelioma: Advances in Pathogenesis, Diagnosis and, Translational Therapies;* Pass, H. I., Vogelzang, N., Carbone, M., Eds.; Springer Science & Business Media, Inc.: New York, 2006; pp 101–111.
123. Murthy, S. S.; Shen, T.; De Rienzo, A; Lee, W. C.; Ferriola, P. C.; Jhanwar, S. C.; Mossman, B. T.; Filmus, J.; Testa, J. R. Expression of GPC3, an X-linked recessive overgrowth gene, is silenced in malignant mesothelioma. *Oncogene.* 2000, 19, 410–416.
124. Nygren, J.; Suhonen, S.; Norppa, H.; Linnainmaa, K. DNA damage in bronchial epithelial and mesothelial cells with and without associated crocidolite asbestos fibers. *Environ. Mol. Mutagen.* 2004, 44, 477–482.
125. Liu, W.; Ernst, J. D.; Broaddus, V. C. Phagocytosis of crocidolite asbestos induces oxidative stress, DNA damage, and apoptosis in mesothelial cells. *Am. J. Respir. Cell Mol. Biol.* 2000, 23, 371–378.
126. Broaddus, V. C.; Yang, L.; Scavo, L. M.; Ernst, J. D.; Boylan, A. M. Asbestos induces apoptosis of human and rabbit pleural mesothelial cells via reactive oxygen species. *J. Clin. Invest.* 1996, 98, 2050–2059.
127. Aljandali, A.; Pollack, H.; Yeldandi, A.; Li, Y.; Weitzman, S. A.; Kamp, D. W. Asbestos causes apoptosis in alveolar epithelial cells: role of iron induced free radicals. *J. Lab. Clin. Med.* 2001, 137, 330–339.
128. Poser, I.; Rahman, Q.; Lohani, M.; Yadav, S.; Becker, H. H.; Weiss, D. G.; Schiffmann, D.; Dopp, E. Modulation of genotoxic effects in asbestos-exposed primary human mesothelial cells by radical scavengers, metal chelators and a glutathione precursor. *Mutat. Res.* 2004, 559, 19–27.
129. Both, K.; Turner, D. R.; Henderson, D. W. Loss of heterozygosity in asbestos-induced mutations in a human mesothelioma cell line. *Environ. Mol. Mutagen.* 1995, 26(1), 67–71.
130. Cortez, B. A.; Machadosantelli, G. M. Chrysotile effects on human lung cell carcinoma in culture: 3-d reconstruction and DNA quantification by Image analysis. *BMC Cancer.* 2008, 8, 181.
131. IARC. *IARC Monographs on the Evaluation of Carcinogenic Risks to Humans, Man- Made Vitreous Fibres;* IARC Monograph: Lyon, France, 2002; Vol. 81.
132. Gordon, G. J.; Jensen, R. V.; Hsiao, L. L.; Gullans, S. R.; Blumenstock, J. E.; Ramaswamy, S.; Richards, W. G.; Sugarbaker, D. J.; Bueno, R. Translation of microarray data into clinically relevant cancer diagnostic tests using gene expression ratios in lung cancer and mesothelioma. *Cancer Res.* 2002, 62, 4963–4967.
133. Arredouani, M. S.; Palecanda, A.; Koziel, H.; Huang, Y. C.; Imrich, A.; Sulahian, T. H.; Ning, Y. Y.; Yang, Z.; Pikkarainen, T.; Sankala, M.; Vargas, S. O.; Takeya, M.; Tryggvason, K.; Kobzik, L. MARCO is the major binding receptor for unopsonized particles and bacteria on human alveolar macrophages. *J. Immunol.* 2005, 175, 6058–6064.
134. Yu, H. B.; Finlay, B. B. The caspase-1 inflammasome: a pilot of innate immune responses. *Cell Host Microbe.* 2008, 4, 198–208.
135. Kim, H. N.; Morimoto, Y.; Tsuda, T.; Ootsuyama, Y.; Hirohashi, M.; Hirano, T.; Tanaka, I.; Lim, Y.; Yun, I. G.; Kasai, H. Changes in DNA 8-hydroxyguanine levels, 8-hydroxyguanine repair activity, and hOGG1 and hMTH1 mRNA expression in human lung alveolar epithelial cells induced by crocidolite asbestos. *Carcinogenesis.* 2001, 22, 265–269.
136. Takeuchi, T.; Morimoto, K. Crocidolite asbestos increased 8-hydroxydeoxyguanosine levels in cellular DNA of a human promyelocytic leukemia cell line, HL60. *Carcinogenesis.* 1994, 15, 635–639.
137. Yamaguchi, R.; Hirano, T.; Ootsuyama, Y.; Asami, S.; Tsurudome, Y.; Fukada, S.; Yamato, H.; Tsuda, T.; Tanaka, I.; Kasai, H. Increased 8-hydroxyguanine in DNA and its repair activity in hamster and rat lung after intratracheal instillation of crocidolite asbestos. *Jpn. J. Cancer Res.* 1999, 90, 505–509.
138. Tashakahi, K.; Pan, G.; Kasai, H.; Hanaoka, T.; Feng, Y.; Liu, N.; Zhang, S.; Xu, Z.; Tsuda, T.; Yamato, H.; Higashi, T.; Okubo, T. Relationship between asbestos exposures and 8-hydroxy deoxyguanosine levels in leukocytic DNA of workers at a Chinese asbestos-material Plant. *Int. J. Occup.Environ. Health.* 1977, 3, 111–119.

139. Marczynski, B.; Czuppon, A. B.; Marek, W.; Reichel, G.; Baur, X. Increased incidence of DNA double-strand breaks and anti-ds DNA antibodies in blood of workers occupationally exposed to asbestos. *Hum. Exp. Toxicol.* 1994, 13, 3–9.

140. Valinluck, V.; Sowers, L. C. Endogenous cytosine damage products alter the site selectivity of human DNA maintenance methyltransferase DNMT1. *Cancer Res.* 2007, 67, 946–950.

141. Haegens, A.; van der Vliet, A; Butnor, K. J.; Heintz, N.; Taatjes, D.; Hemenway, D.; Vacek, P.; Freeman, B. A.; Hazen, S. L.; Brennan, M. L.; Mossman, B. T. Asbestos-induced lung inflammation and epithelial cell proliferation are altered in myeloperoxidase-null mice. *Cancer Res.* 2005, 65, 9670–9677.

142. Christensen, B. C.; Houseman, E. A.; Godleski, J. J.; Marsit, C. J.; Longacker, J. L.; Roelofs, C. R.; Karagas, M. R.; Wrensch, M. R.; Yeh, R. F.; Nelson, H. H.; Wiemels, J. L.; Zheng, S.; Wiencke, J. K.; Bueno, R.; Sugarbaker, D. J.; Kelsey, K. T. Epigenetic profiles distinguish pleural mesothelioma from normal pleura and predict lung asbestos burden and clinical outcome. *Cancer Res.* 2009, 69, 227–234.

143. Huang, S. X.; Partridge, M. A.; Ghandhi, S. A.; Davidson, M. M.; Amundson, S. A.; Hei, T. K. Mitochondria-derived reactive intermediate species mediate asbestos-induced genotoxicity and oxidative stress-responsive signalling pathways. *Environ. Health Perspect.* 2012, 120(6), 840–847.

144. Lohani, M.; Yadav, S.; Schiffmann, D.; Rahman, Q. Diallylsulfide attenuates asbestos-induced genotoxicity. *Toxicol. Lett.* 2003, 143, 45–50.

145. Hei, T. K.; Xu, A.; Huang, S. X.; Zhao, Y. Mechanism of fiber carcinogenesis: from reactive radical species to silencing of the beta igH3 gene. *Inhasl. Toxicol.* 2006, 18, 985–990.

146. Dopp, E.; Yadav, S.; Ansari, F. A.; Bhattacharya, K.; von Recklinghausen, U.; Rauen, U.; Rödelsperger, K.; Shokouhi, B.; Geh, S.; Rahman, Q. ROS-mediated genotoxicity of asbestos-cement in mammalian lung cells in vitro. *Part. Fibre Toxicol.* 2005, 9, 2.

147. Xu, A.; Smilenov, L. B.; He, P.; Masumura, K.; Nohmi, T.; Yu, Z.; Hei, T. K. New insight into intrachromosomal deletions induced by chrysotile in the gpt delta transgenic mutation assay. *Environ. Health Perspect.* 2007, 8(7–92), 115.

148. Nymark, P.; Aavikko, M.; Mäkilä, J.; Ruosaari, S.; Hienonen-Kempas, T.; Wikman, H.; Salmenkivi, K.; Pirinen, R.; Karjalainen, A.; Vanhala, E.; Kuosma, E.; Anttila, S.; Kettunen, E. Accumulation of genomic alterations in 2p16, 9q33.1 and 19p13 in lung tumours of asbestos-exposed patients. *Mol. Oncol.* 2013, 7, 29–40.

149. Borelli, V.; Trevisan, E.; Vita, F.; Bottin, C.; Melato, M.; Rizzardi, C.; Zabucchi, G. J. Peroxidase-like activity of ferruginous bodies isolated by exploiting their magnetic property. *Toxicol. Environ. Health Part A.* 2012, 75, 603–623.

150. Wei, S.; Wang, L. E.; McHugh, M. K.; Han, Y.; Xiong, M.; Amos, C. I.; Spitz, M. R.; Wei, Q. W. Genome-wide gene- environment interaction analysis for asbestos exposure in lung cancer susceptibility. *Carcinogenesis.* 2012, 33(8), 1531–1537.

151. Maeda, M.; Miura, Y.; Nishimura, Y.; Murakami, S.; Hayashi, H.; Kumagai, N.; Hatayama, T.; Katoh, M.; Miyahara, N.; Yamamoto, S.; Fukuoka, K.; Kishimoto, T.; Nakano, T.; Otsuki, T. Immunological changes in mesothelioma patients and their experimental detection. *Clin. Med. Circ. Respir. Pulm. Med.* 2008, 2, 11–17.

152. Allison, A. C.; Harington, J. S.; Badami, D. V. Mineral fibers: chemical, physicochemical, and biological properties. *Adv. Pharmacol. Chemother.* 1975, 12, 291–402.

153. Pooley, F. D. *Evaluation of the Fiber Samples Taken from the Vicinity of Two Villages in Tiekey.* In *Dusts and Disease; Occupational and Environmental Exposure to Selected Particulate and Fibrous Dusts;* Lemen, R., Dement, J., Eds.; Pathotox Publishers: Park Forest South, IL, 1979; pp 41–44.

154. Zoltai, T. Proceedings of Workshop on Asbestos, Definitions and Measurement Methods, National Bureau of Standards Special Publication No 506; Gravatt, C.C., LaFleur, P.D., Heinrich, K. F. G. Ed.; U.S. Government Printing Office, Washington, 1978.

155. Turci, F.; Favero-Longo, S. E.; Tomatis, M.; Martra, G.; Castelli, D.; Piervittori, R.; Fubini, B. A biomimetic approach to the chemical inactivation of chrysotile fibres by lichen metabolites. *Chemistry.* 2007, 13, 4081–4093.

156. Wypych, F.; Adad, L. B.; Mattoso, N.; Marangon, A. A.; Schreiner, W. H. Synthesis and characterization of disordered layered silica obtained by selective leaching of octahedral sheets from chrysotile and phlogopite structures. *J. Colloid Interface Sci.* 2005, 283, 107–112.
157. Daghino, S.; Martino, E.; Vurro, E.; Tomatis, M.; Girlanda, M.; Fubini, B.; Perotto, S. Bioweathering of chrysotile by fungi isolated in ophiolitic sites. *FEMS Microbiol. Lett.* 2008, 285, 242–249.
158. Werner, A. J.; Hochella, M. F.; Guthrie, G. D.; Hardy, J. A.; Aust, A. E.; Rimstidt, J. D. Asbestiform riebeckite (crocidolite) dissolution in the presence of Fe chelators: implications for mineral-induced disease. *Am. Mineral.* 1995, 80, 1093–1103.
159. Lund, L. G.; Aust, A. E. Iron mobilization from asbestos by chelators and ascorbic acid. *Arch. Biochem. Biophys.* 1990, 278, 61–64.
160. Prandi, L.; Tomatis, M.; Penazzi, N.; Fubini, B. Iron cycling mechanisms and related modifications at the asbestos surface. *Ann. Occup. Hyg.* 2002, 46, 140–143.
161. Favero-Longo, S. E.; Girlanda, M.; Honegger, R.; Fubini, B.; Piervittori, R. Interactions of sterile-cultured lichen-forming ascomycetes with asbestos fibres. *Mycol. Res.* 2007, 111, 473–481.
162. Daghino, S.; Martino, E.; Fenoglio, I.; Tomatis, M.; Perotto, S.; Fubini, B. Inorganic materials and living organisms: surface modifications and fungal responses to various asbestos forms. *Chemistry.* 2005, 11, 5611–5618.
163. Chiapello, M.; Daghino, S.; Martino, E.; Perotto, S. Cellular response of Fusarium oxysporum to crocidolite asbestos as revealed by a combined proteomic approach. *J. Proteome Res.* 2010, 9, 3923–3931.
164. Daghino, S.; Turci, F.; Tomatis, M.; Girlanda, M.; Fubini, B.; Perotto, S. Weathering of chrysotile asbestos by the serpentine rock inhabiting fungus *Verticillium leptobactrum*. *FEMS Microbiol. Ecol.* 2009, 69, 132–141.
165. Daghino, S.; Turci, F.; Tomatis, M.; Favier, A.; Perotto, S.; Douki, T.; Fubini, B. Soil fungi reduce the iron content and the DNA damaging effects of asbestos fibers. *Env. Sci. Technol.* 2006, 40, 5793–5798.
166. Martino, E.; Cerminara, S.; Prandi, L.; Fubini, B.; Perotto, S. Physical and biochemical interactions of soil fungi with asbestos fibers. *Environ. Toxicol. Chem.* 2004, 23, 938–944.
167. Fenoglio, I.; Prandi, L.; Tomatis, M.; Fubini, B. Free radical generation in the toxicity of inhaled mineral particles: the role of iron speciation at the surface of asbestos and silica. *Redox Rep.* 2001, 6, 235–241.
168. Martino, E.; Prandi, L.; Fenoglio, I.; Bonfante, P.; Perotto, S.; Fubini, B. Soil fungal hyphae bind and attack asbestos fibers. *Angew. Chem. Int. Ed.* 2003, 42, 219–222.

CHAPTER 13

Eco-Friendly Products as Corrosion Inhibitors for Aluminum–A Review

Rekha N. Nair* and Sharad Bohra

Department of Chemistry, Poornima University, Jaipur, Rajasthan, India
*Email: rekha_124@yahoo.co.in

CONTENTS

ABSTRACT

Metals and their alloys are continuously exposed to acid in the industrial process. Successful enterprises cannot tolerate major corrosion failures and considerable efforts are made in corrosion control at the design stage and in the operational phase. The literature dealing with corrosion of aluminum and possibility of its prevention using eco-friendly inhibitors is examined. The inhibitor effectively secures the metal against corrosion attack. Most inhibitors are organic compounds containing oxygen, nitrogen, and sulfur. These compounds are adsorbed onto the metallic surface blocking the active corrosion sites. This paper discusses the efficiencies of nontoxic eco-friendly inhibitors to reduce the corrosion rate of aluminum.

13.1 INTRODUCTION

Aluminum and its alloys have high technological importance and a wide range of industrial applications. Aluminum alloys are considered as advanced materials in manufacturing of automobile radiators and air conditioners. The use of these materials in lightweight installations is widespread.

It must be noted that various corrosion inhibitors are being used for corrosion inhibition since 19th century. Inhibitors are substances that when added in small concentrations to corrosive media cause decrease in reaction of metal with media or prevent it. Among inhibitors, organic compounds act as good inhibitors due to the presence of hetero atoms like sulfur, nitrogen, and oxygen.[1-7] These atoms coordinate with the corroding metal atom through their electrons. Due to this, protective films are formed on the surface of metal, and thus, corrosion is prevented. But the synthesis of such organic compounds is expensive, and is toxic and hazardous for human beings and the environment. Therefore, the use of chemical inhibitors should be limited and the development of nontoxic, eco-friendly, green inhibitors is regarded as a very important step in the present scenario of environmental threat.

The present paper discusses the efficiencies of nontoxic eco-friendly inhibitors to reduce the corrosion rate of aluminum. The inhibitive property has been found to be very encouraging.

13.2 ECO-FRIENDLY INHIBITORS USED AS CORROSION INHIBITORS FOR ALUMINUM

Several observational studies have been suggested about the applications of eco-friendly inhibitors in corrosive environments. These compounds are nontoxic natural products of plant origin. To investigate the efficiency of these inhibitors, different experimental techniques, such as weight loss method, electrochemical impedance spectroscopy, potentiodynamic polarization, and Fourier transform infrared spectroscopy (FTIR), are being used.

Among acid solutions, hydrochloric acid is one of the most widely used acids. Effects of nontoxic plant materials on the dissolution of aluminum were studied, and the findings are discussed in Table 13.1. From the studies, it is clear that the anticorrosive effect of plant materials varies in their mechanism of action. Natural products as corrosion inhibitors are effective not only in the acid media but also very effective in the basic media too. Because of antiscaling ability, sulfuric acid has been used to clean the metals and its alloys. On the other hand, due to the necessity of corrosion protection of metals in this media, there are many researches in this media also.

TABLE 13.1   Eco-friendly Inhibitors for Corrosion Inhibition of Aluminum and Its Alloys

S No.	Medium	Inhibitor	Methodology	Findings	Ref. No.
1	0.5 M NaOH	*Hibiscus sabdariffa leaves* (AEHSL)	Electrochemical measurements	Mixed-type inhibitor, Langmuir and Dubinin–Radushkevich isotherm	8
2	2 M HCl	*Chromolaena odorata L.* (LECO)	Gasometric and thermometric techniques Temp. 30–60°C	Langmuir adsorption isotherm. Applications in metal surface anodizing and surface coating in industries	9
3	HCl	Ethanolic extract of the leaves of *Ananas sativum*	Weight loss and hydrogen evolution methods	Langmuir adsorption isotherm, activation energies (Ea), activation enthalpy (Δo), and activation entropy (Δo)	10
4	1 M HCl	*Ipomoea involucrata* (IP)	Weight loss technique. Kinetic and thermodynamic techniques. Temp. 30–60°C, KI and KSCN	Langmuir adsorption isotherm	11
5	pH 12	*Hibiscus rosasinensis* (white)	Weight loss method, AC impedance and FTIR methods	Cathodic inhibitor	12
6	0.5 M HCl	*Azadirachia indica* (AZI) plant	Potentiodynamic polarization and impedance techniques	Freundlich adsorption isotherm	13
7	0.5 M NaCl + 2 M sodium hydroxide	Damsissa (*Ambrosia maritime L.*)	Electrochemical techniques, chemical gasometry technique	Mixed-type inhibitor	14
8	0.5 M NaOH and H$_2$SO$_4$	Vigna unguiculata (VU) extract (agricultural waste material)	Weight loss method, electrochemical studies. Temp. 30 and 60°C	Freundlich and Temkin adsorption isotherms, anodic inhibitor	15
9	NaOH	Aqueous extract of garlic	Weight loss method, FTIR	The protective film was analyzed	16

Table 13.1 contd....

S No.	Medium	Inhibitor	Methodology	Findings	Ref. No.
10	HCl	Peepal (*Ficus religeosa*)	Mass loss and thermometric methods	IE dependent upon the concentrations of the inhibitor and the acid	17
11	HCl	*Carica papaya* (CP) and *Azadirachta indica* (AI)	Weight loss, thermometric and hydrogen evolution techniques. Temp. 30–40°C	Freundlich, Temkin and Flory – Huggins adsorption isotherms, Ea, ΔGads and Qads values calculated	18
12	2.0 M HCl	The mucilage extracted from the modified stems of prickly pears (opuntia)	Weight loss, thermometry, hydrogen evolution and polarization techniques	Langmuir adsorption isotherm. Thermodynamic parameters were calculated	19
13	3% NaCl	The third phenolic subfraction of Rosemary leaves extract	Potentiodynamic polarization curves, high-pressure liquid chromatography in the reverse phase (HPLCRP). Temp. 25°C	Cathodic-type corrosion inhibitor, Freundlich adsorption isotherm	20
14	NaOH	*Abrus precatorius*	Weight loss and polarization techniques	Suitable adsorption isotherms were tested graphically	21
15	1 M HCl	Root of ginseng	Weight loss techniques. Temp. 30–60°C.	Freundlich adsorption isotherm, thermodynamic parameters calculated	9
16	0.5 M NaOH and H₂SO₄	*Vigna unguiculata* (VU) extract	Weight loss techniques, electrochemical techniques. Temp. 30 and 60°C.	Freundlich and Temkin adsorption isotherms	22
17	2 M HCl and 2 M KOH	*Sansevieria trifasciata*	Weight loss and hydrogen evolution methods	Physical adsorption, Freundlich isotherm	23
18	2 M NaOH and 0.5 M NaCl	Damsissa	Weight loss, electrochemical method	Mixed-type inhibitor, activation parameters calculated	14
19	5 M HCl	Vanillin	Weight loss measurement, hydrogen evolution method, thermometry and potentiostatic polarization techniques	Langmuir adsorption isotherm	24
20	0.2 N and 0.5 N HCl	*Mangifera indica*	Weight loss method	Langmuir isotherm	1

13.3 RESULTS AND CONCLUSION

The review of literatures has revealed that the use of plant material for inhibition of corrosion of aluminum is an alternate to any other chemicals that harm the environment. Most of the eco-friendly inhibitors have high corrosion inhibition efficiency, but the magnitude of use of extracts of plant materials is more. The reason is that they are rich sources of naturally synthesized chemical compounds that are environmentally acceptable, inexpensive, readily available, and renewable sources of materials, and can also be extracted by simple procedures.[2-7]

The main mechanism of inhibition is found to be adsorption (physical and chemical). Physical adsorption results from electrostatic interaction between the charged centers of molecules and charged metal surface, which results in a dipole interaction of molecules and metal surface. But in chemisorptions, there is adsorption in which the surface atom of a catalyst is bound by electrostatic forces having about the same strength as chemical bonds.[25,26] To detect these mechanisms, different parameters, such as activation energy, Arrhenius plots, surface irregularity, and Langmuir adsorption isotherm, are being studied.

Among corrosion inhibitors, the anticorrosion efficiency of eco-friendly inhibitors is found to be equal to or more than the synthetic inhibitors. Plant extracts can thus be emerged out as corrosion protector for aluminum in coming years that may lower the risk of even environmental pollution.

ACKNOWLEDGMENTS

The author, Rekha N. Nair, expresses her gratitude to her teachers, friends, colleagues, and her family members for their support and guidance and wishes to express sincere thanks to Dr. Alka Sharma, Department of Chemistry for her blessings, incessant support, and encouragement. The author's special thanks are for management of Poornima Foundation for their constant support in exploring new avenues. Rekha N. Nair is thankful to the almighty for giving her so much energy to put in her ideas and knowledge into words in this book. She would also like to thank CRC Press (Taylor & Francis Group) for publishing it in an error-free form.

KEYWORDS

- **Corrosion inhibitors**
- **aluminum**
- **eco-friendly**
- **weight loss**

REFERENCES

1. Nair, R. N.; Kharia, N.; Sharma, I. K.; Verma, P. S.; Sharma, A. Corrosion inhibition study of aluminium in acid media by mango (*Mangifera indica*) leaves as eco-friendly inhibitor. *J. Electro. Chem. Soc. India.* 2007, 56(1/2), 41–47.
2. Umorena, S. A.; Obota, I. B.; Ebensob, E. E.; Obi-Egbedib, N. O. The inhibition of aluminium corrosion in hydrochloric acid solution by exudate gum from *Raphia hookeri. Desalination.* 2009, 247, 561–572.
3. Arora, P.; Kumar, S.; Sharma, M. K.; Mathur, S. P. Corrosion inhibition of aluminium by *Capparis decidua* in acidic media. *Electron. J. Chem.* 2007, 4, 450–456.
4. Nair, R. N.; Sharma, S.; Sharma, I. K.; Verma, P. S.; Sharma, A. Inhibitory efficacy of piper nigrum linn. Extract on corrosion of AA1100 in HCl, RASAYAN. *J. Chem.* 2010, 3(4), 783–795.
5. Sharma, A.; Verma, P. S.; Sharma, I. K.; Nair, R. N. Energy conservation through combating metal-corrosion by means of natural resources. *Proc. World Acad. Sci. Eng. Technol.* 2009, 39, 745–753.
6. Sharma, A.; Nair, R. N.; Sharma, A.; Choudhury, G. Combating aluminium alloy dissolution by employing *Ocimumtenui florum* leaves Extract. *Int. J. Adv. Sci. Tech. Res.* 2012, 6(2), 713–729.
7. Sharma, S.; Pratihar, P. S.; Nair, R. N.; Verma, P. S.; Sharma, A. Influence of ion-additives on inhibitory action of extract of *Trigonella foenum graceums* seeds for AA6063 in Acid Medium. *RASAYAN J. Chem.* 2012, 5, 1.
8. Noor, E. A. Potential of aqueous extract of *Hibiscus sabdariffa* leaves for inhibiting the corrosion of aluminum in alkaline solutions. *J. App. Elec. Chem.* 2009, 39(9), 1465–1475.
9. Obot, I. B.; Obi-Egbedi, N. O. Ginseng root: a new efficient and effective eco-friendly corrosion inhibitor for aluminium alloy of type AA 1060 in hydrochloric acid solution. *J. Elec. Chem. Sci.* 2009, 4(9), 1277–1288.
10. Ating, E. I.; Umoren, S. A.; Udousoro, I. I.; Ebenso, E. E.; Udoh, A. P. Leaves extract of Ananas sativum as green corrosion inhibitor for aluminium in hydrochloric acid solutions. *Green Chem. Lett Rev.* 2010, 3(2), 61–68.
11. Obot, I. B.; Obi-Egbedi, N. O.; Umoren, S. A.; Ebenso, E. E. Synergistic and antagonistic effects of anions and Ipomoea Invulcrata as green corrosion inhibitor for aluminium dissolution in acidic medium. *J. Elec. Chem. Sci.* 2010, 5(7), 994–1007.
12. Rajenderan, S.; Jeyasundari, J.; Usha, P.; Selvi, J. A.; Narayanasamy, B.; Regis, A. P. P.; Renga, P. Corrosion behavior of aluminium in the presence of an aqueous extract of *Hibiscus Rosasinensis. Port. Electrochim. Acta.* 2009, 27(2), 153–164.
13. Arab, S. T.; Al-Turkustani, A. M.; Al-Dhahiri, R. H. Synergistic effect of *Azadirachta Indica* extract and iodide ions on the corrosion inhibition of aluminium in acid media. *J. Kor. Chem. Soc.* 2008, 52(3), 281–294.
14. Abdel-Gaber, A. M.; Khamis, E.; Abo-ElDahab, H.; Adeel, S. H. Inhibition of aluminium corrosion in alkaline solutions using natural compound. *Mater. Chem. Phys.* 2008, 109(2–3), 297–305.
15. Umeron, S. A.; Obet, I. B.; Akpabio, L. E.; Etuk, S. E. Adsorption and corrosive inhibitive properties of *Vigna unguiculata* in alkaline and acidic media. *Pigm. Resin Technol.* 2008, 37(2), 98–105.
16. Priya, S. L.; Chitra, A.; Rajenderan, S.; Anuradha, K. Corrosion behaviour of aluminium in rain water containing garlic extract. *Surf. Eng.* 2005, 21(3), 229–231.
17. Jain, T.; Chowdhary, R.; Arora, P.; Mathur, S. P. Corrosion inhibition of aluminum in hydrochloric acid solutions by Peepal (*Ficus Religeosa*) extracts. *Bull Electrochem.* 2005, 21(1), 23–27.
18. Ebenso, E. E.; Ibok, U. J.; Ekpe, U. J.; Umoren, S.; Jackson, E.; Abiola, O. K.; Oforka, N. C.; Martinez, S. Corrosion inhibition studies of some plant extracts on aluminium in acidic medium. *Trans. SAEST.* 2004, 39(4), 117–123.
19. El-Etre, A. Y. Inhibition of aluminum corrosion using opuntia extract. *Corros. Sci.* 2003, 45(11), 2485–2495.
20. Radosevic, J.; Kliskic, M.; Visekruna, A. Inhibition of corrosion of the Al-2.5Mg Alloy by means of the third acidic phenolic subfraction of aqueous extract of rosemary [Inhibicija korozije Al-2,5Mg slitine

pomoću treće kisele fenolne podfrakcije vodenog ekstrakta ružmarina]. *Kemija U Industriji. J. Chemi. Chem. Eng.* 2001, 50(10), 537–541.

21. Rajalakshmi, R.; Subhashini, S.; Nanthini, M.; Srimathi, M. Inhibiting effect of seed extract of *Abrus precatorius* on corrosion of aluminium in sodium hydroxide. *Oriental J. Chem.* 2009, 25(2), 313–318.
22. Umoren, S. A.; Obot, I. B.; Akpabio, L. E.; Etuk, S. E. Adsorption and corrosive inhibitive properties of *Vigna unguiculata* in alkaline and acidic media. *Pigm. Resin Technol.* 2008, 37(2), 98–105.
23. Oguzie, E. E. Corrosion inhibition of aluminium in acidic and alkaline media by *Sansevieria trifasciata* extract. *Corros. Sci.* 2007, 49, 1527–1539.
24. El-Etre, A. Y. Inhibition of acid corrosion of aluminum using vanillin. *Corros. Sci.* 2001, 43(6), 1031–1039.
25. Roberge, P. R. *Handbook of Corrosion Engineering*; McGraw-Hill: New York, 2000.
26. Raja, P. B.; Sethuraman, M. G. Natural products as corrosion inhibitor for metals in corrosive media – a review. *Mater. Lett.* 2008, 62, 113–116.

CHAPTER 14

Role of Catalyst Particles in the Vertical Alignment of Multiwall Carbon Nanotubes Prepared by Chemical Vapor Deposition

Ved Prakash Arya[1*], V. Prasad[1,2], and P. S. Anil Kumar[1,2]

[1]DESM, Regional Institute of Education, Ajmer, India
[2]Department of Physics, Indian Institute of Science, Bangalore, India
*Email: aryavedp@gmail.com

CONTENTS

ABSTRACT

Aligned multiwall carbon nanotubes were grown on thermally oxidized silicon, quartz, and sapphire substrates without a predeposition of a catalyst. They were grown by chemical vapor deposition at 980?C with benzene as precursor and ferrocene as catalyst and had a length of several tens of microns. It was found that the order in which the precursor and catalyst were introduced during chemical vapor deposition determines the orientation of the nanotubes. Surface elemental analysis shows that the presence of catalyst particles on the substrate is essential for their vertical alignment. Transmission electron microscopy shows that the iron particles are embedded inside the nanotubes.

14.1 INTRODUCTION

Carbon nanotubes (CNTs) have drawn considerable attention due to their unique physical, chemical, and electronic properties.[1-5] To exploit these properties for the realization of nanodevices,[6-9] the alignment of CNTs on substrate materials is an essential requirement. Vertically aligned CNTs can be used for the field electron emitters,[10] as chemical[11] and biological sensors,[12] in power applications,[13] in spin transport studies,[14] etc. The CNTs synthesized by methods using laser evaporation and arc discharge result in randomly oriented CNTs.[15,16] Chemical vapor deposition (CVD)[11,17] has become an important technique for the synthesis of CNTs because it is cheap and gives aligned growth on large areas. The diameter and density of the CNTs can also be tuned in this method. Here, we report a simple thermal-assisted CVD to grow the vertically aligned CNTs on substrates such as quartz, sapphire, and thermally oxidized silicon. The grown CNTs are multiwalled in nature. The vertical alignment of CNTs is associated with the presence of iron particles on the surface of the substrates during the growth. Moreover, when the precursor and the catalyst are introduced simultaneously, we obtain well-aligned tubes on many substrates, whereas when the precursor is introduced first followed by the catalyst, then a misaligned growth takes place.

14.2 EXPERIMENTAL

Sapphire, quartz, and thermally oxidized silicon substrates were used for the growth of CNTs. Each of these substrates was ultrasonically cleaned in isopropyl alcohol, acetone, and deionized water consecutively for 20 min each. After ultrasonication, the substrates were transferred in the CNT formation region of a quartz tube. Here, benzene was used as a source of carbon and ferrocene acted as a catalyst material. The precursor and the catalyst material were taken in the closed end of the quartz tube that was placed inside a horizontal single-zone furnace. The open end of the tube was connected to empty rubber bladder to collect the evolved gases. Argon gas atmosphere was created in the reaction tube when the set temperature was reached. After maintaining the temperature at 980?C for 2 h, it was cooled down to room temperature under ambient conditions.

The morphology and the alignment of the CNTs were observed by scanning electron microscopy (SEM) Quanta 200. The surface elemental analysis was carried out using energy dispersive X-ray (EDX) analysis facility attached with Quanta 200. Acceleration voltage of 25 kV was used for acquiring EDX spectra. High-resolution transmission electron microscopy (HRTEM) FEI Technai F30 operating at 300 kV was used to investigate the structure and crystallinity of the CNTs. The overall crystallinity was characterized by micro-Raman spectroscopy, where 514.5 nm Ar ion laser line was used for excitation.

14.3 RESULTS AND DISCUSSION

14.3.1 SEM Analysis

Figure 14.1(a) and (b) shows the SEM images of the CNTs grown over the sapphire substrate. Very good vertical alignment can be seen here. In order to see the vertical alignment of CNTs, the trenches are made on the deposited film using a sharp tweezer. Some of the amorphous carbon formation can also be seen in these images. The length of the aligned CNTs is ?30 ?m, which is uniform over the whole surface.

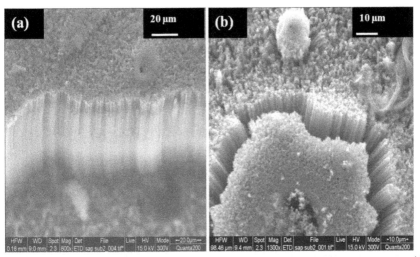

FIGURE 14.1 SEM Images of Vertically Aligned CNTs Grown Over Sapphire Substrate in (a) One-End Region and (b) A Trenched Region.

14.3.2 Raman Study

Figure 14.2 shows the Raman spectra of the CNTs grown over sapphire substrate. Raman spectroscopy is a very useful tool to characterize the CNTs,[18,19] as it tells the overall crystallinity as well as the type of CNTs. This spectrum shows two peaks of the graphitic structure at 1354 and 1579 cm^{-1}. The strong G-line peak at 1579 cm^{-1} is the indication of high-crystalline graphene layers, while the D-line peak at 1354 cm^{-1} tells about the existence of defective graphene layers such as amorphous carbon layers. If the graphene layers had waving structure or buckled structure, the intensity of G-line would become weak. From the prominent G-peak in Figure 14.2, we infer that the overall crystallinity of the CNTs prepared on the sapphire substrate is good, though a partly defective graphite structure is also observed.

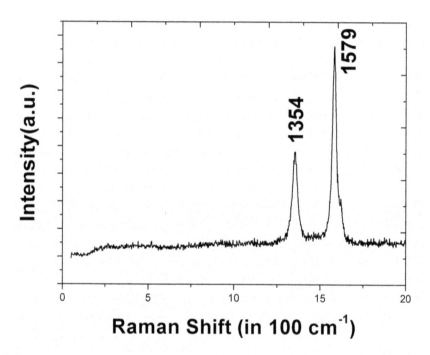

FIGURE 14.2   Raman Spectra of the CNTs Deposited on Sapphire Substrate.

14.3.3   TEM Study

In order to investigate the structural aspects, we have undertaken a TEM study. For this study, some amount of the deposited material from the sapphire substrate was mixed with isopropyl alcohol by ultrasonication for 30 min to get a suspension of the CNTs. One drop from this suspension was dropped on a carbon grid, and the TEM images were taken, which are shown in Figure 14.3(a)–(d). Figure 14.3(a) shows the TEM image of CNTs at low magnification, and it can be seen that the CNTs are having varying range of diameters. The diameter of CNTs is considered to be determined by the catalyst particle size,[20] which means that in the present case also, there is a distribution window of catalyst (Fe) cluster size. Multiwall nature of these CNTs can be seen clearly in the HRTEM image (Fig. 14.3(b)). The HRTEM image in Figure 14.3(c) shows the embedded iron nanocluster inside the CNT. The inset in Figure 14.3(c) is the corresponding fast Fourier transform showing diffused rings for graphitic carbon layers and the diffraction spots for crystalline iron phase. Area shown by a rectangle in Figure 14.3(c) is magnified in Figure 14.3(d), where two kinds of lattice fringes can be seen. The distance between two consecutive lattice fringes in the upper portion of the image is about 0.34 nm corresponding to the graphite (002) planes, whereas the lattice fringe spacing of about 0.2 nm in the lower portion is due to the (110) plane of the bcc-Fe crystal. These two regions are marked as CNT and Fe in Figure 14.3(d). The presence of bcc ?-Fe is observed in the X-ray diffraction pattern also. When catalyst (Fe cluster) gets chemically activated at elevated temperatures, the carbon atoms form a layer on it and the process continues. The growth of the carbon layers decreases the reactivity of the catalyst. Finally, when the catalyst is fully deactivated, further growth is not possible. Phase changes such as carbide formation at catalyst–carbon interface may also contribute to encapsulation as suggested by Pinheiro et al.[21,22]

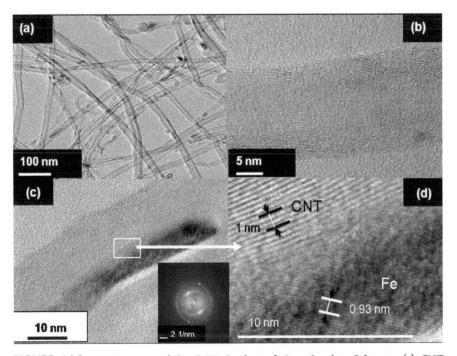

FIGURE 14.3 TEM Images of the CNTs Synthesized Over Sapphire Substrate: (a) CNT Network Having Diameter 10–25 nm, (b) Single MWNT of ?25 nm Diameter Having Many Walls, (c) Embedded Iron Nanocluster in One of the MWCNT (the Inset Is the Fast Fourier Transform), and (d) High Magnification Image of the Area Shown by the Rectangle in (c).

14.3.4 Alignment of CNTs on Substrates

Figures 14.4 and 14.5 show the alignment of CNTs on the quartz and the thermally oxidized silicon substrates. It can be seen that the alignment of CNTs depends on the type of substrates also. It may be due to different adhesion energy of Fe particles on the substrate surfaces and/or different surface diffusion rates of Fe clusters on these substrates. It seems that the vertical alignment is the simplest case if catalyst particles are available all over the substrate. Van der Waals interaction between neighboring CNTs makes them grow straight in the vertical direction to form vertical arrays of CNTs.

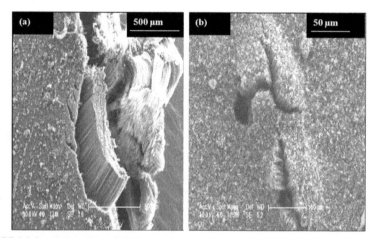

FIGURE 14.4 SEM Images of CNTs Grown Over Quartz Substrate Showing Vertical Alignment in (a) One-End Region and (b) a Trenched Region.

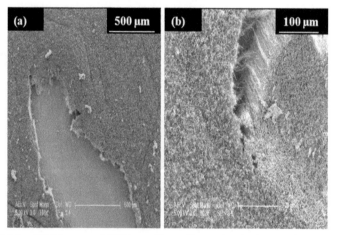

FIGURE 14.5 SEM Images of CNTs on Thermally Oxidized Silicon Substrate Showing (a) Growth Perpendicular to the Surface and (b) Alignment in One of the Trenched Region.

There are various reports on the vertical alignment of CNTs on catalyst predeposited substrates.[23,24] Shin et al.[25] have studied the influence of morphology of Ni catalyst film on vertically aligned CNT growth. They found that the optimum grain size for the vertical alignment of CNTs is 30–60 nm. Kim et al.[26] have examined the dependence of the vertically aligned growth of CNTs on the type of catalysts such as Fe, Co, and Ni nanoparticles. They found that Fe catalyst gave two times higher growth rate than Co and Ni catalysts over the temperature range of 900–1000?C and the growth rate increased by a factor of two for all catalysts, while increasing the temperature from 900 to 1000?C. Report from Ward et al.[27] showed that the patterned growth was strongly influenced by the surface states of substrates such as different materials, roughness, crystallinity, and porosity.

Metal-catalytic growth of CNTs is widely believed to proceed via solvation of carbon vapor into metal clusters, followed by precipitation of excess carbon in the form of nanotubes. Growth of the CNTs by means of the Fe cluster is the result of the surface energy minimization. Graphene formed during the decomposition of the precursors chemically adsorbs to the Fe cluster, reducing its surface energy. Maiti et al.[28] proposed that when metal particles are of the same size or even smaller than the tube diameter, tube closure is prevented by movement of metal particles with the growing tip. This happens due to reactive dangling bonds at the tube tip, which are stabilized by the metal particles, and act as attraction sites for carbon adatoms.

14.3.5   EDX Analysis

In the literature, mostly it is found that the vertically aligned CNTs are grown on predeposited catalyst particles. We have not deposited any catalyst particles prior to CNT deposition and still obtained a good vertical alignment. To check whether it was due to the presence of Fe particles on the surface of the substrates, CNT films were peeled off from the surface of the substrates, and EDX was taken for the identification of elemental composition. After the confirmation of the presence of the Fe on the substrate surface, we tried to investigate the role of the catalyst presence on the substrate for the vertical alignment using two methods of preparation of CNTs. In the first method M_1, both precursors, benzene and ferrocene, are mixed together, in which catalyst particles are readily available in the reaction tube for the growth of CNTs. In the second method M_2, benzene is decomposed first to form a carbon deposit and then ferrocene vapors, along with the precursor, are allowed to enter on the substrate area. CNTs are deposited on the substrate in both cases, but vertical alignment is found in the first case only. Figure 14.6 shows SEM images of the CNTs grown on the sapphire and the quartz substrates by the method M_2. Here, vertical alignment did not take place, but only some random alignment was seen. The EDX spectra (not shown here) of the film surfaces of Figures 14.1 and 14.4 show the presence of carbon and iron, apart from aluminum and silicon, which comes from the contribution of the sapphire and the quartz substrates. Iron present in these samples is in the embedded form and can be seen in the TEM image of Figure 14.3. To identify the elements present on the surface of the quartz substrates (Figs 14.4 and 14.6(b)), CNT film was peeled off from the substrates, and subsequently, EDX spectra at 25 kV were taken from this peeled-off region, which is shown in Figure 14.7.

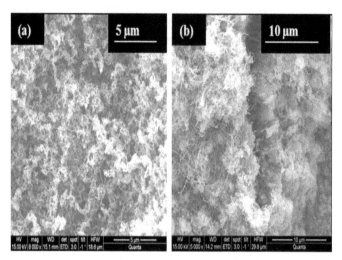

FIGURE 14.6 SEM Images of the CNTs Grown (by M_2 Method) Over (a) Sapphire Substrate and (b) Quartz Substrate.

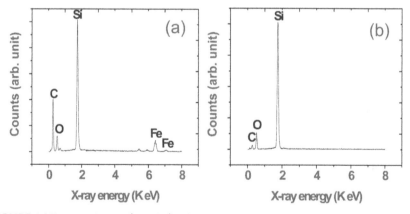

FIGURE 14.7 EDX Spectra (at 25 kV) Taken From the Peeled-Off Portion of Quartz Substrate Sample Grown by (a) M_1 Method (Fig. 14.4) and (b) M_2 Method (Fig. 14.6(b)).

It is clear from these spectra that apart from the silicon and the oxygen peaks arising from the quartz substrate, iron (4.98 wt%) is also present when both ferrocene and benzene vapors are simultaneously available for the CNT growth (Fig. 14.7(a)). On the other hand, iron is not detected when ferrocene vapors are not available initially for the CNT growth (Fig. 14.7(b)). The same trend is observed for other two substrates. It can be seen from Figure 14.6(a) and (b) that the alignment of CNT is not vertical, while a good vertical alignment can be seen in Figures 14.1 and 14.4. It is inferred from the above EDX spectra that the availability of iron particles on the surface of the substrates acts as a catalyst for the CNT deposition, which continues to grow in the vertical direction because of densely available iron particles on the whole substrate. In the second case, when ferrocene vapors are inserted after initial benzene decomposition, carbon particles deposit first on the substrate and then iron vapors enter on the substrate, so vertical alignment is lost, and CNTs are deposited randomly. So, the availability of the catalyst particles in the beginning of the growth is important for the vertical alignment.

14.4 CONCLUSION

We have synthesized vertically aligned CNTs on sapphire, quartz, and thermally oxidized silicon substrates by simple pyrolysis method without any predeposition of the catalyst particles. The grown multiwall CNTs are having well-graphitized layers. We suggest that our method is very simple, economic, and scalable for the realization of vertical alignment of CNTs on various substrates. Further, it is observed that the availability of iron particles in the beginning of the growth of CNTs on the sapphire and quartz substrate gives rise to vertical alignment. We believe that this may be true for other substrates and transition metal catalysts as well, but further investigations are required to confirm this.

ACKNOWLEDGMENTS

The authors are thankful to Prof. S. Umapathy for the Raman spectroscopic studies. Bhawana Singh and Rishikesh Pandey are acknowledged for their help during the experimental work. The authors gratefully thank Sunita for manuscript editing. INI at IISc is acknowledged for SEM and TEM studies. VPA is grateful to UGC, New Delhi, for the financial support.

KEYWORDS

- **Carbon nanotube**
- **chemical vapor deposition**
- **catalyst**
- **SEM**
- **HRTEM**
- **Raman spectroscopy**

REFERENCES

1. Lu, J. P. Elastic properties of carbon nanotubes and nanoropes. *Phys. Rev. Lett.* 1997, 79, 1297.
2. Treacy, M. M. J.; Ebbesen, T. W.; Gibson, J. M. Exceptionally high Young's modulus observed for individual carbon nanotubes. *Nature.* 1996, 381, 678.
3. Kociak, M.; Yu Kasumov, A.; Gueron, S.; et al. Superconductivity in Ropes of Single-Walled Carbon Nanotubes. *Phys. Rev. Lett.* 2001, 86, 2416.
4. Arya, V. P.; Prasad, V.; Anil Kumar, P. S. Magnetic properties of iron particles embedded in multiwall carbon nanotubes. *J. NanoSci. NanoTech.* 2009, 9(9), 5406.
5. Song, S. N.; Wang, X. K.; Chang, R. P. H.; et al. Electronic properties of graphite nanotubules from galvanomagnetic effects. *Phys. Rev. Lett.* 1994, 72, 697.
6. Wong, E. W.; Sheehan, P. E.; Lieber, C. M. Nanobeam mechanics: elasticity, strength, and toughness of nanorods and nanotubes. *Science.* 1997, 277, 1971.
7. Falvo, M. R.; Clary, G. J.; Taylor, R. M.; et al. Bending and buckling of carbon nanotubes under large strain. *Nature.* 1997, 389, 582.
8. Frank, S.; Poncharal, P.; Wang, Z. L.; et al. Carbon nanotube quantum resistors. *Science.* 1998, 280, 1744.
9. Batchtold, A.; Strunk, C.; Salvetat, J.-P.; et al. Aharonov–Bohm oscillations in carbon nanotubes. *Nature.* 1999, 397, 673.
10. Milne, W. I.; Teo, K. B. K.; Chhowalla, M.; et al. Electron emission from arrays of carbon nanotubes/fibres. *Curr. Appl. Phys.* 2002, 2, 509.
11. Kong, J.; Franklin, N. R.; Zhou, C.; et al. Nanotube molecular wires as chemical sensors. *Science.* 2000, 287.
12. Heller, D. A.; Jeng, E. S.; Yeung, T.-K.; et al. Optical detection of DNA conformational polymorphism on single-walled carbon nanotubes. *Science.* 2006, 311, 508.
13. Raffaelle, R. P.; Landi, B. J.; Harris, J. D.; et al. Carbon nanotubes for power applications. *Mater. Sci. Eng. B.* 2005, 116, 233.
14. Tsukagoshi, K.; Alphenaar, B. W.; Wagner, M. Spin transport in nanotubes. *Mater. Sci. Eng. B.* 2001, 84, 26.
15. Thess, A.; Lee, R.; Nikolaev, P.; et al. Crystalline ropes of metallic carbon nanotubes. *Science.* 1996, 273, 483.
16. Bethune, D. S.; Johnson, R. D.; Salem, J. R.; et al. Atoms in carbon cages: the structure and properties of endohedral fullerenes. *Nature.* 1993, 366, 123.
17. Sen, R.; Govindaraj, A.; Rao, C. N. R. Carbon nanotubes by the metallocene route. *Chem. Phys. Lett.* 1997, 267, 276.
18. Eklund, P. C.; Holden, J. M.; Jishi, R. A. Vibrational modes of *carbon* nanotubes; spectroscopy and theory. *Carbon.* 1995, 33, 959.
19. Sveningsson, M.; Morjan, R. E.; Nerushev, O. A.; et al. Properties of CVD-grown carbon nanotube films. *Appl. Phys. A.* 2001, 73, 409.
20. Helveg, S.; Lopez-cartes, C.; Sehested, J.; et al. Atomic-scale imaging of carbon nanofibre growth. *Nature.* 2004, 427, 426.
21. Pinheiro, J. P.; Gadelle, P. Chemical state of a supported iron-cobalt catalyst during CO disproportionation I. Thermodynamic study. *J. Phys. Chem. Solids.* 2001, 62, 1015.
22. Pinheiro, J. P.; Gadelle, P.; Jeandey, C.; et al. Chemical state of a supported iron-cobalt catalyst during CO disproportionation II. Experimental study. *J. Phys. Chem. Solids.* 2001, 62, 1023.
23. Rizzo, A.; Rossi, R.; Signore, M. A.; et al. Effect of Fe catalyst thickness and C_2H_2/H_2 flow rate ratio on the vertical alignment of carbon nanotubes grown by chemical vapour deposition. *Diamond Relat. Mater.* 2008, 17, 1502.
24. Iwasaki, T.; Mejima, S.; Koide, T.; et al. Vertically aligned carbon nanotube growth from Ni nanoparticles. *Diamond Relat. Mater.* 2008, 17, 1443.
25. Shin, Y. M.; Jeong, S. Y.; Jeong, H. J.; et al. Influence of morphology of catalyst thin film on vertically aligned carbon nanotube growth. *J. Crystal Growth.* 2004, 271, 81.

26. Kim, N. S.; Lee, Y. T.; Park, J.; et al. Dependence of vertically aligned growth of carbon nanotubes on catalyst. *J. Phys. Chem. B.* 2002, 106, 9286.
27. Ward, J. W.; Wei, B. Q.; Ajayan, P. M. Substrate effects on the growth of carbon nanotubes by thermal decomposition of methane. *Chem. Phys. Lett.* 2003, 376, 717.
28. Maiti, A.; Brabec, C. J.; Bernholc, J. Kinetics of metal-catalyzed growth of single-walled carbon nanotubes. *Phys. Rev. B.* 1997, 55, R6097.

CHAPTER 15

Electrochemical Process: Review on Research Applications in Machining of Advanced Materials

A. Pandey

Department of Mechanical Engineering, School of Engineering and Technology, Manipal University, Jaipur, Rajasthan, India
Email: anand.pandey@jaipur.manipal.edu

CONTENTS

ABSTRACT

The mechanism of electrochemical process in machining has tremendous applications due to its promising commercial utilization in the manufacturing sector. The process is widely applicable for machining of complicated shapes on difficult-to-machine metals in micron dimensions. The process in commercial terms is referred as electrochemical machining (ECM). ECM is an anodic dissolution method through which workpiece material and tool electrode material are connected as anode and cathode. The process is capable of performing micromachining of harder materials, namely, geometrical features such as groove, microholes, slits, and complex shapes. In the present paper, a review of basic principle working, process capability, and recent applications based on past research works have been discussed.

15.1 INTRODUCTION

Electrochemical machining (ECM) process is one of the noncontact advanced machining processes being widely used to cut difficult-to-machine engineering metals, with a restriction of the workpiece material to be electrically conductive.[1] The cutting process in ECM is possible due to the flow of suitable electrolyte, namely, NaCl, NaClO$_3$, and NaNO$_3$ within the machining zone, which removes the metal ions away from the workpiece material (anode) and tool (cathode). The cavity produced after the machining is the replica of the tool shape in ECM.[2,3] The tool shape is often made of copper, brass, and stainless steel. The electrolytes are pumped at very high rate through the passages within the machining gap. The power supply (DC) in the range of 5–24 V maintains the current density for metal removal.[3–5]

The applications of ECM process make a wide utility in manufacturing of turbine blades, jet-engine parts, engine castings, nozzles, and intricate geometrical features, namely, slit, groove, and holes.[3,6] The process leaves a burr-free machined surface, reducing thermal damages to the workpiece materials of hard materials. The principle mechanism working of ECM can be defined based on the principle of Faradays and Ohm.[7,8] The theory can be described that an electrolyte cell is formed by the anode (workpiece material) and the cathode (tool) in the midst of a following electrolyte. The metal removal takes place by the controlled dissolution of the anode, accrued to the well-known Faradays law of electrolysis. When the electrode is connected to about 20 V electric supply source, flow of current in the electrolyte is established due to positively charged ion being attracted toward cathode.[8]

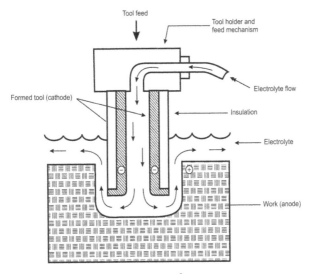

FIGURE 15.1 Schematic Diagram of ECM Process.[2]

Due to electrolysis process at cathode, hydroxyl ions are released, which combine with the metallic ions of anode to form insoluble metal hydroxide. The unwanted metal is thus removed in the form of sludge and precipitated in electrolytic cell.[8,9]

15.2 PROCESS CAPABILITY

ECM process is capable of accomplishing the machining of a wide variety of conductive difficult-to-machine metals,[10] semiconductors, and metal-based composites.[11,12] Geometrical features of miniaturized dimensions can be cut using ECM. Surface finish values of the order of 100 nm can be achieved using ECM-finishing methods.[3,13]

The mechanical components fabricated using ECM are nowadays being widely used CNC systems so that production rate can be increased with high flexibility and parts of close tolerances. In ECM process, the most important parameters, which play an important role for precision machining of harder materials, are described in Section 15.3.

15.3 PROCESS PARAMETERS

15.3.1 Electrolytes

In ECM process, the selection of electrolytes largely depends on the physical, chemical, and mechanical properties of the workpiece material to be machined.[3] The electrolytes selected for machining of metals can be acidic, basic, and neutral aqueous solutions.[14]

15.3.2 Tools

In ECM process, tools are characterized as shaped and nonshaped tools.[3] The basic important properties of ECM tool are good electrical, thermal conductivity, stiffness, and corrosion resistance. Normally, aluminum, copper, brass, and stainless steel are the recommended tool materials.

15.4   VARIANTS PROCESSES

- The variants of ECM process are as follows:[8]
- Electrochemical drilling
- Electrochemical grinding
- Jet electrochemical machining
- Wire electrochemical machining
- Electrochemical discharge machining
- Electrochemical honing
- Ultrasonic-assisted ECM
- Laser-assisted ECM

15.5   PAST WORK

In the present section, the literature review has been subdivided into three major research areas, namely, research based on productivity, research based on tool shape design, and research based on micro-ECM process.

15.5.1   Research Based on Productivity

Researchers reported out some practical applications of ECM micromachining, which appears to be one of the promising and imminent ?-machining method, due to its large number of applications in industries. A ?-tool vibration system was developed, consisting of tool-holding and microtool vibrating units. The author have made best efforts to improve performances, namely, material removal rate (MRR) and dimensional. Holes of smaller sizes were produced on thin copper workpiece materials by ECM with stainless-steel tool.[15]

Some investigations were carried out to study the effects of ECM process parameters on performances, namely, metal removal rate, surface roughness, and overcut. Parameters such as feed rate, electrolyte, flow rate of the electrolyte, and voltage were varied by using sodium chloride and sodium nitrate as electrolytes. The author concluded that feed rate is the most important parameter that affects the MRR.[16]

The author converses the role of NaCl-electrolyte in current-carrying processes in ECM of iron workpiece material. During the investigation, it was revealed that over-voltage calculated with respect to gap and penetration rate indicated that a narrow range of equilibrium gap and penetration rate was permissible.[17] Some works were conferred about the main benefits of the ECM, such as high stock removal rates and damage-free machined surface, which being repeatedly offset by the deprived dimensional control. The state space methodology was applied to transform it into the control model relevant to an ECM control system established on a digital computer. The simulations have been made for the model verification and controller design.[18]

Some investigations were carried out on machining of Al workpiece material using ECM. Sodium chloride was used as an electrolyte at different current densities and was compared with the theoretical values for MRR. The researchers identified that the resistance of the NaCl-electrolyte solution decreased, with increasing current densities.[19] It was conveyed that ECM process has been successfully practical for cutting of quartz through wedge edged tool. Tool has been connected to cathode terminal, whereas workpiece material has been used as a tool. During the process, researchers have cut deep crater on the anode (as a tool) and workpiece interface due to the occurrence of chemical reaction.[20]

An attempt was made to construct a thermal model for the calculation of metal removal rate during ECSM. Temperature distribution within zone of influence of single spark was obtained with the application of finite element method.[21] It has been practically conversed about the MRR of Al workpiece material by ECM. Sodium chloride was chosen as an electrolyte for different current densities. It has been concluded by the author that over-voltage of the system initially increased and attained a saturation value with increased current.[22]

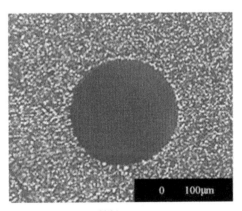

FIGURE 15.2 Hole Making Using ECM.[23,24]

FIGURE 15.3 Microgroove Making Using ECM.[25]

FIGURE 15.4 Microslits Making Using wire ECM.[26]

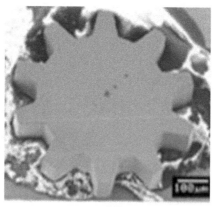

FIGURE 15.5 Gear Making Using Wire ECM.[3,27]

Scientists have suggested that the electrochemical spark machining process has been advantageous for cutting of high-strength metals. The spark energy and the approximate order of hydrogen gas bubble diameter were computed by the proposed valve theory.[28] Rao et al. attempted to investigate the effects of parameters of electrochemical processes, namely, tool feed rate, electrolyte flow velocity, and voltage. Responses such as dimensional accuracy, tool life, MRR, and machining cost were analyzed. The objectives considered are dimensional accuracy, tool life, and the MRR.[29]

Results were shown by emphasizing on those structures of the expansion of mathematical model, which correlates the collaborative and higher-order impacts of various machining parameters, namely, MRR and the overcut phenomena, through response surface methodology. They highlight the mathematical models for analyzing the properties of various parameters on the cutting rate and overcut. The parameters can be used in order to accomplish maximization of the MRR and to minimize the overcut effects of shape features.[30]

15.5.2 Research Based on Tool Shape Design

Some results were deliberated about the actions for cathode as tool design during the ECM. The author developed and tested a new approach that overcomes these difficulties by retaining a finite element method within an optimization formulation. Thus, a least-squares minimization of the deviation of the simulated anode workpiece material shape from that desired has been done.[31]

Some results on the research basis were conferred about the theoretical and experimental investigations, namely, the characteristic shape measurements of anode workpiece surface by the microfeatures of the cathode-tool electrode. This investigation included the study of electrochemical copying of slots, mini-holes, and insulating groove. The author concluded that micromachining capabilities of ECM processes have been increased, and the application of ultra-short pulse current and ultra-small gap size is recommended.[32]

Researchers discussed about the study of electrolyte flow during the ECM process. Researchers attempt to identify the reasons, namely, insulation requirements and machined face considerations that could relate to other ECM components. Researchers observed that by adapting new electrodes for a casting gate increases the metal removal process.[33]

Some investigations were highlighted on accurate prediction of electrode shape for ECM. The author suggests a method using finite element method to design tool. This technique is proficient of designing three-dimensional freeform surface tool from the scanned data of work material. This results in high computing effectiveness, better accuracy, and supple boundary treatment for iterative procedure.[34]

15.5.3   Research Based on μ-ECM

Researchers conferred about the ECM processes afford for drilling miniaturized holes with extremely smooth surface. Applications of current methods include aerospace, electronics, and micromechanics productions. Researchers highlight the expansions, new trends, and the effect of process parameters in influencing the eminence of the holes produced.[23]

Some studies on micro-ECM use ultra-short pulses; 0.1 M sulfuric acid was used as electrolyte. This has been found that to improve productivity, multiple electrodes were applied and multiple structures were machined simultaneously.[35] A practical application of ECM process using low-frequency vibration to the tool electrode and the effect of input parameters and machining conditions on the effectiveness of tool vibration during ECM were analyzed.[36]

The author has described about the application of ultra-short voltage pulses electrochemical reactions, which can be used for nanometer accuracy, and allows for high precision machining of electrochemical active materials. The average potentials of tool electrode and workpiece material, corrosion of the workpiece, and position of the counter reaction of workpiece material dissolution were controlled during the investigation.[37]

It was discussed about the rare application of ECM in micromachining because the electric field is not localized. In this work, ultra-short pulses with tens of nanosecond duration are used to localize dissolution area. The effects of voltage, pulse duration, and pulse frequency on the localization distance were studied. High-quality microhole with 8 ?m diameter was drilled on 304 stainless steel foil with 20 ?m thickness used as workpiece.[38]

15.6   CONCLUSION

ECM process and its variant methods have resulted as the most cost-effective and accurate machining processes in recent days. The ability to machine advanced materials has brought tremendous improvements in surface texture of produced geometrical features, namely, slot, groove, and hole.

ACKNOWLEDGEMENT

The author highly acknowledge to Dr. TanmoyChakraborty and Dr.LalitaLedwani fortheir valuable suggestions and guidance.

REFERENCES

1. McGeough, J. A. *Principles of Electro-Chemical Machining*; Chapman and Hall: London, 1974.
2. Rajurkar, K. P.; Kozak, J.; Wei, B.; McGeough, J. A. Study of pulse electrochemical machining characteristics. *CIRP Ann. Manuf. Technol.* 1993, 42, 231–234.
3. Rajurkar, K. P.; Sundaramb, M. M.; Malshec, A. P. Review of electro chemical and electro discharge machining. *Proc. CIRP.* 2013, 6, 13–26.
4. Kozak, J.; Rajurkar, K. P.; Makkar, Y. Study of pulse electro-chemical micromachining. *J. Manuf. Process.* 2004, 6, 7–14.
5. Schuster, R.; Kirchner, V.; Allongue, P.; Ertl, G. Electrochemical micro-machining. *Science.* 2000, 289, 98–100.
6. Shan, H. S. *Advanced Manufacturing Methods*; Tata McGraw-Hill Publishing. Co.: New Delhi, 2004.
7. Kozak, J.; Rajurkar, K. P. Hybrid Machining Process Evaluation and Development. Second International Conference on Machining and Measurements of Sculptured Surfaces, Krakow, **2000**, pp 501–536.
8. El-Hofy, H. *Advanced Machining Processes*; Mc Graw-Hill: New Delhi, 2005.
9. Rumyantsev, E.; Davydov, A. *Electro-Chemical Machining of Metals*; MIR Publishers: Moscow, 1984.
10. Krauss, W. N.; Holstein, N.; Konys, J. Advanced electrochemical processing of tungsten components for He-cooled divertor application. *Fusion Eng. Design.* 2010, 85(10–12), 2257–2262.
11. Bassu, M. Electro-chemical micromachining as anenabling technology for advanced silicon microstructuring. *Adv. Funct. Mater.* 2012, 22(6), 1222–1228.
12. Senthilkumar, C. Modelling and analysis of electrochemical machining of cast Al/20%SiCp composites. *Mater. Sci. Technol.* 2010, 26(3), 289–296.
13. Mahdavinejad, R.; Hatami, M. On the application of electrochemical machining for inner surface polishing of gun barrel chamber. *J. Mater. Process. Technol.* 2008, 202(1–3), 307–315.
14. Neergat, M.; Weisbrod, K. R. Electro-dissolution of 304 stainless steel in neutral electrolytes for surface decontamination applications. *Corros. Sci.* 2011, 53(12), 3983–3990.
15. Bhattacharyya, B.; Malapati, M.; Munda, J.; Sarkar, A. Influence of tool vibration on machining performance in electrochemical micro-machining of copper. *Int. J. Mach. Tool Manuf.* 2007, 47, 335–342.
16. da Silva Neto, J. C.; da Silva, E. M.; da Silva, M. B. Intervening variables in electrochemical machining. *J. Mater. Process. Technol.* 2006, 179, 92–96.
17. Mukherjee, S. K.; Kumar, S.; Srivastava, P. K. Effect of electrolyte on the current- carrying process in electrochemical machining. *Proc. I Mech. C J. Mech. Eng. Sci.* 2007, 221.
18. Rajurkar, K. P.; Wei, B.; Schnacker, C. L. Monitoring and control of electrochemical machining (ECM). *J. Eng. Ind.* 1993, 115, 217–267.
19. Mukherjee, S. K.; Kumar, S.; Srivastava, P. K. Effect of valance on material removal rate in electrochemical machining of aluminum. *J. Mater. Process. Technol.* 2008, 202, 398–401.
20. Jain, V. K.; Adhikary, S. On the mechanism of material removal in electrochemical spark machining of quartz under different polarity conditions. *J. Mater. Process. Technol.* 2008, 200, 460–470.
21. Bhondwe, K. L.; Yadava, V.; Kathiresan, G. Finite element prediction of material removal rate due to electro-chemical sparks machining. *Int. J. Mach. Tool Manuf.* 2006, 46, 1699–1706.
22. Mukherjee, S. K.; Kumar, S.; Srivastava, P. K.; Kumar, A. Effect of valence on material removal rate in electrochemical machining of aluminum. *J. Mater. Process. Technol.* 2008, 202, 398–401.
23. Sen, M.; Shan, H. S. A review of electrochemical macro- to micro-hole drilling processes. *Int. J. Mach. Tool Manuf.* 2005, 45, 137–152.
24. Yong, L.; Yunfei, Z.; Guang, Y.; Liangqiang, P. Localized electro-chemical micromachining with gap control. *Sens. Actuators.* 2003, 108, 144–148.
25. Wang, K. Electrochemical micro machining using vibratile tungsten wire for high-aspect-ratio micro structures. *Surf. Eng. Appl. Electrochem.* 2010, 46, 395–399.
26. Qu, N. S.; Xu, K.; Zeng, Y. B.; Qia, Y. Enhancement of the homogeneity of micro-slits prepared by wire electro chemical micro machining. *Int. J. Electro Chem. Sci.* 2013, 8, 12163–12171.

27. Shik, H.; Bo Hyun, S. K.; Nam, C. C. Analysis of the side gap resulting from micro electro-chemical machining with a tungsten wire and ultra-short voltage pulses. *J. Micromech. Microeng.* 2008, 18, 075009.

28. Jain, V. K.; Dixit, P. M.; Pandey, P. M. On the analysis of the electrochemical spark machining process. *Int. J. Mach. Tool Manuf.* 1999, 39, 165–186.

29. Rao, R. V.; Pawar, P. J.; Shankar, R. Multi-objective optimization of electrochemical machining process parameters using a particle swarm optimization algorithm. *Proc. I Mech B J. Eng. Manuf.* 2008, 222, 949–958.

30. Bhattacharyya, B.; Sorkhel, S. K. Investigation for controlled electro-chemical machining through response surface methodology-based approach. *J. Mater. Process. Technol.* 1999, 86, 200–207.

31. Yuming, Z.; Jeffrey, J. The cathode design problem. *Chem. Eng. Sci.* 1995, 50, 2679–2689.

32. Kozak, J.; Rajurkar, P. K.; Makkar, Y. Selected problems of microelectrochemical machining. *J. Mater. Process. Technol.* 2004, 149, 426–431.

33. Westley, J. A.; Atkinson, J.; Duffield, A. Generic aspects of tool design for electrochemical machining. *J. Mater. Process. Technol.* 2004, 149, 384–392.

34. Chunhua, S.; Zhu, D.; Zhiyong, L.; Wang, L. Application of FEM to tool design for electrochemical machining freeform surface. *J. Finite Elem. Anal. Design.* 2006, 43, 168–172.

35. Kim, B. H.; Na, C. W.; Lee, Y. S.; Choi, D. K.; Chu, C. N. Micro electrochemical machining of 3D micro structure using dilute sulfuric acid. *Ann. CIRP.* 2005, 54, 191–194.

36. Hewidy, M. S.; Ebeid, S. J.; El-Taweel, T. A.; Youssef, A. J. Modeling the performance of ECM assisted by low frequency vibrations. *J. Mater. Process. Technol.* 2007, 189, 466–472.

37. Kock, M.; Kirchner, V.; Schuster, R. Electrochemical micromachining with ultra short voltage pulses a versatile method with lithographical precision. *Electrochem. Acta.* 2003, 48, 3213/3219.

38. Se Hyun, A.; Shi Hyoung, R.; Deok Ki,C.; Chong Nam, C. Electro-chemical micro-drilling using ultra short pulses. *J. Precis. Eng.* 2004, 28, 129–134.

CHAPTER 16

A Survey of QSAR Studies

Seema Dhail, Tanmoy Chakrborty[*], Lalita Ledwani

Department of Chemistry, Manipal University Jaipur, India
[*]Email: tanmoychem@gmail.com

CONTENTS

ABSTRACT

This report provides a review on quantitative structure–activity relationship (QSAR) modeling, approaches, and its various applications in biological systems. Quantum chemical methods and modeling techniques help in correlating a specific activity for a set of compounds with their structure-derived descriptors by means of a mathematical model. Such models have been widely applied in many fields including chemistry, biology, and environmental sciences. Computational tools or in silico method are cost-effective approaches, which ultimately help in reducing the in vivo experiments on animals also. This study is mainly focused on the method for QSAR modeling as well as the method for systematic comparison of the theoretical molecular descriptors with experimental descriptors.

16.1 INTRODUCTION

In human behavior, one essential feature was observed that relies on comparison and classification for interpreting the observations. Similarly in quantitative structure–activity relationship (QSAR) and quantitative structure–property relationship (QSPR) studies, assumptions are based on the structures of a molecule, such as its geometrical, steric, and electronic properties; it must contain the features that are responsible for its physical, chemical, and biological properties and depending on that we can be able to represent the chemical by more than one numeric descriptors. According to Rouvary, chemistry is like a human behavior in which comparison and classification are related to similarity between the two variables. This shows the ubiquitous nature of the similarity concept. Periodic table is the one of the best examples for grouping the atoms together based on similarity. This statement clearly justifies that the QSAR and QSPR are the best options to examine the similarity between molecules, so that they can be classified and interpreted. According to Herndon and Berz, "Similarity, like beauty, lies in the eyes of the beholder," which means that an organic chemist may also use other concepts to classify the molecules. Molecular descriptors are a molecular property that acts as a concept to describe a molecule. In computational medicinal chemistry, hundreds of such descriptors are used to describe a molecular property over time. Molecular descriptors are simply counting the specific features of a molecule such as the number of hydrogen bond acceptor atoms. Log P is also used as one of the physico-chemical parameters as descriptors, and others as topological and topographical indices, including the Wiener index, the Balaban index, the indices introduced by Randic, the Zagreb index, and the Hosoya index.

16.2 DEVELOPMENT OF QSAR MODEL

For constructing a QSAR/QSPR model (Fig. 16.1), two steps are mainly involved (Chanin Nantasenamat):

i) Molecular description.
ii) Multivariate analysis for correlating the molecular descriptors with the observed activities/properties of selected molecule.

Some intermediate steps involved are

i) Data preprocessing,
ii) Statistical evaluation.

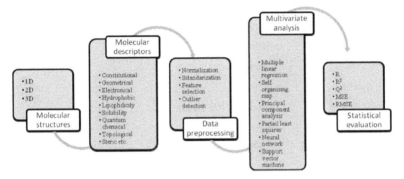

FIGURE 16.1 Schematic Overview of QSAR Process.

16.2.1 Data Understanding

This is the initial and crucial step in QSAR/QSPR studies. First, the researcher must have to become familiar with the nature of the data undergoing the study and QSAR/QSPR model construction, therefore reducing the unnecessary labor and errors that may occur. Before undergoing model construction, a researcher must have the thorough knowledge of literature on the relevant subject, including all its biological or chemical system of interest. First, exploring the data analysis that starts with the observation of the data matrix, particularly, the variables (molecular descriptors), its corresponding data types (characteristics or the kinds of data may be qualitative, i.e., categorical labels, or quantitative, i.e., arithmetic operations), and the data samples (each unique compound) (Chanin Nantasenamat).

16.2.2 Molecular Descriptors

Molecular descriptors are the chemical information that is encoded within the molecular structures for which various sets of algorithms are available. Descriptors are the essential information of a molecule in terms of its constitutional, hydrophobic, quantum chemical, lipophilicity, topological descriptors, etc. Various descriptors used in this study are found in various literatures[1-7] and a more extensive treatment in the encyclopedia *Handbook of Molecular Descriptors.*[8] These descriptors could be calculated using various quantum chemical software given in Table 16.1.

TABLE 16.1

S. No.	Software	Reference
1	Gaussian	Frisch et al.[9]
2	Spartan	Wavefunction[10]
3	GAMESS	Gordon and Schmidt,[11] Schmidt et al.[12]
4	NW Chem	Kendall et al.[13]
5	Jjaguar	Schrödinger[14]
6	MOLCAS	Karlström et al.[15]
7	Q-Chem	Shao et al.[16]
8	Dalton	Angeli et al.[17]
9	MOPAC	Stewart[18]
10	DRAGON	Talete srl,[19] Tetko et al.[20]
11	CODESSA	Katritzky et al.[21]
12	ADRIANA.Code	Molecular Networks GmbH Computerchemie[22]
13	RECON	Sukumar and Breneman[23]

After the calculation of molecular descriptors, it acts as an independent variable for constructing QSAR model further.

16.2.3   Modeled Activities/Properties

Various activities and properties that are calculated can be modeled by QSAR/QSPR as acting as dependent variables for the QSAR model. Calculated dependent variables are assumed to be influenced by various molecular descriptors, that is, independent variables. There are so many biological and chemical properties explored in QSAR studies (Table 16.2).

TABLE 16.2   Biological and Chemical Properties Explored in QSAR Studies (Chanin Nantasenamat)

Biological Properties	Chemical Properties
a. Bioconcentration	• Melting point
b. Carcinogenecity	• Solubility
c. Mutagenecity	• Stability
d. Biodegradation	• Boiling point
e. Permeability	• Dielectric constant
f. Pharmacokinetics	• Diffusion constant
g. Drug metabolism, etc.	• Reactivity, etc.

16.2.4 Data Preprocessing

It is the most important part of data mining process. It helps in ensuring the integrity of the data set before proceeding further with data mining analysis. It is also called as input of useful data and output of unusual data method. For getting a reliable QSAR model, it is essential that data should be handled with care. First, eliminate the invalid character values and errors while collecting the raw data and try to collect full information avoiding incomplete data that may cause trouble for data mining process. For reducing the errors or variability in the range and distribution of each variable in the data set, statistical techniques such as min–max normalization or Z-score standardization were used Minimum and maximum values of each variable are adjusted to uniform range between 0 and 1 by the equation

$$X_{\text{normalized}} = \frac{X_i - X_{\min}}{X_{\max} - X_{\min}}$$

where $X_{\text{normalized}}$ is the min–max normalized value, X_i is the value of interest, $X_{\min}$ is the minimum value, and $X_{\max}$ is the maximum value.

In the case of Z-score standardization, the variable undergoing the study is subjected to statistical operation to achieve mean center and unit variance by using the formula

$$X_{ij}^{\text{stnd}} = \frac{X_{ij} - \overline{X}_j}{\sum_{i=1}^{N}(X_{ij} - \overline{X}_j)^2 / N}$$

where X_{ij}^{stnd} is the standardized value, X_{ij} represents the value of interest, $\overline{X}_j$ represents the mean, and N is the sample size of the data set.

The above equation represents the difference of the value of interest and its mean followed by a division operation with the numerator, which acts as a variance.

16.2.5 Multivariate Analysis

In this approach, the relationship between independent variable (molecular descriptors) and dependent variable (e.g., biological/chemical properties of interest) is quantitatively determined. In the classical approach, linear regression technique is involved by establishing the linear mathematical equation

$$y = a_0 + a_1 X_1 + \cdots + a_n X_n$$

where y is the dependent variable, a_0 is the y-intercept or baseline value for the compound data set, $a_1, \ldots, a_n$ are the regression coefficients calculated from a set of training data in a supervised manner where the independent and dependent variables are known.

This type of linear approach works well for biological/chemical systems, in which the phenomenon of interest is of linear nature. But in some cases, properties are not linear; in that case, nonlinear approaches are used in order to properly model such properties. Artificial neural network (ANN) is one of the most popular techniques used for nonlinear approaches. ANN approach works similarly as inner working of the brain, which is composed of interconnected neurons. It consists of input layer, hidden layer, and output layer. Input layer works for passing the information of the independent variables into the ANN system. Number of neuronal units present in the input layer is equal to the number of independent variables in the data set. Information from the input layer is relayed to the hidden layer for pattern recognition processing and predictions will then be passed from the hidden layer to the output layer. Error is calculated in a back propagation algorithm, which is derived from the difference between the predicted value and the actual value, and if it is acceptable, then the learning process will stop or otherwise signals will be sent backward to the hidden layer for further processing and weight readjustments. This is performed until the solution is reached and finally learning is terminated.

16.2.6 Statistical Evaluation

For construction of a QSAR model, both model validation and statistical parameters are used. In model validation, we divide the data set into training set and testing set. Testing set is used for constructing a predictive model whose predictive performance is evaluated on the testing set. Internal performance can be assessed from predictive performance of the independent testing set that is unknown to the training model. For statistical evaluation, Pearson's correlation coefficient (r) is the commonly used parameter that describes the degree of association between the two variables of interest. The range -1 to $+1$ is taken for the calculation of r value of two different variables of interest; this indicates a negative and positive correlation. Here, r is used to measure the correlation between experimental (x) and predicted (y) values of interest in order to observe the variability that exists between the variables. Calculation of r is carried out by the following equation:

$$r_{xy} = \frac{n\sum xy - \sum x \sum y}{\sqrt{\left(n\sum x^2 - \left(\sum x\right)^2\right)\left(n\sum y^2 - \left(\sum y\right)^2\right)}}$$

where r_{xy} is the correlation coefficient between variables x and y, n is the sample size, x is the individual value of variable x, y is the individual value of variable y, xy is the product of variables x and y, x^2 is the squared value of variable x, and y^2 is the squared value of variable y.

Root mean square (RMS) is used for evaluating the relative error of the QSAR model. It is calculated by the following equation:

$$RMS = \sqrt{\frac{\sum_{i=1}^{n}(x-y)^2}{n}}$$

where x is the experimental value of the activity/property of interest, y is the predicted value of the activity/property of interest, and n is the sample size of the data set.

ANOVA is used for calculating F values. It is the ratio between the explained and the unexplained variance. To compare the performance of multiple QSAR models, the models should have the same set of compounds and descriptors. Each model yields a calculated F value, and the models having the highest value act as the best performing model.

16.2.7   Predictive QSAR Model

Tropsha is one of the commonly used approaches in the field of QSAR.[24] Predictive QSAR model should have the following statistical characteristics:

$q^2 > 0.5$

$R^2 > 0.6$

$\frac{(R^2 - R_0^2)}{R^2} < 0.1$ or $\frac{(R^2 - R_0'^2)}{R^2} < 0.1$

$0.85 < k < 1.15$ or $0.85 < k' < 1.15$

where q^2 is the cross-validated explained variance, R^2 is the coefficient of determination (where R_0^2 represent predicted versus observed activities and $R_0'^2$ represent observed versus predicted activities), and slopes k and k' are the regression lines passing through the origin.

16.2.8   Basic Requirements in QSAR Studies (Tanmoy Chakraborty 2011)

i)   All analogs belong to a congeneric series
ii)   All analogs exert the same mechanism of action
iii) All analogs bind in a comparable manner
iv)   The effects of isosteric replacement can be predicted
v)   Binding affinity is correlated to interaction energies
vi)   Biological activities are correlated to interaction energies
vii) Biological activities are correlated to binding affinity

18.2.9 Specific QSAR Approaches

In QSAR studies, various approaches are found in the literature. Depending on the objectives to be reached by QSAR approaches, QSAR classifies, for example, which type of molecular property is modeled, type of molecular descriptors the model is composed of, and the mathematical method or computational algorithm that is used to estimate the model parameters. Properties such as ADME (absorption, distribution, metabolism, and elimination properties) analysis, environmental QSAR, linear solvation energy relationship (LSER), and binary QSAR are commonly used.

16.2.9.1 *Hansch Approach*

Corwin Hansch is the founder of modern QSAR. According to classic article by a comprehensive model, biological activity for a group of "congeneric" chemicals can be described as follows:

$$\log\left(\frac{1}{C_{50}}\right) = a\pi + b\varepsilon + cS + d$$

where C is the toxicant concentration at which an endpoint is manifested (e.g., 50% mortality or effect) and is related to a hydrophobicity term, π is an electronic term, ε is originally the Hammett substituent constant, S is a steric term (typically Taft's substituent constant, Es), and d is the general additional term depending on the kind of property to be modeled.

Here, the parameter ?, which is the relative hydrophobicity of a substituent, was defined as

$$\pi = \log P_X - \log P_H$$

where P_X and P_H represent the partition coefficients of a derivative and the parent molecule. It acts as a substituent constant denoting the difference in hydrophobicity between a parent compound and a substituted analog and is usually replaced by the more general molecular term the log of the 1 – octanol/water partition coefficient, $\log K_{ow}$ or $\log P$.

Hammett and Taft had contributed for the development of the QSAR paradigm by Hansch and Fujita, which combined the hydrophobic constants with Hammett's electronic constants to yield the linear Hansch equation and its many extended forms.

16.2.9.2 *Free–Wilson Approach*

It is based on the assumption that a biological response can be modeled by various additive substituents. First, a common skeleton for the chemical analogs is defined, then the regression analysis is performed by considering S number of substitution sites R_s (s ? 1, S) and, for each site, a number Ns of different substituents. Hydrogen atoms are also considered as substituents if present in a substitution site of some compounds. Free–Wilson model is defined as

$$y_i = b_0 + \sum_{s=1}^{S} \sum$$

where b_0 is the intercept of the model corresponding to the average biological response calculated from the data set, b_{ks} are the regression coefficients, y acts as biological response, which is usually used in the form $\log(1/C)$, and C is the concentration achieving a fixed effect. The regression coefficients b_{ks} of the Free–Wilson model give the importance of each kth substituent in each sth site in increasing/decreasing the response with respect to the mean response, that is, the activity contribution of the substituent.

16.2.9.3 LSER Approach

LSERs approach is based on the effects of solvent–solute interactions on physico-chemical properties and reactivity parameters. Generally, a property P of a species A in a solvent S can be expressed as

$$P_{AS} = \sum_j \varphi_j(AS)$$

where φ_j are complex functions of both solvents and solutes.

By assuming that these functions can be factorized in two contributions separately depending on solute and solvent, the property can be represented as

$$P_{AS} = \sum_j f_j(A)g_j(S)$$

where f is the function of the solute and g is the function of the solvent.

LSER is based on the possibility to study f and g functions, after a proper choice of the reference systems and properties. In LSER solution, property P is mainly dependent on three factors: a cavity term, a polar term, and a hydrogen bond term:

$$P = \frac{\text{intercept} + \text{cavity term} + \text{dipolarity}}{\text{polarizability term} + \text{hydrogen bond term}}.$$

16.2.10 Quantum Mechanical Descriptors

Quantum chemical molecular descriptors have been actively used in the QSAR studies of various biological systems. Nowadays, both quantum chemical methods and molecular modeling techniques are used in large number of molecular and local quantities, characterizing the reactivity, shape, and binding properties of a complete molecule, as well as the molecular fragments and substituents. Due to well-defined physical information content encoded in large amounts in many theoretical descriptors, their use in the design of a training set in a QSAR study presents two main advantages:

i) Compounds and their substituents can be directly characterized only on the basis of their molecular structure.

ii) Proposed mechanism of action can be directly accounted for in terms of the chemical reactivity of the compounds under study.

Using modern computer hardware, power, and various advanced software enables one to calculate realistic quantum chemical molecular characteristics in a relatively short computational time. On the basis of the size of the molecular system selected for study, either the semiempirical or the ab initio level of algorithms would be applicable. In the case of ab initio model, Hamiltonians represent all nonrelativistic interactions between the nuclei and electrons in a molecule. The Hartree–Fock method employed in basic ab initio calculations is expected to provide better results, when using large basis set. Various ab initio methods beyond Hartree–Fock account for electron correlation in a molecule configuration interaction, multiconfigurational self-consistent field (SCF), correlated pair many-electron theory, and its various coupled-cluster approximations, etc.

Semiempirical quantum chemical methods are more fast and applicable for the calculation of molecular descriptors of long series of structurally complex and large molecules. Various methods are developed within the mathematical framework of molecular orbital theory (SCF MO). The most popularly used semiempirical methods are Austin Model 1 and Parametric Model 3. Apart from them, the methods of computational resources algorithms based on density functional theory (DFT) were used.

Quantum chemically derived descriptors are fundamentally different from experimentally measured quantities. In an experimental measurement, there is an increase in various statistical errors, but in case of quantum chemical calculations, chances of error are less. While using quantum chemistry-based descriptors with a different series of related compounds, the chances of computational error are considered to be approximately constant throughout the series. One of the drawbacks of using quantum chemical descriptors is the failure to directly address bulk effects. Various quantum chemical descriptors are mentioned below according to their various subdivisions:

i) Energy-related descriptors
ii) Electrostatic descriptors
iii) MO-related descriptors
iv) Quantum chemical modeling of empirical descriptors

16.2.10.1 Energy-Related Descriptors

In energy-related descriptors, total energy of the molecule and its different partition-ings can be used as theoretical molecular descriptors. The total energy (E_{tot}) of a molecule is calculated by quantum mechanical method, which is usually referred as quantum mechanical standard state for the energy. According to Born–Oppenheimer approximation, it is written as

$$E_{tot} = \frac{E_{el} + \sum_{A \neq B} ZAZB}{R_{AB}}$$

The net atomic electron attraction energy for a given atomic species in the select-ed molecule can be calculated by the MO method using the elements of the density matrix, the electron repulsion integrals $(\mu\upsilon/\lambda\sigma)$ and the nuclear–electron attraction integrals $(\mu\upsilon\lambda\sigma)$ on the given atomic basis. This energy describes the conformational changes, including various rotational properties and inversion process of a molecule, or detects the atomic reactivity in the molecule by describing the electron–electron repulsion or nuclear–electron attraction driven process. It may also specify the reactiv-ity site or conformational changes in the molecule.

16.2.10.2 Electrostatic Descriptors

The calculation of molecular properties, especially in condensed media, is totally de-pendent on intermolecular interactions. Mostly, electrostatic attractions and repul-sions between two or more molecules or their different parts are due to intermolecular interactions. Such kind of electrostatic descriptors characterizing a molecule would be significantly important in determining its various properties. The charge distribution and the partial charges on atoms in the molecule can also be obtained by using quan-tum chemical calculations. By using Shannon's information theory, a more reliable charge distribution in the molecule can be obtained. For the development of QSAR/QSPR equations, the most negative (minimum) and most positive (maximum) partial charges in the molecule or the minimum and maximum partial charges for particular types of atoms have been used as electrostatic descriptors. By quantum mechanical wave function, electrical moments and their components of a molecule can be calcu-lated. Jurs et al. proposed charged partial surface area descriptors in terms of the sur-face area of the whole molecule or its fragments and in terms of the charge distribution in the molecule.

16.2.10.3 MO-Related Descriptors

Various molecular descriptors, based on physical properties and chemical reactiv-ity of molecules, are available depending on the information available within the molecular orbital formalism. Highly occupied molecular orbital (HOMO) and lowest unoccupied molecular orbital (LUMO) energies belong to quantum chemical descriptors Both the molecular orbitals determine the chemical reactivity of a compound and the possible mechanism of a chemical reaction. The gap between the HOMO and LUMO has also been related to the chemical stability of compounds. Chemical hardness is dependent on HOMO–LUMO energy gap. It is expected that activation hardness is useful in distinguishing between the reaction rates at different sites in the molecule, which is also responsible for predicting orientation effects. Frontier molecular orbital (FMO) theory is based on the concept of superdelocalizability; it is an index characterizing the affinity of occupied and unoccupied orbitals in chemical reactions. Interaction of a compound with the nucleophilic center at the other reactant is considered in case of nucleophilic superdelocalizability, and in case of electrophilic delocalizability, interaction of a compound with the electrophilic center at another reagent is considered. Atomic nucleophilic (N'_A), electrophilic (E'_A), and one-electron (R'_A) extreme values of reactivity indices for a given atomic species in the molecule have often been used as descriptors of molecular reactivity.

Description of donor–acceptor interactions between molecules is dependent on the electron density present in frontier orbitals of atoms. According to this approach, nucleophilic electron density, f_r^N (HOMO), of the donor molecule and electrophilic electron density, f_r^E (LUMO), of an acceptor molecule are mainly responsible for transferring the charges. Frontier electron densities can strictly be used only to describe the reactivity of different atoms in the same molecules.

16.2.10.4 Quantum Chemical Modeling of Empirical Descriptors

These empirical descriptors are very useful in the description of chemical reactivity as well as biological activity in many systems. In order to expand predictive power of the respective QSAR models, it would be beneficial to calculate the empirical descriptors using some quantum mechanical characteristics of molecules. A large number of empirical molecular descriptors involve various solvent effects on different chemical or physical processes. QSAR and QSPR correlation equations are usually parametric, which involve the descriptors reflecting the polarity and the polarizability of the solvent, its ability to act as an acceptor or a donor in a hydrogen bond, and the short-range dispersion and repulsion interactions. Log P, a hydrophobic parameter, can be referred as solvational characteristics because it is directly related to the change of free energy of solvation of a solute in two solvents. Linear solvatochromic relation descriptors have been shown to be successful for calculating a wide range of chemical and physical properties involving solute–solvent interactions as well as biological activities of compounds. In case of theoretical linear solvation energy relationship (TLSER), descriptors have been derived in case of wider selection of solvents. General form of TLSER is

$$\log(\gamma) = c_0 + c_1 V_{mc} + c_2 \pi^* + c_3 \varepsilon_a + c_4 \varepsilon_b + c_5 q^+ + c_6 q^-$$

where V_{mc} is the molecular van der Waals volume, π^* is the polarizability term that is derived from polarization volume of a compound, ε_b is calculated as the difference in energy between LUMO of water and HOMO of the solute; $\bar{q}$ is the most negative atomic charge in the solute molecule, ε_a is the energy difference between the HOMO of water and LUMO of solute, and q^+ is the positive charge of H^+ atom in the solute molecule.

16.2.11 Classification of Molecular Descriptors

Molecular descriptors are mainly subdivided into three different categories:

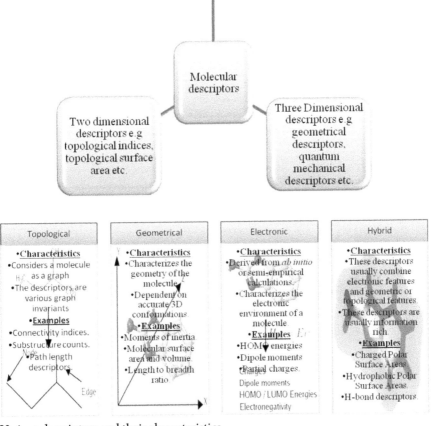

Various descriptors and their characteristics

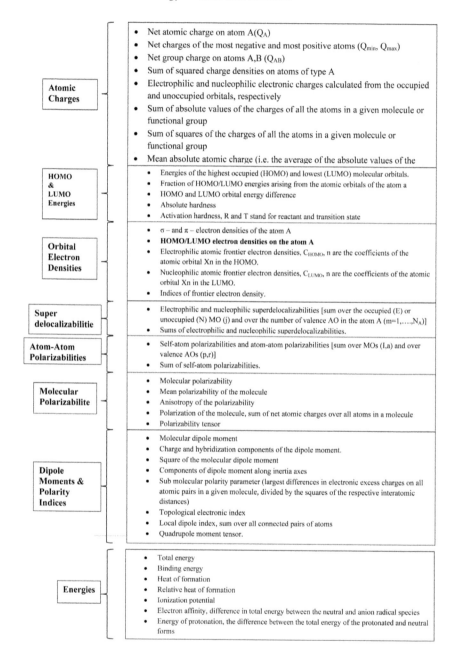

Atomic Charges

- Net atomic charge on atom A(Q_A)
- Net charges of the most negative and most positive atoms (Q_{min}, Q_{max})
- Net group charge on atoms A,B (Q_{AB})
- Sum of squared charge densities on atoms of type A
- Electrophilic and nucleophilic electronic charges calculated from the occupied and unoccupied orbitals, respectively
- Sum of absolute values of the charges of all the atoms in a given molecule or functional group
- Sum of squares of the charges of all the atoms in a given molecule or functional group
- Mean absolute atomic charge (i.e. the average of the absolute values of the

HOMO & LUMO Energies

- Energies of the highest occupied (HOMO) and lowest (LUMO) molecular orbitals.
- Fraction of HOMO/LUMO energies arising from the atomic orbitals of the atom a
- HOMO and LUMO orbital energy difference
- Absolute hardness
- Activation hardness, R and T stand for reactant and transition state

Orbital Electron Densities

- σ – and π – electron densities of the atom A
- **HOMO/LUMO electron densities on the atom A**
- Electrophilic atomic frontier electron densities, C_{HOMO}, n are the coefficients of the atomic orbital Xn in the HOMO.
- Nucleophilic atomic frontier electron densities, C_{LUMO}, n are the coefficients of the atomic orbital Xn in the LUMO.
- Indices of frontier electron density.

Super delocalizabilitie

- Electrophilic and nucleophilic superdelocalizabilities [sum over the occupied (E) or unoccupied (N) MO (j) and over the number of valence AO in the atom A (m=1,....,N_A)]
- Sums of electrophilic and nucleophilic superdelocalizabilities.

Atom-Atom Polarizabilities

- Self-atom polarizabilities and atom-atom polarizabilities [sum over MOs (I,a) and over valence AOs (p,r)]
- Sum of self-atom polarizabilities.

Molecular Polarizabilite

- Molecular polarizability
- Mean polarizability of the molecule
- Anisotropy of the polarizability
- Polarization of the molecule, sum of net atomic charges over all atoms in a molecule
- Polarizability tensor

Dipole Moments & Polarity Indices

- Molecular dipole moment
- Charge and hybridization components of the dipole moment.
- Square of the molecular dipole moment
- Components of dipole moment along inertia axes
- Sub molecular polarity parameter (largest differences in electronic excess charges on all atomic pairs in a given molecule, divided by the squares of the respective interatomic distances)
- Topological electronic index
- Local dipole index, sum over all connected pairs of atoms
- Quadrupole moment tensor.

Energies

- Total energy
- Binding energy
- Heat of formation
- Relative heat of formation
- Ionization potential
- Electron affinity, difference in total energy between the neutral and anion radical species
- Energy of protonation, the difference between the total energy of the protonated and neutral forms

16.3 DENSITY FUNCTIONAL THEORY

In the review of earlier research papers, it was proved that DFT played an important role in the development of quantum chemistry in the past decades. This shows that DFT revolutionized quantum chemistry, offering a computational technique that partly includes electron correlation at a much better quality/cost ratio as compared to the conventional post Hartree–Fock wave function methods.

DFT considers the electron density function $\rho(r)$ as the carrier of all information on the properties of an atom, a molecule, and a cluster instead of the much more complicated wave functions $\psi\,(\chi_1, \chi_2, \ldots, \chi_N)$; it depends on the four spatial and spin coordinates of all N number electrons of the system.

Hohenberg and Kohn had given the ingenious existence proof of the E–E (ρ) functional. This led to the modern DFT. Modern DFT states that each $\rho(r)$ corresponds with an energy value E.

Electron density is one of the most attractive in DFT study. There are two densities for spin polarized systems, $\rho\uparrow(r)$ (spin-up) and $\rho\uparrow(r)$ (spin-down) electrons, as opposed to many-particle wave function, which depends on 3N variables. According to Hohenberg and Kohn (1964), ground-state properties are functional of electron density $\rho(r)$, which is the basic framework for modern density functional methods.

16.3.1 Hohenberg and Kohn Theorems

DFT field was born in 1964 with the publication of Hohenberg and Kohn's paper (1964). In this theory, they proved that the following:

i) Stationary quantum mechanical system of each observable including energy can be calculated in principle exactly from the ground-state density alone.
ii) Using vibrational method involving only density ground state can be calculated.

According to Gross and Kurth (1994), original theorem refers to time-independent ground state, and it will be extended to excited states and time-dependent potentials.

This theorem states that for molecules in a nondegenerative ground state, the ground state molecular energy, wave function, and all other molecular electronic properties are uniquely determined by the ground-state electron probability density, $\rho(x,y,z)$, which is a function of only three variables.

$$E = E_0\,[\rho] \qquad \text{(It gives a functional relationship)}$$

Alternatively, the first Hohenberg–Kohn theorem can be stated as follows: $\rho(r)$ determines the external (due to nuclei potentials), and $v(r)$ and $\rho(r)$ determine N, the total number of electron through its normalization.

$$\int P(r)dr = N$$

N and $v(r)$ determine the molecular Hamiltonian, H_{op}, considering Born–Oppenheimer approximation, neglecting relativistic effects as

$$H_{op}\psi = E\psi$$

where E is the electronic energy, ψ is the wave function, and H_{op} is the Hamiltonian operator.

16.3.2 Kohn–Sham Theorem

If one knows the ground-state electron density $\rho_0(r)$, the Hohenberg–Kohn theorem tells that it is possible in principle to calculate all the ground-state molecular properties from ρ_0. The Hohenberg–Kohn theorem does not tell how to calculate E_0 from ρ_0 or how to find ρ_0 without finding the wave function. A key step to achieve these goals was given by Kohn and Sham (KS). A practical, orbital-state energy within the KS method is given by

$$E_0 = -\frac{1}{2}\Sigma^n < \theta$$

16.3.3 Conceptual DFT

Conceptual DFT mainly follows Parr's dictum. According to Parr's dictum, accurate calculation is not synonymous with useful interpretation. To calculate a molecule is not to understand it.

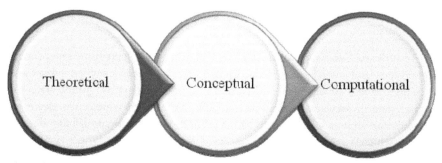

Three directions of DFT

For extracting chemically relevant concepts and principles from DFT, conceptual DFT is mainly used. Many of the chemical concepts and mechanisms have been derived within the scope of sensitivity functional theory. The DFT provides conceptually useful information about chemical reactivity and, most especially, the charge transfer. Some potential density functional reactivity descriptors are chemical potential (?), global hardness (?), and the local softness and Fukui functions. These descriptors are both global and local and have been introduced by Parr et al. Parr et al. discovered a new parameter, chemical potential (?) that has the potential bearing in theoretical chemistry. There are many applications of DFT in chemistry. One is in the calculation of properties of atoms and molecules. Another important use of DFT is in elucidating familiar chemical concepts. Parr et al. have been particularly active in this area. Chemical potential, electronegativity, and hardness are important chemical reactivity parameters that have been extensively used in understanding the molecular structure, properties, reactivity, bonding, and interactions.

16.3.4   Significance

i)   Have a global approach.

ii)   Take advantage of existing data.

iii)   QSAR/QSPR is the scientific methodology in drug discovery and drug research. This scientific approach conserves resources and accelerates the process of development of new molecules for use as drugs, materials, additives, or for any other purpose.

iv)   It establishes a direct relationship between structure and activity/property of bioactive compounds, with some theoretical as well as experimental descriptors.

v)   It predicts the activities of new chemicals.

vi)   Time- and cost-saving potential.

vii)   Used in drug designing.

viii)   Shows the major role in medical sciences and also in pharmacy research.

ix)   It fulfills the demands of a search for new chemical compounds to control various harmful infections/diseases.

x)   It is also useful in catalysis process and protein/DNA functioning.

xi)   Toxicity of a molecule/drug can be detected, which in turn can be detected by their structures.

xii)   Catalyze the development of cost-effective and regulatory acceptable QSAR models in India by leveraging India's IT strengths and technical scientific expertise.

xiii)   Attract entrepreneurship to the area. Promote QSAR models developed in India.

xiv)   Coordination with the GLP Laboratories in the use of QSAR models for decision-making.

v)   Any other IT-based opportunity that may arise including as may be required by other groups.

vi)   It develops a database of experts and expertise on QSAR and other computational techniques available in India.

16.4 QUANTUM QSAR IN BIOLOGICAL SYSTEM

QSAR is the new emerging area in science. Quantum chemical molecular descriptors are actively used in the QSAR studies of biological activities. Some of the active areas are QSAR studies and studies of biological activities, which are discussed below.

16.4.1 In Recognizing the DNA Splice Junction Sites

DNA acts as the genetic material in each living organism. It consists of billions of nucleotides containing numerous genes, which can express several different types of proteins. DNA consists of exons and introns. Exons are the protein-coding regions, and introns are the noncoding regions. DNA undergoes transcription process and gives rise to mRNA; further, it undergoes translation process that ultimately gives rise to specific proteins. Splice junction sites are also present, and they act as boundaries where splicing occurs. Chanin et al. (2009) developed a computational approach for the recognition of DNA splice junction sites. They transformed the DNA sequences to sequences of binary numbers by converting each nucleotide's adenine (A), tyrosine (T), guanine (G), and cytosine (C) as 0001, 0010, 0100, and 1000, respectively. Each entry of the data set describes information surrounding the splice junction site, mainly, 15 nucleotides upstream and 32 nucleotides downstream. Approximately 1424 human DNA sequences data set is made by them that is divided into two portions: i) a training set of 1000 sequences and ii) a testing set of 424 sequences. Various types of predictive models were developed using three different types of learning algorithm, which consists of i) self-organizing map, ii) back propagation neural network, and iii) support vector machine.

16.4.2 Enzymatic Reactions

QSAR is also playing a role in various enzymatic reactions. Quantum chemically calculated electrostatic or MO-related descriptors have been widely used. By using these descriptors, the complex formation between enzyme and substrate was also detected, and the chemical reactivity of the substrate at the site was also detected. According to the study carried out by Koopman and Hudson in 1967, interaction between the drug and receptor is due to either the charge or the orbital control. So, for the electrostatic interactions, atomic charges may be considered. MO-related characteristics characterize the covalent components of the interaction. Various models of electrostatic interactions are applied for testing the several leucine aminopeptidase inhibitors interactions with the enzyme active site. Various electrostatic models together with the charges from electrostatic potentials also give the satisfactory correlation of electrostatic interaction energy with experimental activities of the inhibitors.

Various quantum mechanical molecular electrostatic potentials are also combined with ANNs for predicting the binding energy of several bioactive molecules with their enzyme targets as well as for identifying the quantum mechanical features of inhibitory molecules contributing for the binding.

16.4.3 In Medicinal Activity

In medicinal chemistry, QSAR plays a wider role. Quantum chemical molecular descriptors are extensively used for various pharmacological activities of compounds. Electrostatic descriptors demonstrate the drug–receptor interactions at docking sites. Analysis can be based on the influence of electrostatic interactions on various physicochemical properties, detecting the bioavailability of compounds. Data collected for a particular set of compounds are useful for developing respective QSPR and for comparing with the QSAR for its pharmacological action.

16.4.3.1 Antiviral Activities

For the development of QSARs, antiviral activities using quantum chemical descriptors are also useful. 9-Benzylpurines antirhinoviral activity has been correlated with Huckel MO-generated electronic parameters and substituent constants. The equation for QSAR is

$$-\log(\mathrm{IC}_{50}) = 6.044 + 2.056R_2 + 0.873F_4 - 0.289\pi_4 - 0.094E_\pi^T - 2.323E_{\mathrm{LUMO}}$$

where $R^2 = 0.684$, $S = 0.503$, $n = 50$, and $F = 19$.

The above equation includes LUMO energy and the total π-electron energy (E_π^T) of the compounds as quantum chemical descriptors, where R and F are the Swain–Lupton resonance and field parameters, π is the hydrophobicity substituent constant at a given position in purine ring.

From this, it was interpreted that different serotypes of rhinovirus behave differently in terms of electronic parameters that inhibit their action.

16.4.3.2 Anticancer Activity

In anticancer activity, QSAR plays a very important part. A study has been made on a series of 4-alkynyl and 4-alkenyl-quinazolines and a series of N-4,6-pyrimidine-N-alkyl-N'-phenyl ureas, which are two different series of anticancer tyrosine kinase inhibitors. According to the QSAR study, first series indicate that the activity is controlled by the hydrophobicity of the molecules and molecular connectivity index of the substituent, and in the case of second series of compounds, the activity is found to be controlled by the molecular connectivity index of the substituent and some indicator variables. Various anticancer activities were studied by the QSAR, for example, antiprostatic cancer activity, antibreast cancer activity, and antilung cancer activity.

16.4.3.3   Antibacterial Activity

QSAR studies are also important for the study of antibacterial activities. A series of monocyclic ?-lactam antibiotics, which include atomic charges, bond orders, dipole moment, and others, are studied through QSAR. An in vitro study for antimycobacterial activity against *Mycobacterium tuberculosis* was done by a series of N_2-acyl isonicotinic acid hydrazides was synthesized and tested. The study found that the compound isonicotinic acid N'-tetradecanoyl-hydrazide was more active than the reference compound, isoniazid. The results of antimicrobial activity of the synthesized compounds against *S. aureus, B. subtilis, E. coli, C. albicans,* and *A. niger* indicated that compounds with dichloro, hydroxyl, tri-iodo, and $_2$-tetradecanoyl substituents were the most active ones. The multitarget QSAR model was found to be effective in describing the antimicrobial activity of N_2-acyl isonicotinic acid hydrazides.

16.4.4   Bio and Nonbiodegradation

Various structure-based biodegradation estimation methods have been compared in a recent review of Raymond et al. in 2001. Degradation of various chemical compounds in environment is based on biodegradation. Biodegradation of organic chemicals in natural systems can be classified as primary (structural transformation that alters the molecular integrity), ultimate (conversion to inorganic compounds or normal metabolic products), or acceptable (degradation to the extent that undesirable characteristics are ameliorated). QSAR/QSPR studies presented in the literature focused mainly on primary and ultimate biodegradation. Biodegradability can be expressed in terms of half-lives, diverse biodegradation rates and rate constants, theoretical and biological oxygen demand, etc. In previous literature, primary or ultimate aerobic degradation is the commonly found correlated property. Various models exist for predicting the propensity of a chemical to biodegrade (readily biodegrades or persists) or some quantitative measure of biodegradability such as rate constants.

16.4.5   Toxicity

For the prediction of toxicity, Hansh and Fujita initiated the development of analysis by QSAR analysis. They demonstrated the relationship between biological activities and the hydrophobic, electronic, and steric properties of compounds. The main challenge in the prediction of toxicity is for the development of QSAR for diversified, problematic, and intricate data sets. Konemann et al. gave the classification method based on modes of action, and it was introduced with the concept of "baseline toxicity" during the study of relationship between toxicity and the octanol–water partition coefficient for inert narcotic pollutants. According to their given concept, activity of chemicals with baseline toxicities depends on the hydrophobicity of compounds, and therefore, counted under nonpolar narcosis actors.

16.4.6   General Conclusion

QSAR/QSPR is a new emerging area and will gain significant popularity in different research areas. Due to increased cost in research area mainly to the various experiments involved in research, mainly involving in in vivo studies, together with the increasing power of modern computers and their programs worked together in these direction. Mainly when the modern and complicated programs to precede, from purely empirical selection procedures from greater number of descriptors and to study the structural effects in physically meaningful ways, will be more exploiting.

ACKNOWLEDGMENTS

The authors are grateful to Dr. Sandeep Sancheti, President of Manipal University, Jaipur, India, and Dr. B. K. Sharma, Dean, Faculty of Science, Research, and Innovation of Manipal University, Jaipur, India, for cooperating and providing the facilities for this work.

KEYWORDS

- QSAR modeling
- molecular descriptors
- DFT
- Hartree–Fock
- computational tools

REFERENCES

1. Helguera, A. M.; Combes, R. D.; Gonzalez, M. P.; Cordeiro, M. N. Applications of 2D descriptors in drug design: a DRAGON tale. *Curr. Top. Med. Chem.* 2008, 8, 1628–1655.
2. Karelson, M.; Lobanov, V. S.; Katritzky, A. R. Quantum – chemical descriptors in QSAR/QSPR studies. *Chem. Rev.* 1996, 96(3), 1027–1044.
3. Katritzky, A. R.; Gordeeva, E. V. Traditional topological indices vs electronic, geometrical, and combined molecular descriptors in QSAR/QSPR research. *J. Chem. Inf. Comput. Sci.* 1993, 33, 835–857.
4. Labute, P. A widely applicable set of descriptors. *J. Mol. Graph. Model.* 2000, 18, 464–477.
5. Randić, M. The nature of chemical structure. *J. Math. Chem.* 1990, 4, 157–184.
6. Randić, M.; Razinger, M. *On characterization of 3D molecular structure.* In *From Chemical Topology to Three Dimensional Geometry*; Balaban, A. T., Ed.; Plenum Press: New York, 1997; p 420.
7. Xue, L.; Bajorath, J. Molecular descriptors in chemoinformatics, computational combinatorial chemistry, and virtual screening. *Comb. Chem. High T. Scr.* 2000, 3, 363–372.
8. Todeschini, R.; Consonni, V. *Handbook of Molecular Descriptors*; Wiley-VCH: Weinheim, 2000; Vol. 11.
9. Frisch, M. J.; Trucks, G. W.; Schlegel, H. B.; et al. *Gaussian 03W, Revision C.02*; Gaussian Inc.: Wallingford, 2004.
10. Wavefunction. *Spartan'04*; Wavefunction, Inc: Irvine, CA, 2004.

11. Gordon, M. S.; Schmidt, M. W. *Advances in electronic structure theory: GAMESS a decade later.* In *Theory and Applications of Computational Chemistry: the First Forty Years*; Dykstra, C. E., Frenking, G., Kim, K. S., Scuseria, G. E., Eds.; Elsevier: Amsterdam, 2005; pp 1167–1189.

12. Schmidt, M. W.; Baldridge, K. K.; Boatz, J. A.; Elbert, S. T.; Gordon, M. S.; Jensen, J. H.; Koseki, S.; Matsunaga, N.; Nguyen, K. A.; Su, S.; Windus, T. L.; Dupuis, M.; Montgomery, J. A. General atomic and molecular electronic structure system. *J. Comput. Chem.* 1993, 14, 1347–1363.

13. Kendall, R. A.; Aprà, E.; Bernholdt, D. E.; Bylaska, E. J.; Dupuis, M.; Fann, G. I.; Harrison, R. J.; Ju, J.; Nichols, J. A.; Nieplocha, J.; Straatsma, T. P.; Windus, T. L.; Wong, A. T. High performance computational chemistry: an overview of NWChem a distributed parallel application. *Comput. Phys. Commun.* 2000, 128, 260–283.

14. Schrödinger, Inc. *Jaguar, Version 7.5207*; Portland, OR, 2008.

15. Karlström, G.; Lindh, R.; Malmqvist, P.-Å; Roos, B. O.; Ryde, U.; Veryazov, V.; Widmark, P. O.; Cossi, M.; Schimmelpfennig, B.; Neogrady, P.; Seijo, L. MOLCAS: a program package for computational chemistry. *Comput. Mater. Sci.* 2003, 28, 222–239.

16. Shao, Y.; Molnar, L. F.; Jung, Y.; et al. Advances in methods and algorithms in a modern quantum chemistry program package. *Phys. Chem. Chem. Phys.* 2006, 8, 3172–3191.

17. Angeli, C.; Bak, K. L.; Bakken, V.; et al. *DALTON, A Molecular Electronic Structure Program. Release 2.0*; 2005.

18. Stewart, J. *MOPAC2009*; Colorado, 2009.

19. Talete srl. *DRAGON*; Talete srl: Milano, Italy, 2007.

20. Tetko, I. V.; Gasteiger, J.; Todeschini, R.; Mauri, A.; Livingstone, D.; Ertl, P.; Palyulin, V. A.; Radchenko, E. V.; Zefirov, N. S.; Makarenko, A. S.; Tanchuk, V. Y.; Prokopenko, V. V. Virtual computational chemistry laboratorydesign and description. *J. Comput. Aid. Mol. Des.* 2005, 19, 453–463.

21. Katritzky, A. R.; Karelson, M.; Petrukhin, R. *CODESSA PRO*; Florida, 2005.

22. Molecular Networks GmbH Computerchemie. *ADRIANA.Code*; Molecular Networks GmbH Computerchemie: Erlangen, Germany, 2008.

23. Sukumar, N.; Breneman, C. M. *RECON, Version 5.5*; New York, 2002.

24. Tropsha, A.; Gramatica, P.; Gombar, V. K. The importance of being earnest: validation is the absolute essential for successful application and interpretation of QSPR models. *QSAR Comb. Sci.* 2003, 22, 69–77.

25. Karelson, M. *Quantum – Chemical Descriptors in QSAR*, Chapter 24; Taylor Francis Group LLC: London, 2004.

26. Katritzky, A. R.; Fara, D. C.; Petrukhin, R. O.; Tatham, D. B.; Maran, U.; Lomaka, A.; Karelson, M. *The Present Utility and Future Potential for Medicinal Chemistry of QSAR/QSPR with Whole Molecule Descriptors, Current Topics in Medicinal Chemistry*; Bentham Science Publishers Ltd: Beijing, 2002; Vol. 2, pp 1333–1356.

27. Selassie, C. D. *History of quantitative structure- activity relationships.* In *Burger's Medicinal Chemistry and Drug Discovery*, 6th ed.; Abraham, D. J., Ed.; John Wiley & Sons, Inc: New York, 2003; Vol. 1, pp 187–242.

28. Nantasenamat, C.; Isarankura-Na-Ayudhya, C.; Naenna, T.; Naenna, T.; Prachayasittikul, V. A practical overview of quantitative stricture-activity relationship. *EXCLI J.* 2009, 8, 74–88.

29. Bosse, E.; Roy, J.; Wark, S. *Concepts, Models, and Tools for Information Fusion*; Artech House, Inc: Norwood, MA, 2007.

30. Chen, N.; Lu, W.; Yang, J.; Li, G. *Support Vector Machine in Chemistry*; World Scientific Publishing: Singapore, 2004.

31. Cristianini, N.; Shawe-Taylor, J. *An Introduction to Support Vector Machines and Other Kernel-Based Learning Methods*; Cambridge University Press: Cambridge, 2000.

32. Cros AFA. Action de l'alcool amyliquesur l'organisme. Strasbourg. Thesis, University of Strasbourg, France, 1863.

33. Crum-Brown, A.; Fraser, T. R. On the connection between chemical constitution and physiological action. Pt 1. On the physiological action of the salts of the ammonium bases, derived from strychnia, brucia, thebia, codeia, morphia, and nicotia. *T. Roy. Soc. Ed.* 1868-1869, 25, 151–203.

34. Furusjö, E.; Svenson, A.; Rahmberg, M.; Andersson, M. The importance of outlier detection and training set selection for reliable environmental QSAR predictions. *Chemosphere.* 2006, 63, 99–108.

35. Geladi, P.; Kowalski, B. R. Partial leastsquares regression: a tutorial. *Anal. Chim. Acta.* 1986, 185, 1–17.

36. Goodman, I. R.; Mahler, R. P. S.; Nguyen, H. T. *Mathematics of Data Fusion*; Kluwer Academic Publishers: Dordrecht, Boston, 1997.

37. Hall, D. L.; McMullen, S. A. H. *Mathematical Techniques in Multisensor Data Fusion*; Artech House, Inc.: Boston, MA, 2004.

38. Hammett, L. P. Some relations between reaction rates and equilibrium constants. *Chem. Rev.* 1935, 17, 125–136.

39. Hammett, L. P. The effect of structure upon the reactions of organic compounds. Benzene derivatives. *J. Am. Chem. Soc.* 1937, 59, 96–103.

40. Hansch, C.; Fujita, T. *p-σ-π* analysis. A method for the correlation of biological activity and chemical structure. *J. Am. Chem. Soc.* 1964, 86, 1616–1626.

41. Hansch, C.; Leo, A. *Exploring QSAR*; American Chemical Society: Washington, DC, 1995.

42. Höskuldsson, A. PLS regression methods. *J. Chemometr.* 1988, 2, 211–228.

43. Kim, K. Outliers in SAR and QSAR: 2. Is a flexible binding site a possible source of outliers? *J. Comput. Aid. Mol. Des.* 2007, 21, 421–435.

44. Kim, K. Outliers in SAR and QSAR: Is unusual binding mode a possible source of outliers? *J. Comput. Aid. Mol. Des.* 2007, 21, 63–86.

45. Meyer, H. Zur Theorie der Alkoholnarkose. *Arch. Exp. Path. Pharm.* 1899, 42, 109–118.

46. Nantasenamat, C.; Naenna, T.; Isarankura-Na-Ayudhya, C.; Prachayasittikul, V. Recognition of DNA splice junction via machine learning approaches. *Excli. J.* 2005, 4, 114–129.

47. Nantasenamat, C.; Naenna, T.; Isarankura, N. A.; Ayudhya, C.; Prachayasittikul, V. Quantitative prediction of imprinting factor of molecularly imprinted polymers by artificial neural network. *J. Comput. Aid. Mol. Des.* 2005, 19, 509–524.

48. Nantasenamat, C.; Tantimongcolwat, T.; Naenna, T.; Isarankura-Na-Ayudhya, C.; Prachayasittikul, V. Prediction of selectivity index of pentachlorophenol-imprinted polymers. *Excli. J.* 2006, 5, 150–163.

49. Nantasenamat, C.; Isarankura-Na-Ayudhya, C.; Naenna, T.; Prachayasittikul, V. Quantitative structure-imprinting factor relationship of molecularly imprinted polymers. *Biosens. Bioelectron.* 2007, 22, 3309–3317.

50. Nantasenamat, C.; Isarankura-Na-Ayudhya, C.; Tansila, N.; Naenna, T.; Prachayasittikul, V. Prediction of GFP spectral properties using artificial neural network. *J. Comput. Chem.* 2007, 28, 1275–1289.

51. Nantasenamat, C.; Isarankura-Na-Ayudhya, C.; Naenna, T.; Prachayasittikul, V. Prediction of bond dissociation enthalpy of antioxidant phenols by support vector machine. *J. Mol. Graph. Model.* 2008, 27, 188–196.

52. Overton, C. E. *Studien über die Narkose*; Fischer: Jena, 1901.

53. Richet, M. C. Note sur le rapport entre la toxicité et les propriétes physiques des corps. *Compt. Rend. Soc. Biol. (Paris).* 1893, 45, 775–776.

54. Taft, R. W. *Separation of polar, steric and resonance effects in reactivity*. In *Steric Effects in Organic Chemistry*; Newman, M. S., Ed.; Wiley: New York, 1956; pp 556–675.

55. Torra, V. *Information Fusion in Data Mining*; Springer-Verlag: Secaucus, NJ, 2003.

56. Verma, R. P.; Hansch, C. An approach toward the problem of outliers in QSAR. *Bioorg. Med. Chem.* 2005, 13, 4597–4621.

57. Wang, L. *Support Vector Machines: Theory and Applications*; Springer- Verlag: New York, 2005.

58. Wold, S.; Trygg, J.; Berglund, A.; Antti, H. Some recent developments in PLS modeling. *Chemometr. Intell. Lab.* 2001, 58, 131–150.

59. Worachartcheewan, A.; Nantasenamat, C.; Naenna, T.; Isarankura-Na-Ayudhya, C.; Prachayasittikul, V. Modeling the activity of Furin inhibitors using artificial neural network. *Eur. J. Med. Chem.* 2009, 44, 1664–1673.

CHAPTER 17

Lead Toxicity and Flavonoids

Amrish Chandra* and Deepali Saxena

Amity Institute of Pharmacy, Amity University, Noida, India
*Email: amrish_chandra@yahoo.com

CONTENTS

ABSTRACT

Lead is one of the oldest-established poisons existing and is known to be harmful to living system. Lead exposure continues to be a major health problem in population of both developed and developing countries of Third World Nations. A research on the toxic effects of lead continues, and in the last decade, some new information on the manifold influences of this metal has emerged. These effects can be reduced by use of flavonoids such as Naringenin and Silymarin. Naringenin is considered to have a bioactive effect on human health as antioxidant, free radical scavenger, anti-inflammatory, carbohydrate metabolism promoter, and immune system modulator; and Silymarin has been used medicinally to treat liver disorders, including acute and chronic viral hepatitis, toxin/drug-induced hepatitis, and cirrhosis and alcoholic liver diseases. These flavonoids can be used alone or in combination as dietary supplement to reverse lead poisoning.

17.1 INTRODUCTION

Lead is one of the oldest-established poisons existing and is known to be harmful to living system. Lead exposure continues to be a major health problem in population of both developed and developing countries of Third World Nations. A research on the toxic effects of lead continues, and in the last decade, some new information on the manifold influences of this metal has emerged.

Lead is a naturally occurring bluish-gray metal found in small amounts in the earth's crust. Its atomic number is 82. It has four electrons in its valence shell, and its oxygen state is +2 rather than +4, as only two electrons are easily ionized. Lead can be found in all parts of our environment. It may exist in several species, including metallic, inorganic salt, inorganic oxide, and organic forms.[1] The main source of lead is from human activities, including burning fossil fuels, mining, and manufacturing.[2,3] Lead has many applications, for example, in the production of batteries, ammunition, metal products (solder and pipes), and devices to shield X-rays. Demand for lead is reduced because of health concerns, so lead from gasoline, paints, ceramic products, caulking, and pipe solder has been reduced in recent years. Exposure to low-levels of Pb has been associated with behavioral abnormalities, learning impairment, decreased hearing, and impaired cognitive functions in humans and in experimental animals. Exposure to lead could be due to (a) eating food or drinking water that contains lead. Earlier, water pipes in some older homes may contain lead solder, from which lead can leach out into the water. (b) Spending time in areas where lead-based paints have been used and are deteriorating. Deteriorating lead paint can contribute to lead dust. (c) Working in a job where lead is used or engaged in certain hobbies involving the use of lead such as stained glass, and (d) using health-care products or folk remedies that contain lead. Small children suffer from lead poisoning by eating lead-based paint chips, chewing on objects painted with lead-based paint, or swallowing house dust or soil that contains lead.

Lead may diffuse into the body by various mechanisms such as ingestion through the intestines, inhalation by the lungs through the skin, or by direct swallowing and ingestion. The effects of lead are the same whether it enters the body through breathing or swallowing. Lead can adversely affect almost every organ and system in our body. The main target of lead toxicity is the nervous system, both in adults and in children. Early symptoms of lead neurotoxicity in both adults and children include irritability, fatigue, decreased attention span, memory loss, depression, and low level of cognitive impairment (Table 17.1). It may also cause weakness in fingers, wrists, or ankles. As childhood exposure increases, behavioral symptoms of impulsiveness, inability to follow sequences of directions, decreased play activity, lowered IQ, and poor attentiveness are seen at PbBs of 10–35 g/dL.

TABLE 17.1 Acute and Chronic Symptoms of Lead Toxicity

Earlier Symptoms	Symptoms of Chronic Exposure
• Diffuse muscle weakness	• Abdominal pain/cramping
• General fatigue/lethargy	• Nausea/vomiting
• Myalgia	• Short-term memory loss
• Joint pain/arthritis	• Depression
• Loss of appetite	• Incordination
• Change in taste of food	• Numbness and tingling in extremities
• Headache	• Constipation
• Insomnia	• Inability to concentrate
• Irritability	• Impotence
• Diminished libido	• Somnolence/severe lethargy
• Weight loss of 10 lbs	• Abdominal colic
• Tremulousness	

Lead exposure also causes small increase in blood pressure, particularly in middle-aged and older people, and it can cause anemia. Exposure to high levels of lead can severely damage the brain and kidney in adults or children and ultimately leads to death. In pregnant women, high levels of lead exposure may cause miscarriage. In men, it can damage the organs responsible for sperm production.

Poor nutrition, particularly inadequate intake of calcium and iron, is probably an important risk factor for poisoning. Children's hand-to-mouth activity, increased respiratory rates, and intestinal absorption of lead make them more susceptible than adults to lead exposure. A child who swallows large amounts of lead may develop blood anemia, severe stomachache, muscle weakness, and brain damage. Even at much lower levels such as 10 μg/dL of lead exposure, it can affect a child's mental and physical growth. Lead exposure is more dangerous for young and unborn children. Harmful effects include premature births, learning difficulties, reduced growth in young children, and decreased mental ability in the infant. Some of these effects may persist beyond childhood. Symptoms of lead intoxication include anorexia, apathy, lethargy, anemia, decreased play activity, aggressiveness, and poor coordination.

Blood lead level (μg/dL) is the biologic index most often used as an indicator of recent lead exposure. Atomic absorption spectroscopy is used to measure the blood lead level. In addition to blood lead level, other lead exposure includes free erythrocyte protoporphyrin and zinc protoporphyrin; both are precursors of blood whose levels elevate upon moderate to high exposure to lead. However, these free erythrocyte and zinc protoporphyrins are neither sensitive enough nor specific enough to be used as primary indicators of lead exposure. Lead levels in plasma, urine, bone, and teeth (dentin lead) are less commonly used measures of exposure and body burden.

Lead is a divalent cation and binds strongly to sulfhydryl groups of the many organs affected by lead, and the most vulnerable is central nervous system (CNS). Much of lead's toxicity can be attributed to the alteration in enzymes and structural proteins, but this toxicant has many other targets. Toxic properties of lead are due to its ability to mimic or compete with calcium. At picomolar concentrations, lead competes successfully with calcium, and thereby affects neuronal signaling. It inhibits the entry of calcium into cells. Uncoupled energy metabolism inhibited cellular respiration and altered calcium kinetics.

Attention has also been focused on the haem synthetic pathway, where the number of sites for lead activity is found. δ-Aminolevulinic acid dehydratase (δ-ALAD) is extremely sensitive to lead. Inhibition of this enzyme results in increased circulating aminolevulinic acid (ALA), which may account for some of the behavioral disorders seen in patients with porphyria and perhaps in lead toxicity.

Several mechanisms have been proposed for lead-induced abnormalities, but none have yet been defined explicitly. Disruption of a variety of biochemical processes rather than a single mechanism is responsible for the toxicity.

Lead has three important biochemical properties that contribute to its toxic effects on humans.

- It is an electropositive metal with a high affinity for sulfhydryl groups, and thus inhibits sulfhydryl-dependent enzymes such as δ-aminolevulinic acid dehydratase and ferrochelatase, which are essential for the synthesis of haem.
- Divalent lead acts in a manner similar to calcium and competitively inhibits its actions in important areas such as mitochondrial oxidative phosphorylation. In particular, lead impairs the intracellular messenger system normally regulated

by calcium, and thereby affects endocrine and neuronal function. Lead also changes the vasomotor action of smooth muscle by its effect on Ca-ATPase.

- Lead can affect the genetic transcription of DNA by interaction with nucleic acid-binding proteins with potential consequences for gene regulation.

Oxidative stress has been reported as one of the important mechanism of toxic effect of lead. Lead causes oxidative stress by inducing the generation of reactive oxygen species (ROS) and weakens the antioxidant defense system of cells. Depletion of cell's major sulfhydryl reserves seems to be an important indirect mechanism for oxidative stress induced by lead. Lead irreversibly binds to the sulfhydryl group of proteins, causing impaired function. The disruption of reducing the status of tissue leads to the formation of ROS, which may damage the essential biomolecules such as protein, lipids, and DNA. In vivo generation of highly ROS, such as hydroxyl radical ($HO^{\bullet}$), hydrogen peroxide (H_2O_2), superoxide radical ($O_2^{\bullet}$), and lipid peroxide (LPO), after the lead exposure, may result in systematic mobilization and depletion of the cells' intrinsic antioxidant defenses. At high levels, these ROS can be toxic to cells and may contribute to cellular dysfunction and poisoning. A significant decrease in the activity of superoxide dismutase (SOD), a free radical scavenger, and metalloenzyme (zinc/copper) occurs due to an increase in lead concentration, thereby reducing the disposal of superoxide radicals. Significant depletion of other antioxidant enzymes, such as catalase, GPx, and glutathione S-transferase (GST), also occurs. Catalase is an efficient decomposer of H_2O_2 and known to be susceptible to lead toxicity. Lead-induced decrease in GPx activity may arise as a consequence of impaired functional groups, such as GSH and NADPH- or selenium-mediated detoxification of toxic metals. While antioxidant enzyme GST is known to provide protection against oxidative stress, the inhibition of this enzyme on lead exposure might be due to the depletion in the status of tissue thiol moiety. These enzymes are important for maintaining critical balance in the glutathione redox state. The concentration of malondialdehyde (MDA), which is a reflection of endogenous lipid oxidation level, gets increased on lead exposure.

Lead-induced disruption of the pro-oxidant/antioxidant balance in lead-exposed tissue contributes to tissue injury via oxidative damage to critical biochemical variables, such as lipids, proteins, and DNA. Since oxidative stress is implicated in lead toxicity, a therapeutic strategy to increase the antioxidant capacity of cells may fortify the long-term effective treatment of its poisoning. This may be accomplished either by reducing the possibility of metal interacting with critical biomolecules and inducing oxidative damage or by bolstering the cells' antioxidant defenses through endogenous supplementation of antioxidant molecules.

The most important initial aspect of management of lead poisoning is the removal of the patient from the source of exposure. One of the major focuses is to address the issue of recovery in altered biochemical variables in two major target sites (hematopoietic and neurological disorders), reducing body lead burden adopting few newer strategies with antioxidants.

Flavonoids are polyphenolic compounds that are ubiquitous in nature. Over 4000 flavonoids have been identified, many of which occur in fruits, vegetables, and beverages (tea, coffee, beer, wine, and fruit drinks). The flavonoids have aroused considerable interest recently because of their potential beneficial effects on human health, and they have been reported to have antiviral, antiallergic, antiplatelet, anti-inflammatory, antitumor, and antioxidant activities.

Antioxidants are compounds that protect cells against the damaging effects of ROS, such as singlet oxygen, superoxide, peroxyl radicals, hydroxyl radicals, and peroxynitrite. An imbalance between antioxidants and ROS results in oxidative stress, leading to cellular damage. Flavonoids may help provide protection against these diseases by contributing, along with antioxidant vitamins and enzymes, to the total antioxidant defense system of the human body. Epidemiological studies have shown that flavonoid intake is inversely related to mortality from coronary heart disease and to the incidence of heart attacks. The capacity of flavonoids to act as antioxidants depends upon their molecular structure. The position of hydroxyl groups and other features in the chemical structure of flavonoids are important for their antioxidant and free radical scavenging activities.[4]

"Naringenin", a flavanoid found in grapefruit juice, has been shown to have an inhibitory effect on the human cytochrome P450 isoform CYP1A2 that can change pharmacokinetics in a human (or orthologous) host of several popular drugs in an adverse manner, even resulting in carcinogens of otherwise harmless substances.[5] Naringenin (Fig. 17.1) is considered to have a bioactive effect on human health as antioxidant, free radical scavenger, anti-inflammatory, carbohydrate metabolism promoter, and immune system modulator. This substance has also been shown to reduce oxidative damage to DNA in vitro. Scientists exposed cells to 80 µmol of naringenin per liter, for 24 h, and found that the amount of hydroxyl damage to the DNA was reduced by 24% in that very short period of time. A full glass of orange juice will supply about enough naringenin to achieve a concentration of about 0.5 µmol/L.

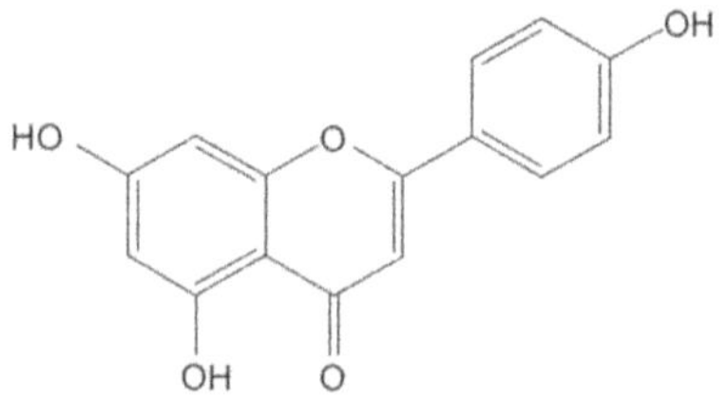

FIGURE 17.1   Chemical Structure of Naringenin.

"Silymarin", a flavonolignan from the seeds of "milk thistle" (*Silybum marianum*), has been widely used from ancient times because of its excellent hepatoprotective action. It is a mixture of mainly three flavonolignans: silybin, silidianin, and silychristine, with silybin being the most active. Silymarin has been used medicinally to treat liver disorders, including acute and chronic viral hepatitis, toxin/drug-induced hepatitis, and cirrhosis and alcoholic liver diseases. Its mechanism of action includes inhibition of hepatotoxin binding to receptor sites on the hepatocyte membrane; reduction of glutathione oxidation to enhance its level in the liver and intestine; antioxidant activity; and stimulation of ribosomal RNA polymerase and subsequent protein synthesis, leading to enhanced hepatocyte regeneration.[6]

Silymarin (Fig. 17.2) is extracted using 95% ethanol, from the seeds of the milk thistle. The plant consists of approximately 70–80% of the silymarin flavonolignan and approximately 20–30% of a chemically undefined fraction, comprising mostly polymeric and oxidized polyphenolic compounds. The most prevalent component of the silymarin complex is silybin (50–60% of silymarin), which is the most active photochemical and is largely responsible for the claimed benefit of the silymarin. Besides silybin, which is a mixture of two diastereomers (A and B) in approximately 1:1 proportion, considerable amounts of other flavonolignans are present in the silymarin complex, namely, silychristin (20%), silydianin (10%), isosilybin (5%), and dehydrosilybin, and a few flavonoids, mainly taxifolin. The seeds also contain betaine, trimethylglycine, and essential fatty acids that may contribute to silymarin's hepatoprotective and anti-inflammatory effects.[7–9]

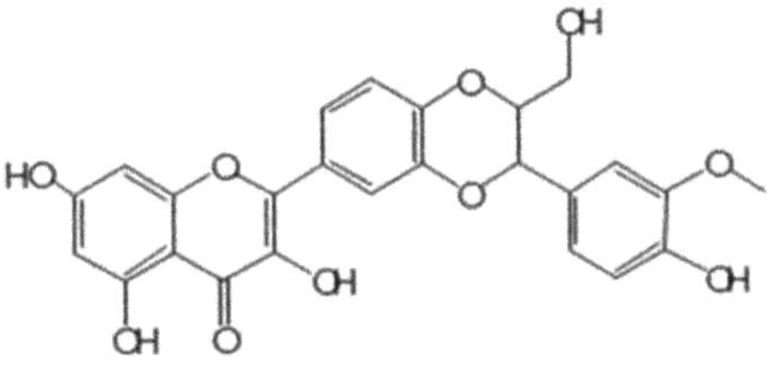

FIGURE 17.2   Chemical Structure of Silymarin.

17.2 LITERATURE REVIEW

Lead is a metal of antiquity and is detectable in practically all phases of the inert environment and in all biological systems, having widespread industrial applications. It is a highly toxic metal, the clinical manifestations of its toxicity are termed as "plumbism", have been known since ancient times. Significant exposure to lead is an environmental threat to optimal health and to physical development in young children that affects all socioeconomic groups.[1] Lead is widely distributed in nature. It is usually associated with other metals, particularly silver and zinc. Although, mined and used for centuries, galena still remains the principal source of lead today. Lead, a ubiquitous environmental toxin, induces a broad range of physiological, biochemical, and behavioral dysfunction. Its poisoning is thus an environmental disease, but it is also a disease of lifestyle. It is known to affect the structure and function of various organs and tissues.

17.2.1 Physical and Chemical Properties

Lead is a member of Group IVB of the Periodic Table, together with tin, germanium, carbon, and silicon. The chemical symbol is Pb, from the Latin name *plumbum*. Other characteristics are given in Table 17.2.

TABLE 17.2 Physical and Chemical Properties of Lead

Name	Lead
Symbol	Pb
Atomic number	82
Atomic mass	207.2 amu
Melting point	327.5°C (600.65 K, 621.5°F)
Boiling point	1740.0°C (2013.15 K, 3164.0°F)
Number of protons/electrons	82
Number of neutrons	125
Crystal structure	Cubic
Density at 293 K	11.34 g/cm^3
Color	Bluish

17.2.2 Chemical Forms of Lead

Lead is released into the environment in many different chemical forms, which determines its water solubility, types of chemical reactions that occur in the atmosphere, water, and soil, and the extent to which lead binds in soils.[10]

17.2.2.1 Inorganic Lead Compounds

Inorganic lead is without doubt one of the oldest occupational toxins, evidence of which can be found dating back to Roman times. The majority of compounds detected consist of the lead halides: lead bromide ($PbBr_2$), lead chloride ($PbCl_2$), lead bromochloride (PbBrCl), the alpha and beta forms of the double salt lead bromochloride ammonium chloride ($2PbBrCl \cdot NH_4Cl$), lead sulfide (PbS), lead sulfate ($PbSO_4$), and elemental lead.[10] Another form emitted from mining operations and smelters is lead oxide.[11] Once these lead compounds are present in the atmosphere, they are converted through chemical reactions into a large number of additional lead compounds. Some of the lead compounds identified in the atmosphere are $2PbBrCl \cdot NH_4Cl$, $PbSO_4$, lead carbonate ($PbCO_3$), $PbBr_2$, $PbCl_2$, lead oxide (PbOx), lead hydroxychloride (Pb(OH)Cl), and lead ($2PbO \cdot PbBrCl$).[10] Inorganic lead compounds are also found in water and soil. The amount of lead in surface water depends on the pH and the dissolved salt content. In the environment, the divalent form (Pb^{2+}) is the stable ionic form of lead.

17.2.2.2 Organic Lead Compounds

The tetra alkyl lead compounds, specifically tetraethyl lead (TEL) and tetramethyl lead (TML), are the primary organic lead compounds that were used as automotive gasoline additives. The phase-out of TEL and TML from automotive gasoline was initiated in the 1970s. These two compounds no longer are present in large amounts in the atmosphere; however, their degradation products still exist in the atmosphere. These compounds decompose rapidly to trialkyl and dialkyl lead compounds when exposed to sunlight and eventually degrade to inorganic lead oxides.[11]

In water, the TEL compounds are subject to photolysis and volatilization. The more volatile compounds are lost to the atmosphere by evaporation. In air, the degradation process consists of trialkyl lead compounds degrading to dialkyl lead and finally to inorganic lead. Triethyl and trimethyl lead are more water soluble than are TEL or TML and therefore are more often detected in aquatic environments.[11] Another source of organic lead in water is the conversion of inorganic lead to TML by microorganisms living in anaerobic lake sediments.[12] However, if the water over the sediments is aerobic, the TML will be oxidized, resulting in release of lesser amounts of TML to the water.

In soil, organic lead compounds such as TML and TEL may be converted to highly water-soluble compounds, such as the trialkyl lead oxides. These compounds could be subject to leaching from the soil.

17.2.3 Lead in the Environment

17.2.3.1 Occurrence

The level of lead in the earth's crust is about 20 mg/kg. In the environment, it may be derived from either natural or anthropogenic sources. Most of the lead present was buried in subsurface deposits composed of a relatively inert (insoluble) form. As a consequence, humans and other living species have no known use for lead. The amount of lead on the earth's crust is larger than might be predicted. One reason is that it was concentrated during the earth-forming process and a second is that it is the "sink" for radioactive decay of uranium and thorium. A lead sulfide containing ore found at the surface of the earth's crust undergoes weathering to the mineral $PbSO_4$ or "anglesite". Further weathering may result in "cerrusite", $PbCO_3$. Lead and iron sulfide ore at the interface with unweathered ore, high in clay, is known as "jarosite". Most lead in ore bodies is in the form of galena or cerrusite, both of which are attractive minerals. Lead has four common isotopes: lead-204, lead-206, lead-207, and lead-208. The last three forms of lead result from the radioactive decay of thorium and two different isotopes of uranium. Lead is found with silver, and due to this reason, it is extensively mined. Lead is also found in conjunction with many other trace elements, especially with antimony.[3]

17.2.3.2 Environmental Sources

Although lead is ubiquitous in the environment of industrialized nations, the contribution of natural sources of lead to concentrations in the environment is low compared to the contribution from human activities.[4] Through human activities such as mining, smelting, refining, manufacturing, and recycling, lead finds its way into the air, water, and surface soil. Lead-containing manufactured products (gasoline, paint, printing inks, lead water pipes, lead-glazed pottery, lead-soldered cans, battery casings, etc.) also contribute to the lead burden. Lead in contaminated soil and dust can find its way into the food and water supply.

The source of greatest concern is old housing, specifically houses once painted with products containing lead as a pigment. The chips and dust from peeling or cracking leaded paint remain highly toxic. Sanding, scraping, or heating painted doors, windows, stairs, or fences can release leaded dust into the air, where children and adults may breathe it in. Even vacuuming, sweeping, or walking can circulate the dust, which eventually gathers on the floor where it is accessible to infants and toddlers engaging in hand-to-mouth activity.[5]

Air

Lead is released to air by natural processes such as volcanic activity, forest fires, crustal weathering, and radioactive decay from radon. These natural contributions are of relatively minor consequence. The vast majority of lead in the atmosphere results from human activity. The overall emissions of lead to air dropped significantly beginning in the 1970s and continuing until a complete ban on lead in fuel went into effect in 1996. When organic lead compounds are exposed to sunlight, they decompose rapidly to trialkyl and dialkyl lead compounds. These compounds eventually decompose to inorganic lead oxides, through a combination of photolysis and reactions with hydroxyl radicals and ozone. The half-lives of TEL and TML in the summer are 2 and 9 h, respectively, while in winter, both compounds have half-lives up to several days.[11]

The release of lead from paint to the local environment is an important source of environmental exposure. Lead carbonate was the most frequently used lead pigment, and lead oxide and lead chromate also were common. Flaking or peeling of aging paint can be a major point source of environmental lead exposure, as well as sanding of painted surfaces during home renovation. By these processes, lead can become mobilized in dust to the air and soil. In a pooled analysis of 12 epidemiological studies,

Lanphear *et al.*[13] confirmed that lead-contaminated house dust was the major source of lead exposure to children.

Water

Lead enters groundwater from natural weathering of rocks and soil, indirectly from atmospheric fallout and directly from industrial sources. In water, organic lead compounds undergo photolysis and volatilization. Degradation processes convert trialkyl lead to dialkyl lead and then to inorganic lead compounds.[11] An additional and distinct hazard to the water supply is lead piping or lead solder in older plumbing systems. Areas with a supply of soft (acidic) water are more susceptible to release of lead from plumbing, which can result in levels of lead in drinking water high enough to have significant effects on human health.[14]

Soil

The largest amount of lead released into the environment is released to land, predominantly to landfill sites. Lead-containing wastes result from ore production, household renovation, and remediation of lead paint, use of lead in ammunition, solder, weights, and bearing metals, and production of iron and steel. Although lead is now banned in motor fuels in the United States and several other countries, organic lead compounds continue to be present or actively deposited in the soil.

17.2.3.3 Occupational Exposure

The most common route of occupational exposure to lead is inhalation of lead fumes or leaded dusts in air and absorption of lead through the respiratory system. Lead also may be ingested and absorbed in the gastrointestinal tract. Absorption through the skin occurs with organic lead[15] and possibly also with the more soluble species of inorganic lead.[16] The National Institute for Occupational Safety and Health has estimated that more than 3 million Americans potentially are occupationally exposed to some form of lead.[17] Occupational exposure to lead may occur during the production of lead-acid batteries, in which grids are formed either by melting lead blocks and pouring molten lead into molds or by feeding rolled sheets of lead through punch presses. A lead oxide paste also is applied into grid spaces. Leaded glassware is made by combining lead oxide compounds with molten quartz. This process results in lead fumes and dusts, and glassblowing is an additional avenue for potential contact with lead. Production of pigments can involve lead oxide, lead carbonate, and lead chloride. Lead miners and mine workers are involved in the extraction, crushing, grinding, and concentration of lead.

Workers in the plastics and vinyl industries may be exposed to lead when it is used as a stabilizing or coloring agent. Powdered pigments, such as lead chromate, are blended with plastic pellets and heated to form some plastic products. Similarly, lead has frequently been used in compounding rubber. Greater health awareness has led to reduction in applications of lead in plastics and rubber manufacture. Automobile-repair workers may be exposed to lead through work around batteries and other parts, engine reconditioning, solder, and, until recently in the United States and several other countries, leaded gasoline. Lead continues to find application in the production of cable sheathing because of its workability, durability, and resistance to corrosion.

17.2.4   Lead Toxicity – Global Scenario

Despite being a dangerous toxin, lead also forms one of man's most valuable commodities in present scenario. Occurring naturally in the environment, the metal is mined and processed in about 60 countries. The usage continues to increase and has risen from 4 million tons per year. Of this, nearly 2 million tons per year are produced in Asia. Secondary production or recycling is now widely practiced and currently accounts for about 50% of usage worldwide. The main producers of lead mineral are China, Australia, USA, Peru, Canada, and Mexico. These six countries produce three-quarters of world output. All industrialized nations use lead. The USA is by far the greatest consumer, most of it being used for batteries. Other major consumers are China, UK, Germany, Japan, Republic of Korea, France, and Italy, while Spain, Mexico, and Brazil use lesser amounts.[18]

In India, about 75% of total demand is from the domestic battery industries. Demand is growing at the rate of 6–7% per annum and will continue to grow in the near future.[18] Annual demand for lead is nearly 1.60 lakh tones. When an ore body occurs at some depth below the surface, it must be mined by underground methods; in India, lead mines are found at Zawar, Rajpura –Dariba, and Rampura – Agucha, which are highly mechanized. The first two are underground mines. Rampura – Agucha is an opencast mine. There are small lead mines at Sargipalli and Agnigundala. In India, lead mines were commissioned in May 1991 together with a smelter at Chandeliya, which proved to be a significant step in reducing dependence on imports in the case of both the metals. China is a major world producer and important supplier of refined lead, ranking first in mine production followed by the United States, Canada, Germany, the United Kingdom, and Japan. In China, lead is produced mainly as a co-product of zinc.

Ten million people are estimated to be at risk of lead metal toxicity at identified sites. Lead continues to be a significant public health problem especially in developing countries, where there are considerable variations in the sources and pathways of exposure.[18] For example, in many Latin American countries, leaded paint is not a significant source of recurrent exposure, whereas lead-glazed ceramics are such a source. Exposure attributable to miscellaneous sources may be even more significant than universal exposure associated with leaded petrol, especially for people living in poverty. The situation is similar in other countries where industrialization is occurring. In Dhaka, lead concentrations in airborne particulate matter averaged 453 ng/ m^3 during the low rainfall season of November to January.[19]

Environmental levels of lead have been increasing for hundreds of years and are only just starting to decrease in response to greater awareness of its harmful effects. Today, much of the lead in circulation exists in car batteries, also called used lead-acid batteries (ULAB). Of the 6 million tons of lead that are used annually, approximately three-quarters go into the production of lead-acid batteries in which 97% are eventually recycled to regain the lead.[20] Acute lead poisoning commonly results from people inhaling lead particles in dust or through ingestion of lead-contaminated dirt. This was the case around the Haina ULAB recycling facility in the Dominican Republic, where at least 28% of children required immediate treatment for lead exposure and 5% had blood lead levels that put them at risk for neurological damage.[21] The extraordinary danger that lead poses was recently highlighted by a catastrophe in Dakar, Senegal, where between November 2007 and March 2008, 18 children died from acute lead poisoning due to lead dust and soil exposure from ULAB recycling. Until the contamination was discovered, the main economic activity in the Dakar community of Thiaroye Sur Mer was ULAB recycling. Initial tests of children living in the area found an average blood lead level of 129.5 μg/dL, drastically exceeding the US Centres for Disease Control (CDC) and Prevention action level of 10 μg/dL.[19] Once lead is on the ground, it can remain in the upper layer of soil for many years. Lead can migrate into ground water supplies, particularly in areas that receive acidic or "soft" rainwater. Furthermore, levels of lead can build up in plants and animals when the surrounding environment is contaminated.

Many developing countries have been actively engaged in lead reduction programs, particularly in respect of leaded petrol, "the mistake of the twentieth century". Bangladesh, China, Egypt, Haiti, Honduras, Hungary, India, Kuwait, Nicaragua, Malaysia, and Thailand have, for example, made dramatic efforts to phase out leaded petrol in recent years. Success in this endeavor requires government commitment, incentive policies, a broad consensus among stakeholders, and public understanding, acceptance, and support.

17.2.5 Biological Aspects of Lead Metabolism

17.2.5.1 *Absorption*

Lead may be absorbed into the body by ingestion,[6,7] inhalation,[8] or through skin.[16] The absorption of lead from different sources is dependent on many factors, such as amount of lead presented to portals per unit time and the physical and chemical states in which lead is presented. It is also influenced by factors such as age and physiological status.

Most of the studies of ingested lead indicate that not more than 10% lead is absorbed from the gastrointestinal tract[9] and maximum passes out in the feces. Intestinal absorption of lead by children is much greater than in the adults. Nutritional iron deficiency enhances lead toxicity, thereby giving concern that pregnant women and young children may be more susceptible to lead toxicity.22,23

17.2.5.2 Distribution and Retention

Blood contains lead in a nondiffusible form bound to erythrocytes and in a diffusible form in plasma. Plasma occupies a central position in the distribution equilibrium and would be expected to reflect the concentration of lead in all the body tissues. In the absence of significant previous exposure, lead within red blood cells (RBCs) was found bound primarily to hemoglobin.[24] Lead in blood is primarily bound to RBC (99%) rather than plasma.[25] Within the cell, 50% is bound to hemoglobin and other to proteins.[26] Iron deficiency is well documented to increase susceptibility to lead intoxication.[5]

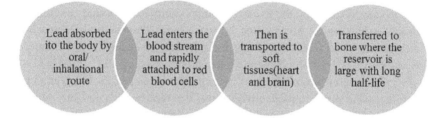

FIGURE 17.3   Lead Metabolism.

Among body organs, the greatest percentage of lead is taken into the kidney followed by the liver and the other soft tissues such as heart and brain, but the lead in the skeleton represents the major body fraction.[27] Within the skeleton, lead is incorporated into the mineral in place of calcium. There it accounts for about 94–95% of the total body burden in adults and about 70% in young children.[28] Lead is readily transferred across the placenta, and the concentration of lead in the blood of new-born children is similar to that of their mothers, suggesting that lead come into equilibrium between the mother and fetus and the increased mobilization of bone lead during pregnancy may also continue to increase.[7,29,30] Lead also crosses the blood–brain barrier (BBB), and a study by Antonio et al.[31] showed that the plasma lead fraction is in equilibrium with that in the cerebrospinal fluid (CSF). Assuming that the BBB is functioning normally, plasma lead concentration is a good indicator of CSF lead concentration, and therefore, this parameter is a reliable index of potential lead transfer to brain tissues.[31] A number of factors that are known to increase bone turnover, such as pregnancy, lactation, chemotherapy, tumor infiltration of the bone, or postmenopausal osteoporosis, may be associated with the mobilization of lead in stores, leading to chronic lead toxicity. Hypothyroidism too is known to increase bone turnover; it has rarely been implicated in the pathogenesis of lead poisoning.

17.2.5.3 Excretion

Inorganic lead is not metabolized, and the excretion is low primarily through urinary tract. Absorbed lead is eliminated primarily via the kidney in the urine (about 76%) and to a lesser extent by the gastrointestinal tract (about 16%) through biliary secretion.[32] Other routes for elimination (hair, nails, sweat, and exfoliated skin) account for approximately 8%.[33] Lead is also excreted in milk in concentrations of up to 12 μg/L. In general, lead is excreted very slowly from the body with its biological half-life estimated at 10 years, thus facilitating accumulation in the body (Fig. 17.3).

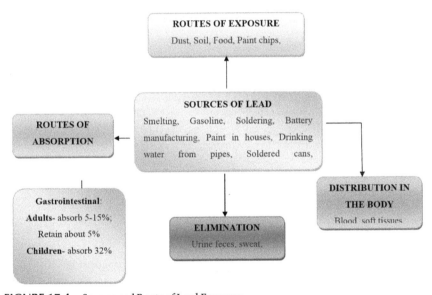

FIGURE 17.4 Sources and Route of Lead Exposure.

17.2.6 Clinical Effects of Lead Toxicity

Lead enters the system through various sources (Fig. 17.4) Lead is known to cause acute, subchronic, and chronic toxicity. The most commonly used biological marker is the concentration of lead in blood. The concentration of lead in plasma is very low and thus is not recommended.

17.2.6.1 Acute Toxicity

Acute poisoning is uncommon. It results from inhalation of large quantities of lead due to occupational exposure among industrial workers and in children through ingestion of large oral dose from lead-based paint on toys. The clinical symptoms of acute poisoning are characterized by metallic taste, abdominal pain, vomiting, diarrhea, anemia, oliguria, collapse, and coma. Blood lead levels: Acute lead toxicity occurs at blood levels of 100–120 µg/dL in adults and 80–100 µg/dL in children. The various symptoms are given in Table 17.3.

TABLE 17.3 General Signs and Symptoms of Lead Toxicity

Mild	Moderate	Severe
Myalgias	Headache	Encephalopathy
Irritability	Tremor	Motor neuropathy
Paresthesias	Vomiting	Seizures
Mild fatigue	General fatigue	Coma
Intermittent	Diffuse-abdominal pain	Abdominal colic
Abdominal pain	Weight loss	Lead lines
Lethargy	Loss of libido	Oliguria
	Constipation	

Adapted from case studies in environmental medicine: lead toxicity. Atlanta: Agency for Toxic Substances and Disease Registry, 1992.

17.2.6.2 Chronic Toxicity

This is more common and can be described in three stages of progression:

- The "early stage" is characterized by loss of appetite, weight loss, constipation, irritability, occasional vomiting, fatigue, weakness, gingival lining on gums, and anemia;
- The "second stage" is marked by intermittent vomiting, irritability, nervousness, tremors, and sensory disturbances in the extremities, most often accompanied by stippling of RBCs; and
- A "severe stage" of toxicity is characterized by persistent vomiting, encephalopathy, lethargy, delirium, convulsions, and coma.

17.2.7 Biochemical and Toxicological Effects

17.2.7.1 Impairment of Normal Metabolic Pathway

Lead is distributed in all cells. The biochemical basis for its toxicity is its ability to bind the biologically important molecules, thereby interfering with their function by a number of mechanisms. Enzymes may bind lead resulting in an adverse function. It binds to sulfhydryl and amide groups frequent components of enzymes, altering their configuration and diminishing their activities. It may also compete with essential metallic cations for binding sites, inhibiting enzyme activity or altering the transport of essential cations such as calcium. At the subcellular level, the mitochondrion appears to be the main target organelle for toxic effects of lead in many tissues.[34] It has been reported to impair normal metabolic pathways in children at very low blood levels.[35,36] At least three enzymes of the haem biosynthetic pathway are affected by lead, and at high blood lead levels, the decreased haem synthesis that results leads to decreased synthesis of hemoglobin. There is some evidence that accumulation of one of the intermediates, δ-ALA, exerts toxic effects on neural tissues through interference with the activity of the neurotransmitter gamma-amino-butyric acid (GABA).[37]

17.2.7.2 Toxic Effects of Lead on Organ Systems

Lead is a poison that affects virtually every system in the body. Children are more vulnerable to lead exposure than adults because of the frequent hand-to-mouth activity and a higher rate of intestinal absorption and retention. Lead is a cumulative poison. It produces a range of effects, primarily on the hematopoietic system, the nervous system, and the kidneys.

17.2.7.3 Hematopoietic System

The relationship between lead and haem pathway enzymes has been well studied (Fig. 17.5). This pathway is found in all cells. Three of the seven enzymes in the production of haem are downregulated by lead, resulting in the dose-dependent diminished production of haem and in the accumulation of precursor molecule. The hematological effects of lead can be attributed to the combined effect of (a) inhibition of hemoglobin synthesis and (b) shortened life spans of circulating erythrocytes. These effects may result in anemia, which may be mild, hyper, chronic, and, sometimes, microcytic anemia. Anemia is also associated with shortened RBCs life span. Basophilic stippling of RBC is a feature of lead-induced anemia. The adverse effects of lead appear even with blood concentration as low as 10 μg/dL.[38,39]

Lead inhibits many stages in the pathway of haem synthesis. δ-ALAD catalyzes the formation of prophobilinogen from δ-ALA and ferrochelatase, which incorporates iron into protoporphyrin.[40] It is suggested that the inhibition of ALAD can occur at blood lead as low as 5 µg/dL. At higher concentrations, ALAD inhibition is more pronounced (90% at blood lead 55 µg/dL). ALA in urine has been used for many years as indicator of exposure, inhibition of hematopoiesis among industrial workers, and the diagnosis of lead poisoning. A significant correlation coefficient between lead in blood (and lead in urine) and ALA-U or ALA-D has been suggested.[40,41]

Ferrochelatase is the enzyme that catalyzes the incorporation of iron into the porphyrin ring. If as a result of lead toxicity, the enzyme is inhibited and its pathway is interrupted, or if adequate iron is not available, zinc is substituted for iron, and zinc protoporphyrin concentration is increased.[42] The critical target, however, seems to be the enzyme's haem synthesis, essential for the insertion of iron into the precursor, protoporphyrin IX. The major consequences of this effect, which have been evaluated in both adults and children, are reduction of hemoglobin and the inhibition of cytochrome P450-dependent phase-I metabolism. Lead inhibits normal hemoprotein function in both respects, which results in basophilic stippling of erythrocytes related to clustering of ribosome and microcytosis. The threshold for this effect in children is approximately 15 µg/dL.[38,43]

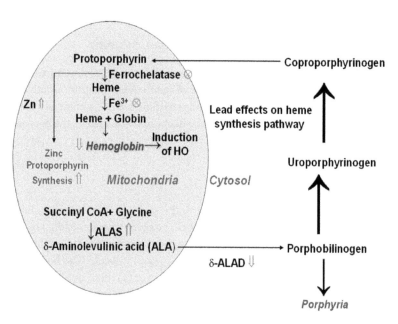

FIGURE 17.5 Effects of Lead on Haem Synthesis Pathway (Source: Kelada *et al.*, Am. J. Epidemiol., 154, 1–13, 2001).

17.2.7.4 Renal Effects

Lead-induced chronic renal insufficiency may result in gout. A direct effect on the kidney of long-term lead exposure is nephropathy. Impairment of proximal tubular function manifests in aminoaciduria, glycosuria, and hyperphosphaturia (a Fanconi-like syndrome). Overt effects of lead on the kidney in man and experimental animals, particularly the rat and mouse, begin with acute morphological changes consisting of nuclear inclusion bodies or lead protein complexes and ultrastructural changes in organelles, particularly mitochondria. Dysfunction of proximal renal tubules (Fanconi syndrome) as manifested by glycosuria, aminoaciduria, and hyperphosphaturia in the presence of hypophosphatemia and rickets was first noted in acute lead poisoning.[44,45] Long-term exposure to lead may give rise to the development of irreversible functional and morphological renal changes. This includes intense interstitial fibrosis, tubular atrophy, and dilation and the development of functional as well as ultrastructural changes in renal tubular mitochondria.[46,47]

The symptoms resulting from lead poisoning are subtle, and often the patients remain asymptomatic until significant reductions of renal function have occurred. Chronic exposure to high levels of lead results in irreversible changes in the kidney, including interstitial fibrosis, tubular atrophy, glomerular sclerosis, and ultimately renal failure. In general, individuals with blood lead levels of 60 µg/dL or more are at a definite risk of developing renal failure. Although, recently it has become evident that blood lead levels as low as 10 µg/dL, previously considered to be safe, may also be associated with renal functional abnormalities.

17.2.7.5 Neurological and Neurobehavioral Effects

The most vulnerable target of lead poisoning is the nervous system (Fig 17.6). In children, neurological deficits have been documented at exposure levels once thought to cause no harmful effects. In addition to the lack of a precise threshold, childhood lead toxicity may have permanent effects.[48,49] The effect of lead on cognitive and behavioral development of the CNS is the critical effect on infants and children and is the focus for current prevention strategies.[1]

- Effects in children generally occur at lower blood lead levels than in adults.
- The developing nervous system in children can be affected adversely at blood lead levels of less than 10 g/dL.
- Neurological deficits, as well as other effects caused by lead poisoning, may be irreversible.

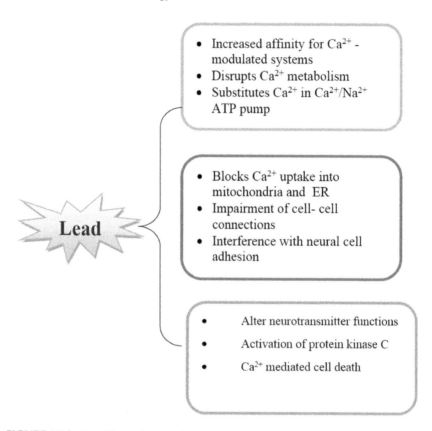

FIGURE 17.6 Possible Mechanism of Lead-Induced Neurological Effects.

Generally recognized effect of lead on the CNS is encephalopathy. Headache, poor attention spam, irritability, loss of memory, and dullness are the early symptoms of the effects of lead exposure on the CNS. The developing nervous system of the child is more sensitive to lead-induced impairment. The most serious manifestation of lead poisoning is acute encephalopathy, the symptoms of which include persistent vomiting, ataxia, seizures, impaired consciousness, and coma. Survivors suffered a number of neurological complications such as mental retardation, deafness, blindness, and convulsions. A chronic form of encephalopathy has also been described in which progressive mental retardation, loss of motor skills, and behavioral disorders occur rather than the more precipitous symptoms seen in acute encephalopathy.[33,50,51] The primary effects of lead on the peripheral nervous system are reduced motor activity, muscular weakness, especially of the exterior muscles, tremor fatigue, and lack of muscular coordination.

17.2.7.6 Reproductive Toxicity

Lead exerts its toxic effect directly on the developing fetus after gestation begins and indirectly on paternal or maternal physiology before and during the reproduction process.[52] Several studies have been conducted on the association between lead exposure and sterility, abortion, still births, and neonatal deaths, but little evidence is available as to whether subtoxic levels of lead affect fertility or cause fetal injuries in human.[53,54] Increase in blood lead reduces semen volume, semen density, and counts of total motile and viable sperm, percentage of progressively motile sperm, levels of zinc, acid phosphatase, and citric acid and increases the percentage of pathological sperm.[55,56]

Abortions, miscarriages, and still births have also been reported among women working in lead industries. Prenatal exposure to lead has been associated with toxic effects on the fetus. These include reduced gestational age, birth weight, and adversely delayed cognitive development. Recent studies have suggested that a significant amount of bone lead is mobilized, enters the circulation during pregnancy and lactation, and crosses the placenta.

17.2.7.7 Effects on Bone

Lead in bone is of interest for two reasons. Bone is the largest depository of the body burden of lead, and second, it is now recognized that lead may in fact have an effect on bone metabolism. Current concern that lead in bone may be mobilized during a number of physiological and pathological conditions, such as age, endocrine status, osteoporosis, renal disease, and, in particular, during pregnancy and lactation[57] has suggested that the dose/rate of lead exposure influences location and concentration of lead in different sites in bone, which may influence its availability for mobilization. Other factors are maternal age, gestation age, parity, and race. The major determinant for each of these factors is nutritional status, particularly dietary calcium. Gulson et al.[58] have reviewed the possible mechanisms where by lead may directly or indirectly alter several aspects of bone cell function.[57]

17.2.7.8 Carcinogenic Effects

There are a number of experimental evidences where exposure to lead has been shown to cause renal tumors. Incidences of renal adenocarcinoma have been reported in lead-exposed animals depending upon length and severity of exposure.[59] It is believed that males are more susceptible than females to lead-induced carcinomas. Not many epidemiological studies and no conclusive evidences, however, are available in the literature for the possible association between lead exposure and increased incidence of cancer. One of the possible mechanisms suggested for such effects could be related to transformation of disordered renal epithelial cell growth to renal cyst. Cells lining cysts become transformed and proliferate abnormally presumably in response to increased intracystic fluid volume. It can thus be suggested that following lead-induced chronic renal failure contributes to an increase in renal adenocarcinomas.

17.2.7.9 Cardiovascular Disorders

Lead poisoning is associated with a number of morphological and biochemical changes in the cardiovascular system. A number of factors seem to indicate that the association between hypertension and blood lead is causal but doubts still exist on this point, in particular since epidemiological has had major shortcomings, i) these studies are mainly cross-sectional and very few prospective studies available in the literature and ii) these studies do not provide information on a number of relevant confounders. The pathophysiological mechanisms by which lead could induce vascular diseases are not clear. However, it could be attributed to i) inhibition of cytochrome P450 leading to the accumulation of lipids in vessel wall and ii) inhibition of SOD resulting in the elevation of serum of LPO levels, which is a serious risk factor for heart diseases. LPOs are also known to promote the adherence and aggregation of platelets in the vessels of the CNS, thereby initiating blood clotting and iii) lead-induced carcinogenicity. The clinical features of lead poisoning are depicted (Fig. 17.7).

17.2.8 Mechanism of Lead-Induced Toxicity

17.2.8.1 Oxidative Stress

It has been observed that oxygen is both life-sustaining and life-threatening inhalant. During the past two decades, the evidence supporting the deleterious effects of oxygen-free radicals in many pathological processes has grown considerably.[60] Free radicals may play an important role in several pathological conditions of the CNS where they directly injure tissue and their formation may also be a consequence of tissue injury. Recently, attention has also been focused on the contribution of oxygen-free radicals to brain dysfunction and brain cell death after brain injury such as cerebral ischemia and head trauma.[61]

Free radicals produce tissue damage through multiple mechanisms, including ex-citotoxicity, metabolic dysfunction, and disturbance of intracellular calcium homeostasis.[62] A free radical is defined as an atom or molecule in a particular state with an unpaired electron in its outer orbit. The one-, two-, and three-electron reduction of molecular oxygen results in the formation of the superoxide radical $(O_2^{\cdot-})$, hydrogen peroxide (H_2O_2), and the hydroxyl radical, respectively.[63]

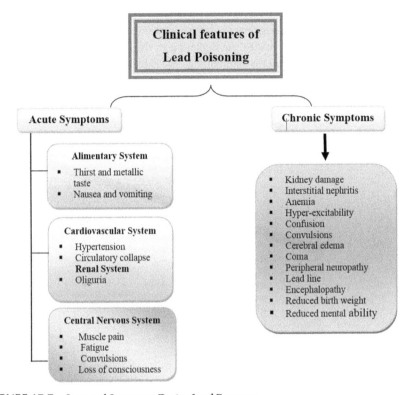

FIGURE 17.7   Signs and Symptoms During Lead Exposure.

Lead-induced oxidative stress

Oxidative stress has been reported as one of the important mechanisms of toxic effect of lead. Lead causes oxidative stress (Fig. 17.8) by inducing the generation of ROS and weakens the antioxidant defense system of cells.[64,65] Depletion of cell's major sulfhydryl reserves seems to be an important indirect mechanism for oxidative stress induced by lead. Lead irreversibly binds to the sulfhydryl group of proteins, causing impaired function. The disruption of reducing status of tissue leads to the formation of ROS, which may damage essential biomolecules such as protein, lipids, and DNA.

In vivo generation of highly ROS, such as hydroxyl radical $(HO^{\bullet})$, hydrogen peroxide (H_2O_2), superoxide radical $(O_2^{\bullet})$, and LPO, the aftermath of lead exposure, may result in systematic mobilization and depletion of the cells' intrinsic antioxidant defenses. At high levels, these ROS can be toxic to cells and may contribute to cellular dysfunction and poisoning.[66] Significant depletion of other antioxidant enzymes, such as catalase, GPx, and GST, also occur. Catalase is an efficient decomposer of H_2O_2 and known to be susceptible to lead toxicity.[67] Lead-induced decrease in GPx activity may arise as a consequence of impaired functional groups such as GSH and NADPH- or selenium-mediated detoxification of toxic metals. While, antioxidant enzyme GST is known to provide protection against oxidative stress and the inhibition of this enzyme on lead exposure might be due to the depletion in the status of tissue thiol moiety. These enzymes are important for maintaining critical balance in the glutathione redox state. The concentration of MDA, which is a reflection of endogenous lipid oxidation level, gets increased on lead exposure.

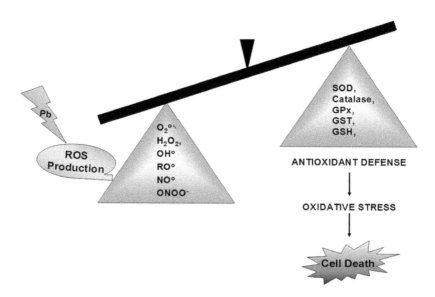

FIGURE 17.8   Lead-Induced Oxidative Stress and Cell Death.

Production of GSH is considered to be the first line of defense against oxidative injury and free radical generation, where GSH functions as a scavenger and a co-factor in metabolic detoxification.[68] GSH has carboxylic groups, an amino group, a sulfhydryl group, and two peptide linkages as sites for the reaction of lead. Its functional group, –SH, plays an important role in lead binding.

17.2.8.2 Ionic Mechanism

Interaction with essential elements has been assigned to be one of the possible mechanisms of lead-induced toxicity. Various studies have shown that lead has an inhibitory effect on the peripheral nervous system through stimulus-coupled or calcium-dependent release of acetylcholine.[69] Lead has diverse and complex action on calcium messenger system, emphasizing the importance of this pathway as a key molecular and cellular target of lead toxicity. Lead substitutes for calcium in affecting the activity of second messengers. Calmodulin, activated by calcium, stimulates several protein kinases, cyclic AMP, and phosphodiesterase and affects potassium channels.[70] Lead also has its effect on sodium concentrations. Synaptosomes prepared from rodents exposed to lead in vivo demonstrate inhibition of sodium-dependent high-affinity uptake of choline, dopamine, and GABA.[71]

17.2.8.3 Apoptosis

Apoptosis (programmed cell death) can be induced by a variety of stimuli. Apoptosis occurs when a cell activates an internally encoded suicide program as a response to either intrinsic or extrinsic signals. One of the better characterized apoptotic cascade pathways has mitochondrial dysfunction as its initiator. Mitochondrial dysfunction initiated by the opening of the mitochondrial transition pore leads to mitochondrial depolarization, release of cytochrome C, activation of a variety of caspases, and cleavage of downstream death effect or proteins and ultimately results in apoptotic cell death. While, a variety of stimuli can trigger opening of the mitochondrial transition pore and cause apoptosis, a sustained intracellular increase in Ca^{2+} is one of the better-known triggers; accumulation of lead is another. Lead disrupts calcium homeostasis, causing a marked accumulation of calcium in lead-exposed cells.[50] Lead, in nanomolar concentrations, also induces mitochondrial release of calcium thus initiating apoptosis.[72] A summary of the different mechanisms involved in lead induced toxicity is presented in Fig. 17.9.

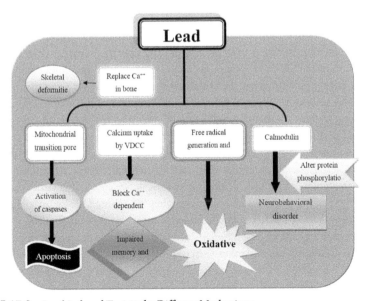

FIGURE 17.9 Lead-Induced Toxicity by Different Mechanisms.

17.2.9   Clinical Biochemical Indicators

Inhibition of blood δ-ALAD activity, increased uptake of blood lead, and enhanced urinary excretion of lead and δ-ALA are the most essential diagnostic tests of lead poisoning. Other tests including blood hemoglobin, basophilic granulation of RBCs, and blood zinc protoporphyrin can also provide some essential information. Lead in urine is less often used as an indicator of exposure. The urinary excretion of lead increases immediately after exposure and thus could be considered as an indicator for recent lead intake. Urinary lead excretion depends on various factors and also varies with time besides no direct conclusion about the pattern of exposure could be drawn from urinary lead excretion. A total of 95% of the absorbed lead is incorporated in the bone and skeleton. Half-time of the bone lead is more than 10 years and thus could be a useful tool for estimating long-term integrated lead exposure. However, performing bone biopsies is difficult, and thus, this test is not generally used for large surveys.

Radiological examination could be a tool for diagnosing acute lead poisoning as it may reveal radiographic evidence of paint chip ingestion in children but this could not be a sensitive method. Neurobehavioral testing could also be useful as this may reflect changes in perceptual motor speed and memory deficits that are characteristic of lead toxicity.

17.2.10   Therapies and Prevention Against Lead Toxicity

17.2.10.1   Role of Flavonoids

Antioxidants are compounds that protect cells against the damaging effects of ROS, such as singlet oxygen, superoxide, peroxyl radicals, hydroxyl radicals, and peroxynitrite. An imbalance between antioxidants and ROS results in oxidative stress, leading to cellular damage. Oxidative stress has been linked to cancer, aging, atherosclerosis, ischemic injury, inflammation, and neurodegenerative diseases (Parkinson's and Alzheimer's). Flavonoids may help provide protection against these diseases by contributing, along with antioxidant vitamins and enzymes, to the total antioxidant defense system of the human body. Epidemiological studies have shown that flavonoid intake is inversely related to mortality from coronary heart disease and to the incidence of heart attacks. However, recent studies have demonstrated that flavonoids found in fruits and vegetables may also act as antioxidants.[73]

Flavonoid is any member of a class of widely distributed biological natural products containing aromatic heterocyclic skeleton of flavan (2-phenylbenzopyran). Generally, flavonoids are biological pigments providing colors to the plants. Flavonoids are a class of secondary plant phenolics with significant antioxidant and chelating properties. In the human diet, they are most concentrated in fruits, vegetables, wines, teas, and cocoa. Flavonoids, such as quercetin, silymarin, hesperetin, naringenin, and epicatechin, have been proposed to exert beneficial effects in a multitude of disease states including cancer, cardiovascular disease, and neurodegenerative disorders. Many of the biological actions of flavonoids have been attributed to their antioxidant properties. Flavonoids are ideal scavengers of peroxyl radicals due to their favorable reduction potentials relative to alkyl peroxyl radicals, and thus, they are effective inhibitors of lipid peroxidation. Due to the hydrogen (electron)-donating ability of a flavonoid molecule, they act to scavenge a reactive radical species and are primarily associated with the presence of a B-ring catechol group.[74] One important structural feature that is partly responsible for the antioxidant properties of flavonoids involves the presence of 2, 3 unsaturation in conjugation with a 4-oxo group in the C-ring. In addition, the presence of functional groups involving both hydroxyl groups of ring-B and the 5-hydroxy group of ring-A are all important contributors in the ability of flavonoids to chelate redox-active metals, and thus prevent catalytic breakdown of hydrogen peroxide.[63] In addition to free radical scavenger, flavonoids can also act as carcinogen inactivators, inhibition of tumor cell growth, modulators of DNA repair, and inducers of apoptosis.[75]

17.2.10.2 Naringenin

Naringenin is a flavanone, a type of flavonoid, that is considered to have a bioactive effect on human health as antioxidant, free radical scavenger, anti-inflammatory, carbohydrate metabolism promoter, and immune system modulator. It is the predominant flavanone in grapefruit.[76] Naringenin is a natural flavonoid, aglycone of naringin, and is widely distributed in citrus fruits, tomatoes, cherries, grapefruit, and cocoa. Naringenin has also been extensively investigated for its pharmacological activities, including antitumor,[77] anti-inflammatory, and hepatoprotective effects.[76,78]

17.2.10.3 Silymarin

Silymarin is a standardized mixture of antioxidant flavonolignans extracted from the medicinal plant S. marianum, commonly known as milk thistle. It has been known since ancient times and recommended in ancient European and Asiatic medicine mainly for the treatment of various diseases.[79] Silymarin is extracted from the dried seeds of milk thistle plant, where it is present in higher concentrations. Silymarin is a complex mixture of four flavonolignan isomers, namely silybin, isosilybin, silydianin, and silychristin. Among the isomers, silybin is the major and most active component and represents about 60–70%. It is a free radical scavenger and a membrane stabilizer that prevents lipoperoxidation and hepatotoxicity. The cytoprotective effects of silymarin are mainly attributable to its antioxidant and free radical scavenging properties. It can prevent the absorption of toxins into the hepatocytes by occupying the binding sites as well as inhibiting many transport proteins at the membrane. Oral effectiveness, good safety profile, availability in India, and, most importantly, an affordable price make silymarin a suitable candidate for the treatment of metal-induced toxicity.[80]

The possible mechanism underlying the protective properties of silymarin include the prevention of GSH depletion,[81] maintenance of hepatic protein synthesis via RNA activation,[82] and preservation of mitochondrial transport function. Silymarin was found to be able to counteract lead-induced oxidative stress.[83] Jain reported therapeutic efficacy of silymarin and naringenin in reducing a lead-induced hepatic damage in young rats.[84]

Silymarin also prevents epithelial and other types of cancer, such as prostate, bladder, lung, ovarian, and breast cancers.[85,86] Silymarin induces the increase of glutathione content in the cell and the inhibition of lipid peroxidation. Recent studies[87] have shown that silymarin has antioxidant effect in the CNS, being able to enter the CNS through the BBB.

17.3   CONCLUSION

The present review suggests that administration of Silymarin and Naringenin with lead has some protective efficacy against lead-induced oxidative stress by reducing ROS levels and activating antioxidant defense enzymes in the system. This strategy of supplementation of these flavonoids in diet may have beneficial effects during lead toxicity. Use of antioxidants brings another option to the therapeutic intervention. Because of the rebound effect of chelators, chelation therapy could not be effective when the patient is in continuous contact with lead exposure. Antioxidants, however, are recognized as safe molecules along with the chelation property. Consequently, experiments are needed to show the effects of antioxidants on the cells or animals that are treated concomitantly for lead exposure. Detailed mechanistic studies are also required to understand the mechanisms underlying the beneficial effects of some antioxidants and to explore the optimum dosage and duration of treatment to obtain better clinical recoveries in lead intoxication cases.

KEYWORDS

- **Lead**
- **flavonoids**
- **naringenin**
- **silymarin**

REFERENCES

1. CDC. Centers for Disease Control and Prevention. *Preventing Lead Poisoning in Young Children: A Statement by the Centers for Disease Control*; US Dept of Health and Human Services: Atlanta, GA, 1991.

2. Norman, J.; Kucharik, C.; Gower, S.; Baldocchi, D.; Crill, P.; Rayment, M.; Striegl, R. A comparison of six methods for measuring soil-surface carbon dioxide fluxes. *J. Geophys. Res.*, 1997, 102, 28771-77.

3. Needleman, H. L.; Bellinger, D. Studies of lead exposure and the developing central nervous system: a reply to Kaufman. *Arch. Clin. Neuropsychol.* 2001, 16, 359–374.

4. Breen, J. J.; Stroup, C. R. *Lead Poisoning: Exposure; Abatement; Regulation*; CRC Press: Boca Raton; FL, 1995; pp 3–12.

5. McElvaine, M. D.; DeUngria, E. G.; Matte, T. D. Prevalence of radiographic evidence of paint chip ingestion among children with moderate to severe lead poisoning; St. Louis; Missouri; 1989 through 1990. *Pediatrics.* 1992, 89, 740–742.

6. Markowitz, M. Lead poisoning: a disease for the next millennium. *Curr. Probl. Pediatr.* 2000, 30, 62–69.

7. Gulson, B. L.; Jameson, C. W.; Mahaffey, K. R. Pregnancy increased mobilisation of lead from maternal skeleton. *J. Lab. Clin. Med.* 1997, 130, 51–62.

8. Hodgkins, D. G.; Rogins, T. G.; Hinkamp, D. L. The effect of airbone lead particle size on worker blood-lead levels: an empirical study of battery workers. *Br. J. Ind. Med.* 1991, 49, 241–248.

9. Philip, A. T.; Gerson, B. Lead poisoning – part II effects and assay. *Clin. Lab. Med.* 1994, 14, 651–670.

10. NSF. *Physical and Chemical Characteristics of Environmental Lead – in Lead in the Environment*; National Science Foundation, 1977.

11. ATSDR US. *Toxicological Profile for Lead. Agency for Toxic Substances and Disease Registry*; U.S. Public Health Service: Atlanta; GA, 1999.

12. Craig P.J.; Emsley J.; Faulkner D.J.; Huang P.M.; Paul E.A.; Schidlowski M.W.; Stumm J.C.G.; Walker P.J.; Wangersky J.; Westall A.J.B.; Zehnder Zinder S.H. The natural environment and the biogeochemical cycles. *Hutzinger, O. ed. Springer-Verlag*, 1980, 185-197.

13. Lanphear, B. P.; Matte, T. D.; Rogers, J.; Clickner, R. P.; Dietz, B.; Bornschein, R. L.; Succop, P.; Mahaffey, K. R.; Dixon, S.; Galke, W.; Rabinowitz, M.; Farfel, M.; Rohde, C.; Schwartz, J.; Ashley, P.; Jacobs, D. E. The contribution of lead contaminated house dust and residential soil to children's blood lead levels a pooled analysis of 12 epidemiologic studies. *Environ. Res.* 1998, 79, 51–68.

14. Lee, R. G.; Becker, W. C.; Collins, D. W. Lead at the tap: sources and control. *Am. Water Works Assoc. J.* 1989, 81, 52–62.

15. Bress, W. C.; Bidanset, J. H. Percutaneous in vivo and in vitro absorption of lead. *Vet. Hum. Toxicol.* 1991, 33, 212–214.

16. Stauber, J. L.; Florence, T. M.; Gulson, B. L. Percutaneous absorption of inorganic lead compounds. *Sci. Total Environ.* 1994, 145, 55–70.

17. Staudinger, K. C.; Roth, V. S. Occupational lead poisoning. *Am. Fam. Physician.* 1998, 57, 719–726.

18. NMCE. Report on lead. National Multi-Commodity Exchange of India Limited, 2004.

19. Haefliger, P.; Mathieu-Nolf, M.; Lociciro, S.; Ndiaye, C.; Coly, M.; Diouf, A. Mass lead intoxication from informal used lead acid battery recycling in Dakar; Senegal. *Environ. Health Perspect.* 2009, 117, 1535–1540.

20. Robar. (2003). – Reference to be changed
International Lead and Zinc Study Group (2009). End uses of lead [web site]. Lisbon, International Lead and Zinc Study Group (http://www.ilzsg. org/static/enduses.aspx?from=1, accessed 19 November 2009)

21. Kaul, B.; Mukerjee, H. Elevated blood lead and erythrocyte protoporphyrin levels of children near a battery-recycling plant in Haina; Dominican Republic. *Int. J. Occup. Environ. Health.* 1999, 5, 4.

22. Bruner, A. B.; Joffe, A.; Duggan, A. K.; Casella, J. F.; Brandt, J. Randomised study of cognitive effects of iron supplementation in non-anaemic iron-deficient adolescent girls. *Lancet.* 1996, 348, 992–996.

23. McGregor, G. S.; Ani, C. A review of studies on the effect of iron deficiency on cognitive development in children. *J. Nutr.* 2001, 131, 649S–666S.

24. Cheng, Y.; Willett, W. C.; Schwartz, J.; Sparrow, D.; Weiss, S.; Hu, H. Relation of nutrition to bone lead and blood lead levels in middle-aged to elderly men. The Normative Aging Study. *Am. J. Epidemiol.* 1998, 147, 1162–1174.

25. Cake, K. M.; Bowins, R. J.; Vaillancourt, C. Partition of circulating lead between serum and red blood cells is different for internal and external sources of exposure. *Am. J. Ind. Med.* 1996, 29, 440–445.

26. Al-Modhefer, A. J.; Bradbury, M. W. B.; Simons, T. J. B. Observations on the chemical nature of lead in human blood serum. *Clin. Sci.* 1991, 81, 823–829.

27. Grobler, S. R.; Theunissen, F. S.; Kotze, T. J. The relation between lead concentrations in human dental tissues and in blood. *Arch. Oral. Biol.* 2000, 45, 607–609.

28. Gulson, B. L.; Mizon, K. J.; Korsch, M. J.; Howarth, D. Non-orebody sources are significant contributors to blood lead of some children with low to moderate lead exposure in a major lead mining community. *Sci. Total Environ.* 1996, 191, 299–301.

29. Lagerkuist, B. J.; Ekesrydh, S.; Englyst, Y. Increased blood lead and decreased calcium levels during pregnancy: a prospective study of Swedish women living near a smelter. *Am. J. Public Health.* 1996, 86, 1247–1252.

30. Abadin, H. G.; Hibbs, B. F.; Pohl, H. R. Breast feeding exposure of infants to cadmium, lead, and mercury: a public health view point. *Toxicol. Ind. Health.* 1997, 15, 1–24.

31. Antonio, M. T.; Corpas, I.; Leret, M. L. Neurochemical changes in newborn rat's brain after gestational cadmium and lead exposure. *Toxicol. Lett.* 1999, 104, 1–9.

32. Saryan L. A.; Zenz C. Lead and its compounds. In: Zenz C, Dickerson OB, Horvath EP, editors. Occupational Medicine. 3. St Louis: Mosby; 1994. pp. 506–41.

33. Al-Saleh, I. A. S. The biochemical and clinical consequences of lead poisoning. *Med. Res. Rev.* 1994, 14, 415–486.

34. Cory-Slechta, D. A.; Schaumburg, H. H. *Lead; Inorganic.* In *Experimental and Clinical Neurotoxicology*; Spencer, P. S., Schaumburg, H. H., Ludolph, A. C., Eds.; Oxford University Press: New York, 2000; Vol. 2, pp 708–720.

35. Finkelstein, Y.; Markowitz, M.; Rosen, J. Low level lead induced neurotoxicity in children: an update on central nervous system effects. *Brain Res Brain Res Rev.* 1998, 27, 168–176.

36. Mendelsohn, A. L.; Dreyer, B. P.; Fierman, A. H.; Rosen, C. M.; Legano, L. A.; Kruger, H. A. Low-level lead exposure and behavior in early childhood. *Pediatrics.* 1998, 101, E10.

37. Fitsanakis, V. A.; Aschner, M. The importance of glutamate; glycine; and gamma-aminobutyric acid transport and regulation in manganese; mercury and lead neurotoxicity. *Toxicol. Appl. Pharmacol.* 2005, 204, 343–354.

38. Lanphear, B. P.; Dietrich, K.; Auinger, P.; Cox, C. Cognitive deficits associated with blood lead concentrations <10μg/dl in US children and adolescents. *Public Health Rep.* 2000, 115, 521–529.

39. Bergdahl, L. A.; Gerhardsson, L.; Schutz, A.; Desnick, R. J.; Wetmur, J. G.; Sassa, S.; Skerfying, S. Lead binding to δ-ALAD in human erythrocytes. *Pharmacol. Toxicol.* 1997, 81, 153–158.

40. Jaffe, E. K. Porphobilnogen synthase; the first source of heme asymmetry. *J. Bioenerg. Biomembr.* 1995, 27, 169–179.

41. Wetmur, J. G. Influence of the common human δ-aminolevulinate dehydratase polymorphism on lead body burden. *Environ. Health Perspect.* 1994, 102, 215–219.

42. Gurer, H.; Ozgunes, H.; Neal, R.; Spitzand, D. R.; Ercalv, N. Antioxidant effects of N-acetyl cystein and succimer in red blood cells from lead exposed rat. *Toxicology.* 1998, 128, 181–189.

43. Schwartz, B. S.; Lee, B. K.; Stewart, W.; Ahn, K. D.; Springer, K.; Kelsey, K. Associations of δ-aminolevulinic acid dehydratase genotype with plant; exposure duration; and blood lead and zinc protoporphyrin levels in Korean lead workers. *Am. J. Epidemiol.* 1995, 142, 738–745.

44. Al-Saleh I.A.S. The biochemical and clinical consequences of lead poisoning. *Med. Res. Rev.,* 1994, 14, 415-486.

45. Sanchez-Fructuoso, A. I.; Torralbo, A.; Arroyo, M. Occult lead intoxication as a cause of hypertension and renal failure. *Nephrol. Dial. Transplant.* 1996, 11, 1775–1780.

46. Vyskocil, A.; Semecky, V.; Fiala, Z.; Cizkova, M.; Viau, C. Renal alterations in female rats following subchronic lead exposure. *J. Appl. Toxicol.* 1995, 15, 257–262.

47. Damek-Poprawa, M.; Sawicka-Kapusta, K. Histopathological changes in the liver; kidneys; and testes of bank voles environmentally exposed to heavy metal emissions from the steelworks and zinc smelter in Poland. *Environ. Res.* 2004, 96, 72–78.

48. Anderson, A. C.; Pueschel, S. M.; Linakis, J. G. *Pathophysiology of Lead Poisoning*. In *Lead Poisoning in Children*; Pueschel, S. M., Linakis, J. G., Anderson, A. C., Eds.; P.H. Brookes: Baltimore, MD, 1996; pp 75–96.

49. Tong, S. P.; Baghurst, A.; McMichael, M. S.; Mudge, J. Lifetime exposure to environmental lead and children's intelligence at 11-13 years: The Port Pirie cohort study. *Br. Med. J.* 1996, 312, 1569.

50. Bressler, J.; Kim, K. A.; Chakraborti, T.; Goldstein, G. Molecular mechanisms of lead neurotoxicity. *Neurochem. Res.* 1999, 24, 595–600.

51. Brown, L. L.; Schneider, J. S.; Lidsky, T. I. Sensory and cognitive functions of the basal ganglia. *Curr. Opin. Neurobiol.* 1997, 7, 157–163.

52. Ronis, M. J. J.; Badger, T. M.; Shema, S. J. Endocrine mechanism underlying the growth effects of developmental lead exposure in rat. *J. Toxicol. Environ. Health.* 1998, 54, 101–120.

53. Mitchell J.W. Occupational medicine forum: lead toxicity and reproduction. *J. Occup. Med.,* 1987, 29: 397-399.

54. Li, P. J.; Sheng, Y. Z.; Wang, Q. Y.; Gu, L. Y.; Wang, Y. L. Transfer of lead via placenta and breast milk in human. *Biomed. Environ. Sci.* 2000, 13, 85–89.

55. Alexander, B. H.; Checkoway, H.; Van Netten, C. Semen quality of men employed at a lead smelter. *Occup. Environ. Med.* 1996, 53, 411–416.

56. Lin, S.; Hwang, S.; Marshall, E. G. Fertility rates among lead workers and professional bus drivers: a comparative study. *Ann. Epidemiol.* 1996, 6, 201–208.

57. Needleman H.L.; Riess J.A.; Tobin M.J.; Biesecker G.E.; Greenhouse J.B. Bone lead levels and delinquent behaviour. *JAMA* 1996, 275: 363–369.

58. Gulson B.L.; Pounds J.G.; Mushak P.; Thomas B.J.; Gray B.; Korsch M.J. Estimation of cumulative lead release (lead flux) from the maternal skeleton during pregnancy and lactation. *J. Lab. Clin. Med.,* 1999, 134: 631–640.

59. Goyer, R. A. *Toxic Effects of Metals*. In *Casarett & Doull's Toxicology: The Basic Science of Poison*; Klaassen, C., Ed.; McGraw-Hill: New York, 1996; pp 691–737.

60. Halliwell, B.; Gutteridge, J. M. C. Role of free-radicals and catalytic metal-ions in human-disease – an overview. *Methods Enzymol.* 1990, 186, 1–85.

61. Savolainen, K. M.; Loikkanen, J.; Eerikainen, S. Interactions of excitatory neurotransmitters and xenobiotics in excitotoxicity and oxidative stress: glutamate and lead. *Toxicol. Lett.* 1998, 103, 363–367.

62. Choi, Y. J.; Park, J. W.; Suh, S. I.; Mun, K. C.; Bae, J. H.; Song, D. K.; Kim, S. P.; Kwon, T. K. Arsenic trioxide-induced apoptosis in U937 cells involve generation of reactive oxygen species and inhibition of Akt. *Int. J. Oncol.* 2002, 21, 603–610.

63. Valko, M.; Rhodes, C. J.; Moncol, J.; Izakovic, M.; Mazur, M. Free radicals; metals and antioxidants in oxidative stress-induced cancer. *Chem. Biol. Interact.* 2006, 160, 1–40.

64. Sandhir, R.; Julka, D.; Gill, K. D. Lipoperoxidative damage on lead exposure in rat brain and its implications on membrane bound enzymes. *Pharmacol. Toxicol.* 1994, 74, 66–71.

65. Flora, S. J. S. Nutritional components modify metal absorption; toxic response and chelation therapy. *J. Nutr. Environ. Med.* 2002, 12, 53–67.

66. Saxena, G.; Flora, S. J. S. Lead induced oxidative stress and hematological alterations and their response to combined administration of calcium disodium EDTA with a thiol cheator in rats. *J. Biochem. Mol. Toxicol.* 2004, 18, 221–233.

67. Sandhir, R.; Gill, K. D. Effect of lead on lipid peroxidation in liver of rats. *Biol. Trace Elem. Res.* 1995, 48, 91–97.

68. Bray, T. M.; Taylor, C. G. Tissue glutathione; nutrition and oxidative stress. *Can. J. Physiol. Pharmacol.* 1993, 71, 746–775.

69. Reddy, G. R.; Riyaz Basha Md; Devi, C. B.; Suresh, A.; Baker, J. L.; Shafeek, A.; Heinz, J.; Chetty, C. S. Lead induced effects on acetylcholinesterase activity in cerebellum and hippocampus of developing rat. *Int. J. Dev. Neurosci.* 2003, 21, 347–352.

70. Seino, S.; Miki, T. Physiological and pathophysiological roles of ATP-sensitive K+ channels. *Prog. Biophys. Mol. Biol.* 2003, 81, 133–176.

71. Lasley, S. M.; Green, M. C.; Gilbert, M. E. Influence of exposure period on in vivo hippocampal glutamate and GABA release in rats chronically exposed to lead. *Neurotoxicology.* 1999, 20, 619–629.

72. Silbergeld, E. K.; Sauk, J.; Somerman, M. Lead in Bone: Storage site; exposure source; and target organ. *Neurotoxicology.* 1993, 14, 225–236.

73. Briskin, Donald P. Medicinal Plants and Phytomedicines. Linking Plant Biochemistry and Physiology to Human Health. *Plant Physiology,* 2000, 124(2), 507-14.

74. Youdim, K. A.; Spencer, J. P.; Schroeter, H.; Rice-Evansc. Dietary flavonoids as potential neuroprotectants. *Biol. Chem.* 2002, 388(3–4), 503–519.

75. Duthie, S. J.; Dobson, V. L. Dietary flavonoids protect human erythrocyte DNA from oxidative attack in vitro. *Eur. J. Nutr.* 1999, 38(1), 28–34.

76. Lee, M. H.; Yoon, S.; Moon, J. O. The flavonoid naringenin inhibits dimethylnitrosamine-induced liver damage in rats. *Biol. Pharm. Bull.* 2004, 27, 72–76.

77. Kanno, S.; Tomizawa, A.; Hiura, T.; Osanai, Y.; Shouji, A.; Ujibe, M.; Ohtake, T.; Kimura, K.; Ishikawa, M. Inhibitory effects of naringenin on tumor growth in human cancer cell lines and sarcoma S-180-implanted mice. *Biol. Pharm. Bull.* 2005, 28, 527–530.

78. Hirai, S.; Kim, Y. I.; Goto, T.; Kang, M. S.; Yoshimura, M.; Obata, A.; Yu, R.; Kawada, T. Inhibitory effect of naringenin chalcone on inflammatory changes in the interaction between adipocytes and macrophages. *Life Sci.* 2007, 81, 1272–1279.

79. Radjabian, T.; Huseini, H. F. Anti-hyperlidemic and anti-atherosclerotic.activities of silymarin from cultivated and wild plants of *Silybum marianum* L with different content of flavonolignans. *Iran. J. Pharmacol. Ther.* 2010, 9, 63–67.

80. Grange, L.; Wang, M.; Watkins, R.; Ortiz, D.; Sanchez, M.; Konst, J.; Lee, C.; Reyes, E. Protective effects of the flavonoid mixture, silymarin, on fetal rat brain and liver. *Journal of Ethnopharmacology,* 1999, 65(1), 53-61.

81. Campos, R.; Garrido, A.; Guerra, R.; Valenzuela, A. Silybin dihemisuccinate protects against glutathione depletion and lipid peroxidation induced by acetaminophen on rat liver. *Planta Med.* 1989, 55, 9–417.

82. Conti, M.; Malandrino, S.; Magistretti, M. J. Protective activity silipide on liver-damage in rodents. *Jpn. J. Pharmacol.* 1992, 60, 315.

83. Bongiovanni, G. A.; Soria, E. A.; Eynard, A. R. Effects of the plant flavonoids silymarin and quercetin on arsenite-induced oxidative stress in CHO-K1 cells. *Food Chem. Toxicol.* 2007, 45, 971.

84. Jain, A.; Yadav, A.; Bozhkov, A. I.; Padalko, V. I.; Flora, S. J. Therapeutic efficacy of silymarin and naringenin in reducing arsenic-induced hepatic damage in young rats. *Ecotoxicol. Environ. Saf.* 2010, 74, 607–614.

85. Shalan, M. G.; Mostafa, M. S.; Hassouna, M. M.; Hassab El-Nabi, S. E.; El-Refaie, A. Amelioration of lead toxicity on rat liver with Vitamin C and silymarin supplements. *Toxicology.* 2005, 206, 1–15.

86. Kaur, M.; Agarwal, R. Silymarin and epithelial cancer chemoprevention: how close we are to bedside? *Toxicol. Appl. Pharmacol.* 2007, 224, 350–359.

87. Nencini, C.; Giorgi, G.; Micheli, L. Protective effect of silymarin on oxidative stress in rat brain. *Phytomedicine.* 2007, 14, 129–135.

CHAPTER 18

A Theoretical Analysis of Bimetallic Ag–Au$_n$ (N = 1–7) Nanoalloy Clusters Invoking DFT-Based Descriptors

Prabhat Ranjan[1], Srujana Venigalla[2], Ajay Kumar[3], and Tanmoy Chakraborty[2*]

[1]Department of Electronics and Communication Engineering, Manipal University Jaipur, Dehmi-Kalan, Jaipur, India
[2]Department of Chemistry, Manipal University Jaipur, Dehmi-Kalan, Jaipur, India
[3]Department of Mechatronics Engineering, Manipal University Jaipur, Dehmi-Kalan, Jaipur, India
*Email: tanmoy.chakraborty@jaipur.manipal.edu; tanmoychem@gmail.com

CONTENTS

ABSTRACT

Due to many fold applications in the field of engineering and science, nowadays bimetallic nanoalloy clusters are extensively popular. A deep insight is required to explore the properties of such compounds. Among such nanoclusters, the compound formed between Ag and Au has gained a considerable interest because they possess unique optical, electronic and magnetic, and mechanical properties, which have extensive application in the field of radiation medicine, biophysics, and nanoscience. Density functional theory (DFT) is a new paradigm of quantum mechanics to study the electronic properties of materials. Recently, conceptual DFT-based descriptors have been turned to the indispensable tools for studying the experimental properties of compounds. In this venture, we have analyzed the experimental properties of the Ag–Au$_n$ ($n = 1$–7) nanoalloy clusters invoking DFT methodology. A nice correlation has been made between properties of aforesaid nanoclusters with our evaluated theoretical descriptors. This study is probably the first attempt to establish such type of correlation.

18.1 INTRODUCTION

Since last few years, nanomaterials and nanotechnology have emerged as a new branch in the research domain of science and technology.[1] Due to the existence of a large number of quantum mechanical and electronic effects, nanoparticles possess various unique physicochemical properties.[2–4] The classification of nanoparticles is done in terms of size range of 1–100 nm. That particular size range exists between the levels of atomic/molecular and bulk material.[5] But, there are still some instances of nonlinear transition of certain physical properties, which may vary depending on their size, shape, and composition.[6,7] A large number of scientific reports are available for describing the effects of size and structure to change the optical, electronic, magnetic, chemical, and other physical properties of nanoparticles.[1,3,4] A deep insight into the research of nanoparticles with well-defined size and structure may lead to some other alternatives for better performance.[8] The nanoparticles, due to its vast applications in the areas of biological labeling, photochemistry, catalysis, information storage, magnetic device, optics, sensors, photonics, optoelectronics, nanoelectronics, etc., have got immense importance.[1,3,9–11]

The noble metals can be extensively applied in several technological areas due to its superior catalytic, magnetic, and electronic properties.[12–18] There are a number of examples, where it has been observed that the combination of two or more noble metals enhances the above-mentioned properties.[13,19,20] Nowadays, different compositions of nanoalloys are being utilized for the advancement of methodologies and characterization techniques.[13,19,21] A deep study of core–shell structure of nanocompounds is very much popular because its properties can be tuned through the proper control of other structural and chemical parameters. Group 11 metal (Cu, Ag, and Au) clusters exhibit the filled inner d-orbitals having one unpaired electron in the valence shell.[22] This electronic arrangement is responsible to reproduce exactly similar shell effects,[23-27] which are experimentally observed for the alkali metal clusters.[28-30] As a result, bimetallic noble metal clusters follow to exhibit similar physicochemical trends with bimetallic alkali metal clusters.[31-35] Among the nanoclusters of Group 11 elements, the compound formed between Ag and Au is very much popular due to its large-scale applications. Fabbi et al.[36] already reported the importance of AgAu dimer using dispersed fluorescence spectroscopy. A number of theoretical observations have been made about the physicochemical properties and importance of AgAu dimer.[37,38] The location of silver has a controlling effect on the optical properties of such particles because optical properties are governed by plasmon resonance frequency of silver, which is also dependent on its structural environment.[39] It has been already established that gold–silver bimetallic nanoclusters as catalysts can enhance the reaction efficiency and selectivity.[40,41] Though a number of experimental studies have been done on this particular type of compounds, a theoretical analysis invoking density functional theory (DFT) is still unexplored.

DFT is a new branch of quantum mechanics to explore the electronic properties of materials in terms of quantitative descriptors. As for the larger systems electron density is more manageable as compared to wave function, DFT is very much popular to study the many-body systems.[8] Superconductivity of metal-based alloys,[42] magnetic properties of nanoalloy clusters,[43,44] quantum fluid dynamics,[45] molecular dynamics,[46] and nuclear physics[47,48] can be extensively studied by DFT methodology. Recently, we have established the importance of DFT-based descriptors in the domain of drug-designing process.[49,50] The study of DFT has been broadly classified into three subcategories, namely, theoretical, conceptual, and computational.[51-54] The conceptual DFT is highlighted following Parr's dictum "Accurate calculation is not synonymous with useful interpretation. To calculate a molecule is not to understand it."[55]

In this venture, we have successfully studied some bimetallic nanoclusters containing Ag and Au, in terms of DFT-based global descriptors, namely, hardness, highest occupied molecular orbital (HOMO)–lowest unoccupied molecular orbital (LUMO) gap, softness, and electrophilicity index. An attempt has been made to correlate the properties of instant compounds with their computational counterparts.

18.2   COMPUTATION DETAILS

In this study, we have made an analysis on the bimetallic nanoalloy clusters of Ag–Au$_n$, where $n = 1$–7. Three-dimensional modeling and structural optimization of all the compounds have been performed using ADF software package[56] within DFT framework. For optimization purpose, the local density approximation (LDA) has been adopted. Although LDA functionals are not so much complex, but its effectiveness is already established in many fold applications, particularly for solid-state physics,[57] where accurate phase transitions in solids[58] and liquid metals[59,60] are predicted, and for lattice crystals, in which 1% precision are successfully achieved.[61] The used computation methodology in this paper is based on the molecular orbital approach, using linear combination of atomic orbitals. Energy minimization of all the bimetallic alloys has been performed without imposing any restriction on molecular spin. Z-axis has been chosen for the spin polarization axis. In this process, the symmetrized fragment orbitals are combined with auxiliary core functions to ensure orthogonalization on the (frozen) core orbitals. For the optimization of the structures, double zeta basis sets have been used along with frozen core approximation. The quadrupole moment of the molecule is calculated in terms of analytical integration methodology.

Invoking Koopmans' approximation,[62] we have calculated the ionization energy (I) and electron affinity (A) of all the nanoalloys using the following ansatz:

$$I = e\varepsilon_{\text{HOMO}} \tag{18.1}$$

$$A = -e\varepsilon_{\text{LUMO}} \tag{18.2}$$

Thereafter, using I and A, the conceptual DFT-based descriptors, namely, electronegativity (χ), global hardness (η), molecular softness (S), and electrophilicity index (ω), have been computed. The equations used for such calculations are as follows:

$$c\chi = -m\mu = \frac{I+A}{2} \tag{18.3}$$

where μ represents the chemical potential of the system.

$$h\eta = \frac{1}{2}(I - A) \tag{18.4}$$

$$S = \frac{1}{2 \times h\eta} \tag{18.5}$$

$$w\omega = \frac{m\mu^2}{2h\eta} \tag{18.6}$$

In this connection, it is important to note that molecular polarizability can be correlated in terms of the molecular softness as these two properties always run hand in hand.

18.3 RESULTS AND DISCUSSION

In this study, a detailed theoretical analysis of Ag–Au bimetallic nanoclusters has been performed in terms of electronic structure theory. The orbital energies in the form of

HOMO–LUMO gap along with computed DFT-based descriptors for instant nano-clusters, namely, molecular electronegativity, global hardness, global softness, and global electrophilicity index, have been reported in Table 18.1. Table 18.1 reveals that HOMO–LUMO gaps of the Ag–Au nanoclusters are maintaining direct relationship with their evaluated global hardness values. As the frontier orbital energy gap increases, their hardness value increases. This trend is expected on the basis of experimental point of consideration. As the molecule possesses the highest HOMO–LUMO gap, it will be least prone to response against any external perturbation. Table 18.1 clearly shows that AgAu is the least reactive species, whereas $AgAu_6$ will exhibit maximum response. Though there are no such quantitative data of optical properties of aforesaid clusters, we can assume that there must be a direct qualitative relationship between op-tical properties of AgAu nanoclusters with their computed HOMO–LUMO gap. The assumption is based on the fact that optical properties of materials are interrelated with flow of electrons within the systems, which in turn depend on the difference between the distance of valence and conduction band. There is a direct linear relationship be-tween HOMO–LUMO gaps with the difference in the energy of valence–conduction band.[63] On that basis, we may conclude that optical properties of instant bimetallic nanoclusters increase with the increase of their hardness values. Similarly softness data exhibit an inverse relationship toward the experimental optical properties. The lin-ear correlation between HOMO–LUMO gaps along with their computed hardness is lucidly plotted in Figure 18.1. The high value of correlation coefficient $(R^2 = 1)$ ob-served in Figure 18.1 validates our predicted model. From the obtained correlation coefficients of several DFT-based descriptors and their HOMO–LUMO gaps, it can be concluded that the best linear relationship is observed in the case of hardness and the least one for electrophilicity index $(R^2 = 0.268)$ of these nanoclusters.

TABLE 18.1 Computed DFT-Based Descriptors of $AgAu_n$, $n1$–7, Nanoclusters

Species	HOMO–LUMO gap (eV)	Electronegativity (eV)	Hardness (eV)	Softness (eV)	Electrophilicity Index (eV)
AgAu	2.017	4.318	1.009	0.495	2.159
$AgAu_2$	0.394	4.654	0.197	2.534	2.327
$AgAu_3$	1.395	5.289	0.696	0.717	2.645
$AgAu_4$	0.255	5.423	0.127	3.909	2.712
$AgAu_5$	1.432	4.272	0.716	0.698	2.136
$AgAu_6$	0.134	5.269	0.067	7.501	2.635
$AgAu_7$	0.686	4.307	0.342	1.458	2.153

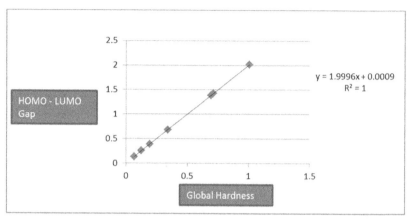

FIGURE 18.1   A Linear Correlation Between Global Hardness versus HOMO–LUMO Gap.

18.4   CONCLUSION

In recent times, bimetallic nanoalloy clusters have got immense importance for its diverse nature of applications. A marked optical property is observed in the case of nanocluster-containing Group 11 metals, namely, silver and gold. The theoretical study of engineering properties of any nanomaterial invoking DFT descriptors is an active field of research. In this paper, we have studied for the first time about the system of $AgAu_{n=1-7}$ in terms of conceptual DFT-based descriptors, namely, global hardness, electronegativity, softness, and electrophilicity index. In this analysis, it is observed that the computed HOMO–LUMO gap runs hand in hand along with its evaluated global hardness. Because of the absence of any quantitative benchmark, the optical property of Ag–Au nanocluster has been assumed to be exactly equivalent to its HOMO–LUMO gap. Here, our evaluated data reveal that the optical property of these compounds maintains a direct relationship with hardness and inverse relationship with softness. This trend is expected from the consideration of other experimental facts. The high value of linear correlation between hardness and HOMO–LUMO gap successfully supports our predicted model.

ACKNOWLEDGEMENT

The authors are grateful to Dr. Sandeep Sancheti, President of Manipal University Jaipur, India and Dr. B K Sharma, Dean Faculty of Science, Research and Innovation of Manipal University Jaipur, India for cooperating and providing the facilities for this work)

KEYWORDS

- **Density functional theory**
- **bimetallic nanoalloy**
- **hardness**
- **softness**
- **electrophilicity index**

REFERENCES

1. Zabet-Khosousi, A.; Dhirani, A.-A. Charge transport in nanoparticle assemblies. *Chem. Rev.* 2008, 108, 4072–4124.
2. Daniel, M. C.; Astruc, D. Gold nanoparticles: assembly, supramolecular chemistry, quantum-size-related properties, and applications toward biology, catalysis, and nanotechnology. *Chem. Rev.* 2004, 104, 293–346.
3. Ghosh, S. K.; Pal, T. Interparticle coupling effect on the surface plasmon resonance of gold nanoparticles: from theory to applications. *Chem. Rev.* 2007, 107, 4797–4862.
4. Ghosh Chaudhuri, R.; Paria, S. Core/shell nanoparticles: classes, properties, synthesis mechanisms, characterization, and applications. *Chem. Rev.* 2012, 112, 2373–2433.
5. Alivisatos, A. P. Semiconductor clusters, nanocrystals, and quantum dots. *Science.* 1996, 271, 933–937.
6. Kastner, M. A. Artificial atoms. *Phys. Today.* 1993, 46, 24–31.
7. Haruta, M. Catalysis of gold nanoparticles deposited on metal oxides. *CATTECH.* 2002, 6, 102–115.
8. Ismail, R. Theoretical Studies of Free and Supported Nanoalloy Clusters, Ph.D Thesis, 2012, 20–38.
9. Roucoux, A.; Schulz, J.; Patin, H. Reduced transition metal colloids: a novel family of reusable catalysts? *Chem. Rev.* 2002, 102, 3757–3778.
10. Munoz-Flores, B. M.; Kharisov, B. I.; Jimenez-Perez, V. M.; Elizondo Martinez, P.; Lopez, S. T. Recent advances in the synthesis and main applications of metallic nano-alloys. *Ind. Eng. Chem. Res.* 2011, 50, 7705–7721.
11. Murray, R. W. Nanoelectrochemistry: metal nanoparticles, nanoelectrodes, and nanopores. *Chem. Rev.* 2008, 108, 2688–2720.
12. Teng, X.; Wang, Q.; Liu, P.; Han, W.; Frenkel, A. I.; Wen, N.; Marinkovic, J. C.; Hanson, J. A. Rodriguez, formation of Pd/Au nanostructures from Pd nanowires via galvanic replacement reaction. *J. Am. Chem. Soc.* 2008, 130, 1093–1101.
13. Ferrando, R.; Jellinek, J.; Johnston, R. L. Nanoalloys: from theory to applications of alloy clusters and nanoparticles. *Chem. Rev.* 2008, 108, 845–910.
14. Henglein, A. Physicochemical properties of small metal particles in solution: "microelectrode" reactions, chemisorption, composite metal particles, and the atom-to-metal transition. *J. Phys. Chem.* 1993, 97, 5457–5471.
15. Davis, S. C.; Klabunde, K. J. Unsupported small metal particles: preparation, reactivity, and characterization. *Chem. Rev.* 1982, 82, 153–208.
16. Lewis, L. N. Chemical catalysis by colloids and clusters. *Chem. Rev.* 1993, 93, 2693–2730.
17. Schmid, G. Large clusters and colloids. Metals in the embryonic state. *Chem. Rev.* 1992, 92, 1709–1727.
18. Schon, G.; Simon, U. A fascinating new field in colloid science: small ligand-stabilized metal clusters and possible application in microelectronics. *Colloid. Polym. Sci.* 1995, 273, 101–117.
19. Oderji, H. Y.; Ding, H. Determination of melting mechanism of Pd24 Pt14 nanoalloy by multiple histogram method via molecular dynamics simulations. *Chem. Phys.* 2011, 388, 23–30.

20. Liu, H. B.; Pal, U.; Medina, A.; Maldonado, C.; Ascencio, J. A. Structural incoherency and structure reversal in bimetallic Au-Pd nanoclusters. *Phys. Rev. B.* 2005, 71, 075403–075409.

21. Baletto, F.; Ferrando, R. Structural properties of nanoclusters: energetic, thermodynamic, and kinetic effects. *Rev. Mod. Phys.* 2005, 77, 371–423.

22. Alonso, J. A. Electronic and atomic structure, and magnetism of transition-metal clusters. *Chem. Rev.* 2000, 100, 637–677.

23. Katakuse, I.; Ichihara, T.; Fujita, Y.; Matsuo, T.; Sakurai, T.; Matsuda, H. Mass distributions of copper, silver and gold clusters and electronic shell structure: I. *Int. J. Mass Spectrom. Ion Processes.* 1985, 67, 229–236.

24. Katakuse, I.; Ichihara, T.; Fujita, Y.; Matsuo, T.; Sakurai, T.; Matsuda, H. Mass distributions of negative cluster ions of copper, silver and gold: I. *Int. J. Mass Spectrom. Ion Processes.* 1986, 74, 33–41.

25. de Heer, W. A. The physics of simple metal clusters: experimental aspects and simple models. *Rev. Mod. Phys.* 1993, 65, 611–676.

26. Gantefor, G.; Gausa, M.; Meiwes-Broer, K.-H.; Lutz, H. O. Photoelectron spectroscopy of silver and palladium cluster anions. Electron delocalization *versus*, localization. *J. Chem. Soc. Faraday Trans.* 1990, 86, 2483–2488.

27. Leopold, D. G.; Ho, J.; Lineberger, W. C. Photoelectron spectroscopy of mass-selected metal cluster anions. I. Cu_n^-, n=1-10. *J. Chem. Phys.* 1987, 86, 1715–1726.

28. Lattes, A.; Rico, I.; de Savignac, A.; Ahmad-Zadeh Samii, A. Formamide, a water substitute in micelles and microemulsions [xxx] structural analysis using a Diels-Alder reaction as a chemical probe. *Tetrahedron.* 1987, 43, 1725–1735.

29. Chen, F.; Xu, G.-Q.; Hor, T. S. A. Preparation and assembly of colloidal gold nanoparticles in CTAB-stabilized reverse microemulsion. *Mater. Lett.* 2003, 57, 3282–3286.

30. Taleb, A.; Petit, C.; Pileni, M. P. Optical properties of self-assembled 2D and 3D superlattices of silver nanoparticles. *J. Phys. Chem. B.* 1998, 102, 2214–2220.

31. Deshpande, M. D.; Kanhere, D. G.; Vasiliev, I.; Martin, R. M. Density-functional study of structural and electronic properties of Na_nLi and Li_nNa (1<n<12) clusters. *Phys. Rev. A.* 2002, 65, 033202.

32. Dhavale, A.; Shah, V.; Kanhere, D. G. Structure and stability of Al-doped small Na clusters: Na_nAl (n=1,10). *Phys. Rev. A.* 1998, 57, 4522.

33. Zhao, G. F.; Sun, J. M.; Liu, X.; Guo, L. J.; Luo, Y. H. Geometries and electronic properties of Na_mSi_n(m+n>7) clusters. *J. Mol. Struct. THEOCHEM.* 2008, 851, 348–352.

34. Zhang, X. R.; Gao, C. H.; Wu, L. Q.; Tang, H. S. The theory study of electronic structures and spectram properties of WnNim(n+m≤7; m=1, 2) clusters. *Acta Phys. Sin.* 2010, 59, 5429.

35. Wang, X. T.; Guan, Q. F.; Qiu, D. H.; Cheng, X. W.; Li, Y.; Peng, D. J.; Gu, Q. Q. Defect microstructures in polycrystalline pure copper induced by high-current pulsed electron beam – the vacancy defect clusters and surface micropores. *Acta Phys. Sin.* 2010, 59, 7252.

36. Fabbi, J. C.; Langenberg, J. D.; Costello, Q. D.; Morse, M. D.; Karlsson, L. Dispersed fluorescence spectroscopy of jet-cooled AgAu and Pt2. *J. Chem. Phys.* 2001, 115, 7543.

37. Weis, P.; Welz, O.; Vollmer, E.; Kappes, M. M. Structures of mixed gold-silver cluster cations (AgmAun+,m+n<6): ion mobility measurements and density-functional calculations. *J. Chem. Phys.* 2004, 120, 677–684.

38. Bonačić-Koutecký, V. Density functional study of structural and electronic properties of bimetallic silver-gold clusters: comparison with pure gold and silver clusters. *J. Chem. Phys.* 2002, 117, 3120–3131.

39. Langlois, C.; Wang, Z. W.; Pearmain, D.; Ricolleau, C.; Li, Z. Y. HAADF-STEM imaging of CuAg core-shell nanoparticles. *J. Phys.* 012047, 241(2010), 012043–.

40. Jankowaik, J. T. A.; Barteau, M. A. Ethylene epoxidation over silver and copper-silver bimetallic catalysts: I. kinetic and selectivity. *J. Catal.* 2005, 236, 366–378.

41. Piccinin, S.; Zafeiratos, S.; Stampfl, C.; Hansen, T. W.; Hävecker, M.; Teschner, D.; Bukhtiyarov, V. I.; Girgsdies, F.; Knop-Gericke, A.; Schlögl, R.; Scheffler, M. Alloy catalyst in a reactive environment: the example of Ag-Cu particles for ethylene epoxidation. *Phys. Rev. Lett.* 2010, 104, 035503.

42. Wacker, O. J.; Kummel, R.; Gross, E. K. U. Time-dependent density-functional theory for superconductors. *Phys. Rev. Lett.* 1994, 73, 2915–2918.

43. Illas, F.; Martin, R. L. Magnetic coupling in ionic solids studied by density functional theory. *J. Chem. Phys.* 1998, 108, 2519–2527.

44. Gyorffy, B.; Staunton, J.; Stocks, G. In *Fluctuations in Density Functional Theory: Random Metallic Alloys and Itinerant Paramagnets*; Gross, E., Dreizler, R., Eds.; Plenum: Plattsburgh, 1995; p 461.

45. Kümmel, S.; Brack, M. Quantum fluid-dynamics from density functional theory. *Phys. Rev. A.* 2001, 64, 022506.

46. Car, R.; Parrinello, M. Unified approach for molecular dynamics and density. Functional theory. *Phys. Rev. Lett.* 1985, 55, 2471–2474.

47. Koskinen, M.; Lipas, P.; Manninen, M. Shapes of light nuclei and metallic clusters. *Nucl. Phys. A.* 1995, 591, 421–434.

48. Schmid, R. N.; Engel, E.; Dreizler, R. M. Density functional approach to quantum hadrodynamics: local exchange potential for nuclear structure calculations. *Phys. Rev. C.* 1995, 52, 164–169.

49. Chakraborty, T.; Ghosh, D. C. Correlation of the drug activities of some anti-tubercular chalcone derivatives in terms of the quantum mechanical reactivity descriptors. *Int. J. Chemoinf. Chem. Eng.* 2011, 1, 53–65.

50. Chakraborty, T.; Ghosh, D. C. Correlation of the drug activities and identification of the reactive sites in the structure of some anti-tubuculour juglone derivatives in terms of molecular orbital and the density functional descriptors. *Int. J. Chem. Model.* 2012, 4, 413–440.

51. Parr, R. G.; Yang, W. Density-functional theory of the electronic structure of molecules. *Annu. Rev. Phy. Chem.* 1995, 46, 701–728.

52. Kohn, W.; Becke, A. D.; Parr, R. G. Density functional theory of electronic structure. *J. Phys. Chem.* 12980, 100(1996), 12974.

53. Liu, S.; Parr, R. G. Second-order density-functional description of molecules and chemical changes. *J. Chem. Phys.* 1997, 106, 5578–5586.

54. Ziegler, T. Approximate density functional theory as a practical tool in molecular energetics and dynamics. *Chem. Rev.* 1991, 91, 651–667.

55. Geerlings, P.; De Proft, F. Chemical reactivity as described by quantum chemical methods. *Int. J. Mol. Sci.* 2002, 3, 276–309.

56. ADF. *SCM, Theoretical Chemistry*; Vrije Universiteit: Amsterdam, The Netherlands, 2013.

57. Jones, R. O.; Gunnarsson, O. The density functional formalism, its applications and prospects. *Rev. Mod. Phys.* 1989, 61, 689–746.

58. Zupan, A.; Blaha, P.; Schwarz, K.; Perdew, J. P. Pressure-induced phase transitions in solid Si, Sio$_2$, and Fe: performance of LSD and GGA density functionals. *Phys. Rev. B.* 1998, 58, 11266–11272.

59. Theilhaber, J. Quantum-molecular-dynamics simulations of liquid metals and highly degenerate plasmas. *Phys. Fluids B.* 1992, 4, 2044–2051.

60. Stadler, R.; Gillan, M. J. First-principles molecular dynamics studies of liquid tellurium. *J. Phys. Condens. Matter.* 2000, 12, 6053–6061.

61. Argaman, N.; Makov, G. Density functional theory: an introduction. *Am. J. Phys. Condens. Matter.* 2000, 68, 69–79.

62. Parr, R. G.; Yang, W. *Density Functional Theory of Atoms and Molecules*; Oxford University Press: Oxford, 1989.

63. Xiao, H.; Tahir-Kheli, J.; Goddard, W. A. III. Accurate band gaps for semiconductors from density functional theory. *J. Phys. Chem. Lett.* 2011, 2, 212–217.

CHAPTER 19

Study of Pesticide Residue in Vegetables and Fruits in India: A Review

Sanjan Choudhary and Nitu Bhatnagar*

Department of Chemistry, Manipal University Jaipur, India
*Email: nitu.bhatnagar@jaipur.manipal.edu

CONTENTS

ABSTRACT

This review focuses on the concentration of multi-pesticide residues in commonly consumed species of vegetable and fruit samples based on survey carried out by various institutions from different parts of India during the period 2008 to 2013 and the methods that have been adopted for their analysis. Organophosphorus pesticide residues have been detected in most of the vegetable samples, such as potato, tomato, onion, okra, brinjal, cabbage, and cauliflower, and fruit samples, such as apple, banana, mango, and grapes. On the basis of the survey, it was observed that a number of methods have been used for the analysis of pesticide residues in vegetables and fruits, but these methods are unable to identify targeted compounds and specific pesticide residues and also are time consuming. These problems associated with traditional analytical techniques can be overcome by the use of ultra-performance liquid chromatography–time-of-flight mass spectrometry that has high sensitivity and high selectivity toward multi-pesticide residue analysis.

19.1   INTRODUCTION

Pesticides have played a very important role in constantly boosting agriculture and have been routinely used in most countries of the world to control harmful pests. But, on the other hand, contamination with pesticides may be introduced into natural water sources via direct application for control of aquatic weeds and/or indirectly by drainage from agricultural lands. As a result, high levels of these chemicals may cause contamination in both irrigation and drainage water. In recent years, numerous countries including India are currently initiating programs to reduce pesticide usage in conventional agriculture.[1,2] Although pesticides have played a very important role in constantly boosting agriculture, but the hazards that come along with them are dangerous to food safety and human health. As a result, this issue has increasingly become the focus of world attention. Therefore, different countries have initiated various programs to reduce pesticide usage in conventional agriculture. The presence of pesticide residues in vegetables, fruits, milk, meat, fish, egg, soft drink, and water is an important concern not only for consumers but also for government authorities as they are daily used in everyday meal of population. Monitoring of pesticide residues in food community helps to assess the potential risk of these products to consumers' health and provides information on the pesticides that have been used on the field crops. In recent years, there has been an increase in violation rates of the maximum residue limits (MRLs) and incidents of misuse of pesticides. As a result, consumers are exposed to pesticides, usually in minute quantities, in several food groups including fruits, juices, vegetables, sweetened carbonated drinks, and many agricultural products.[3–7] To detect the pesticide residue in several food groups, a number of methods have been developed for their analysis. Pesticide residues have been traditionally monitored by gas chromatography (GC)-based multi-residue methods. But, several new polar and ionic pesticides cannot be determined directly by this method due to their poor thermal stability or volatility.[8-12] Liquid chromatography (LC) coupled with tandem mass spectrometry (MS) has become a dominating technique in multiple residues analysis.[13-15] High sensitivity and selectivity of pesticide residues detection can be achieved by tandem mass analyzers operating in a selective reaction monitoring mode, enabling us to optimize the parameters for each target analyte. This approach, however, is not able to identify nontargeted compounds. A novel approach represented by LC–time-of-flight MS (LC–TOF-MS) has been introduced for target and nontarget analysis of pesticide residues in food.[16,17] Various studies on pesticides in food and water have been appreciated for the unique ability of accurate mass to identify both target and nontarget compounds by LC–TOF-MS.[18-20] The ultra-performance (UP) LC–TOF-MS instrumentation provides sensitive full-scan acquisition, identification, and confirmation of target and nontarget analytes of pesticide residues in vegetable and fruit samples within short run time. The use of sub-2 ?m UPLC column provides better chromatographic resolution and more sensitivity. The objective of the present review is to generate awareness about the presence of pesticide residues in vegetable and fruit samples and rapid multi-residue method, which are used for the analysis of pesticide residues in vegetables and fruits using UPLC–TOF-MS analysis.

19.2 PREVIOUS STUDIES ON PESTICIDE RESIDUES IN VEGETABLE AND FRUIT SAMPLES

India currently uses about 60,000 tons of pesticides, but worldwide, there has been a 44% increase in the use of herbicides over the past decade, with a concomitant reduction in insecticides by 30%. For determining the extent of pesticide contamination in the food stuffs, including vegetables and fruits, a program entitled "Monitoring of Pesticide Residues in Products of Plant Origin in the European Union" started to be established in the European Union (EU) since 1996. In 1996, seven pesticides (acephate, chlopyriphos, chlopyriphos-methyl, methamidophos, iprodione, procymidone, and chlorothalonil) and two groups of pesticides (benomyl group and maneb group, i.e., dithiocarbamates) were analyzed in apples, tomatoes, lettuce, strawberries, and grapes.[21] On the basis of report generated by ministry of health and family welfare, India, May 21, 2013, a total of 6441 vegetable samples of brinjal, okra, tomato, cabbage, capsicum, chilli, and cauliflower collected from the various Agriculture Produce Market Committee (APMC)/wholesale/retail markets located at different parts of the country were analyzed during April 2011–March 2012 by 17 participating laboratories. In 208 (3.2%) samples, the residues were found above MRL levels. In Delhi, a total of 1120 vegetable samples were collected from APMC markets and retail outlets. The pesticide residues above MRL were detected in 60 (5.3%) samples. The common residues , namely, chlorpyriphos, profenophos, triazophos, imidacloprid, acephate, ethion, phorate, quinalphos, and cypermethrin, were detected in the vegetable samples that included unapproved pesticides. Also, a total of 2170 samples of fruits (apple, banana, grapes, orange, pomegranate, guava, and mango) were analyzed by 14 laboratories. Residues found above MRL were detected in 12 (0.5%) samples. Frequently detected pesticides were chlorpyriphos, EDBC, cypermethrin, profenophos, and lambda-cyhalothrin. Residues of some unapproved pesticides, which were commonly detected, were quinalphos, profenophos, and cypermethrin.

Many researchers have also estimated the pesticide residues in various vegetables and fruits, including potato, tomato, onion, okra, brinjal, cabbage, and cauliflower and banana, mango, apple, watermelon, grape, orange, lemon, pear, pineapple, strawberry, kiwi fruit, beet, papaya, and litchi, respectively.[22–25] The occurrence of pesticide residues is reported to be even more than MRL values as recommended by EU, World Health Organization (WHO), and Food and Agricultural Organization. Literature reveals that in India, vegetables consume 14% of the total pesticides used, in which the share of different types of pesticides in Indian agriculture market shows that organophosphorus (50%) ranked first, followed by pyrethroids (19%), organochlorines (18%), carbamates (4%), and biopesticides (1%).[26] Thus, although pesticide application is a necessary step for coping with pest-related problems, it is very important to assess their residues in vegetables and fruits at the same time. Recently, many countries have established regular monitoring programs for quantitative determination of residues in food products[27] as pesticide residues above the maximum tolerance limits at harvest time are a subject of great concern both globally and nationally. Surveys carried out by institutions spread throughout the country indicate that 50–70% of vegetables are contaminated with insecticide residues.[28] In general, food is the main exposure route. Exposure to pesticide residues through the diet is assumed to be of five orders of magnitude higher than other exposure routes, such as air and drinking water.[29] According to WHO (2003), fruits and vegetables constitute on an average 30% (based on mass) of food consumption, and they are the most frequently consumed food group (WHO, 2003). Furthermore, because fruits and vegetables are mainly consumed raw or semi-processed, they contain high pesticide residue levels compared to other food groups such as bread and other foodstuffs that are based on cereal processing.[24]

19.3 RISK INVOLVED DUE TO MULTI-PESTICIDE RESIDUES IN VEGETABLE AND FRUIT SAMPLES

Vegetables and fruits are important components of human diet; however, they are infested by various insect pests. To control the damage, different group of pesticides are used. But pesticides are rapidly applied by farmers during entire period of vegetable farming including fruiting stage and also on fruit to control damage. The indiscriminate use of pesticides and the nonadoption of waiting period lead to accumulation of pesticide residue in vegetables and fruits. These pesticides enter into food chain either from the surrounding environment or from diet. Since they are toxic in nature, they are problematic for human beings in the course of the food chain. Table 19.1 shows the incidence of multi-pesticide residues in fruit and vegetable samples analyzed from different parts of India according to a survey conducted by ministry of health and family welfare, Govt. of India. The result of the studies reveals (only selected vegetable and fruit samples) that 30–80% of sample contains more than one pesticide residue.[25,30] Hence, periodic monitoring of pesticide is necessary to minimize the risk of human health problem due to consumption of vegetables and fruits.

TABLE 19.1 Incidence of Multi-pesticide Residues in Fruits and Vegetables from India

Name of Sample	Scientific Name	% of Sample Containing One or More Residue
Potato	*Solanum tuberosum*	60
Tomato	*Lycopersicon esculentus*	67
Cabbage	*Brassica oleracea*	83
Onion	*Allium cepa*	58
Okra	*Hibiscus esculentus*	50
Mango	*Mangifera indica*	30
Banana	*Musa sapientum*	30
Apple	*Malus domestica*	25
Orange	*Citrus sinensis*	25
Grape	*Vitis vinifara*	30

19.4 SAMPLING PROCEDURE

The pesticide residues are present in extremely small quantity in heterogeneous materials including vegetables and fruits; therefore, good sampling technique and analytical method used for analysis are extremely important to find the trace level of pesticide residue. The crucial or challenging part in pesticide residue analysis consists of two main stages: the isolation of the pesticides from the matrix (sample treatment) and the analytical method for the determination. Sample treatment, which involves both the extraction of the pesticides and the purification of the sample extract obtained, still remains as the bottleneck of the entire procedure, despite much progress on automation has been accomplished. Different studies have been described in the literature about the sample preparation and chromatographic determination of pesticide residues in food and feedstuffs.[31] The effect of sample matrix depends on sample composition. Food samples can differ significantly in their physical–chemical properties and type of compounds present. The factor that mainly affects the extraction of trace level of pesticides from food depends on their polarity and the type of matrix. The common method of sample preparation consists of homogenization of the sample with an organic solvent alone or mixed with water or pH adjusted, using an ultrasonic bath, a blender, or a homogenizer. In most cases, although the analytes of interest are isolated from the bulk matrix, several contaminants may also be co-extracted, as well as part of the matrix, that could interfere in the determination step of the analysis compounds in the presence of extractable major sample components, such as lipids, that offer special problems in the subsequent determination by GC or LC.

19.5 CHALLENGES INVOLVED IN QUANTIFYING TRACE LEVEL PESTICIDE RESIDUE IN VEGETABLES AND FRUITS

Traditional methods for the determination of pesticide residues in food samples, especially in vegetables and fruits, are laborious, time consuming, and usually involve large amounts of solvents, which are expensive, generate considerable waste, contaminate the sample, and can enrich it for analytes. In addition, usually more than one cleanup stage prior to detection is required .[16] Therefore, finding the trace level of pesticide residues in vegetable and fruit samples are is a challenging job. The modern analytical techniques have been developed to overcome the drawbacks of the traditional approaches and also were able to quantify the trace level of pesticide residue. Growing concern over food safety necessitates more rapid and automated procedures to take into account the constant increase in the number of samples to be tested, so interest in procedures that are fast, accurate, precise, inexpensive, and amenable to automation for online treatment is ongoing. In this review article, the modern technique that can analyze the number of samples in less time using liquid chromatography ionization time-of-flight mass spectrometer (LC–TOF-MS) is reported. The LC–TOF-MS, using most advanced technique, has several advantages over most traditional methods of analysis in the following ways: i) a good separation and high sensitivity were achieved by LC–TOF-MS method for all pesticides, ii) the classical procedure that involves extraction with 1% acetic acid in acetonitrile cleanup with primary secondary amine and magnesium sulfate showed an efficient removal of interferences, providing a simple, rapid, and reliable analysis of pesticides in all matrices; iii) for most of the pesticides assayed, the performance characteristics obtained within validation study were acceptable, within the quality control requirements, iv) high recoveries are achieved for a wide polarity and volatility range of pesticides, v) solvent usage and waste are very small. This method is useful for detection of pesticide residue present in almost all type of vegetables and fruits and also it is most effective and widely acceptable in terms of accuracy and reliability.

19.6 CONCLUSION

The present review article serves as a reference document and is thus helpful in taking necessary and timely investigation to monitor the level of pesticide residue in vegetable and fruit samples. It also gives idea that the consumption of vegetable is safe or not from consumer's point of view as residues of all the pesticides was far below/above their MRLs. The review reveals that multi-residue analysis method using UPLC–TOF-MS is helpful for sensitive identification and quantification of multi-pesticide residues in vegetable and fruit samples. The merit of this method is less run time and less consumption of mobile phase for the determination of multi-pesticide residues in vegetables and fruits.

ACKNOWLEDGMENTS

The authors gratefully acknowledge the support provided by Manipal University, Jaipur, in carrying out this study. The technical support and guidance provided by Prof. Man Singh, Dean, School of Chemical Sciences, Central University of Gujarat, is also duly acknowledged.

KEYWORDS

- **Multi-pesticide residue**
- **fruit and vegetable**
- **ultra-performance liquid chromatography–time-of-flight mass spectrometry (UPLC–TOF-MS)**

REFERENCES

1. Abilash, P. C.; Singha, N. Pesticides use and application: an Indian scenario. *J. Hazard Mater.* 2009, 15, 165.
2. CIBRC Central Insecticides Board, Insecticides/Pesticides Registeredunder section 9(3) of the Insecticides Act, 2012.
3. Sinha, S. N.; Vasudev, K.; Rao, M. V. V. Quantification of organoposphate insecticides and herbicides in vegetable samples using the "quick easy cheap effective rugged and safe" (QuEChERS) method and a high-performance liquid chromatography-electrospray ionisation-mass spectrometry (LC-MS/MS) technique. *Food Chem.* 2012, 132, 1574.
4. Stan, H. J. Pesticide residue analysis in foodstuffs applying capillary gas chromatography with mass spectrometric detection. State-of-the-art use of modified DFG-multimethod S19 and automated data evaluation. *J. Chromatogr. A.* 2000, 347, 892.
5. Hernando, M. D.; Agüera, A.; Fernández-Alba, A. R.; Piedra, L.; Contreras, M. Gas chromatographic determination of pesticides in vegetable samples by sequential positive and negative chemical ionization and tandem mass spectrometric fragmentation using an ion trap analyser. *Analyst.* 2001, 126, 46.
6. Anastassiades, M.; Lehotay, S. J.; Stajnbaher, D.; Schenck, F. J. Fast and easy multiresidue method employing acetonitrile extraction/partitioning and "dispersive solid-phase extraction" for the determination of pesticide residues in produce. *J. AOAC. Int.* 2003, 86, 412.
7. Yang, X.; Zhang, H.; Liu, Y.; Wang, J.; Zhang, Y. C.; Dong, A. J.; Zhao, H. T.; Sun, C. H.; Cui, J. Determination of herbicide propisochlor in soil, water and rice by quick, easy, cheap, effective, rugged and safe (QuEChERS) method using by UPLC-ESI-MS/MS. *Food Chem.* 2011, 127, 855.
8. Fillion, J.; Sauve, F.; Selwyn, J. Multiresidue method for the determination of residues of 251 pesticides in fruits and vegetables by gas chromatography/mass spectrometry and liquid chromatography with fluorescence detection. *J. AOAC. Int.* 2000, 83, 698.

9. Podhorniak, L. V.; Negron, J. F.; Griffith, F. D. Gas chromatography with pulsed flame photometric detection multiresidue method for organophosphate pesticide and metabolite residues at the parts-per-billion level in representatives commodities of fruits and vegetable crop groups. *J. AOAC. Int.* 2001, 84, 873.

10. González-Rodríguez, R. M.; Rial-Otero, R.; Cancho-Grande, B.; Simal-Gándara, J. Determination of 23 pesticide residues in leafy vegetables using gas chromatography-ion trap mass spectrometry and analyte protectants. *J. Chromatogr. A.* 2008, 100, 1196–1197.

11. Cajka, T.; Hajslova, J.; Lacina, O.; Mastovska, K.; Lehotay, S. J. Rapid analysis of multiple pesticide residues in fruit-based baby food using programmed temperature vaporiser injection-low-pressure gas chromatography-high-resolution time-of-flight mass spectrometry. *J. Chromatogr. A.* 2008, 1186, 281.

12. Pihlstrom, T.; Blomkvist, G.; Friman, P.; Pagard, U.; Osterdahl, B. G. Analysis of pesticide residues in fruit and vegetables with ethyl acetate extraction using gas and liquid chromatography with tandem mass spectrometric detection. *Anal. Bioanal. Chem.* 2007, 389, 1773.

13. Lacina, O.; Urbanova, J.; Poustka, J.; Hajslova, J. Identification/quantification of multiple pesticide residues in food plants by ultra-high-performance liquid chromatography-time-of-flight mass spectrometry. *J. Chromatogr. A.* 2010, 648, 1217.

14. Hernandez, F.; Pozo, O. J.; Sancho, J. V.; Bijlsma, L.; Barreda, M.; Pitarch, E. Critical review of the application of liquid chromatography/mass spectrometry to the determination of pesticide residues in biological samples. *J. Chromatogr. A.* 2006, 242, 1109.

15. Pozo, O. J.; Barreda, M.; Sancho, J. V.; Hernandez, F.; Lliberia, J.; Cortes, M. A.; Bago, B. Multiresidue pesticide analysis of fruits by ultra-performance liquid chromatography tandem mass spectrometry. *Anal. Bioanal. Chem.* 2007, 389, 1765.

16. Kovalczuk, T.; Lacina, O.; Jech, M.; Poustka, J.; Hajslova, J. Novel approach to fast determination of multiple pesticide residues using ultra-performance liquid chromatography-tandem mass spectrometry (UPLC-MS/MS). *Food Addit. Contam. A Chem. Anal. Control. Expo. Risk Assess.* 2008, 4, 444.

17. Ferrer, I.; Thurman, E. M. Multi-residue method for the analysis of 101 pesticides and their degradates in food and water samples by liquid chromatography/time-of-flight mass spectrometry. *J. Chromatogr. A.* 2007, 24, 1175.

18. Gilbert-Lopez, B.; Garcia-Reyes, J. F.; Ortega-Barrales, P.; Molina-Diaz, A.; Fernández-Alba, A. R. Multi-residue determination of pesticides in fruit-based soft drinks by fast liquid chromatography time-of-flight mass spectrometry. *Rapid Commun. Mass Spectrom.* 2007, 21, 2059.

19. Lacorte, S.; Fernandez-Alba, A. R. Time of flight mass spectrometry applied to the liquid chromatographic analysis of pesticides in water and food. *Mass Spectrom. Rev.* 2006, 25, 866.

20. Gilber-Lopez, B.; Garcia-Reyes, J. F.; Ortega-Barrales, P.; Molina-Diaz, A.; Fernandez-Alba, A. R. Large scale pesticide multiresidue methods in food combining liquid chromatography—time-of-flight mass spectrometry and tandem mass spectroscopy. *Rapid Commun. Mass Spectrom.* 2007, 21, 2059.

21. Mezcua, M.; Malato, O.; Garcia-Reyes, J. F.; Molina-Diaz, A.; Fernandez-Alba, A. R. Accurate-mass databases for comprehensive screening of pesticide residues in food by fast liquid chromatography time-of-flight mass spectrometry. *Anal. Chem.* 2009, 81, 913.

22. Gilbert-Lopez, B.; Garcia-Reyes, J. F.; Mezcua, M.; Ramos-Martos, N.; Fernandez-Alba, A. R.; Molina-Diaz, A. Multi-residue determination of pesticides in fruit-based soft drinks by fast liquid chromatography time-of-flight mass spectrometry. *Talanta.* 2010, 81, 1310.

23. European Commission. *Method Validation and Quality Control Procedure Foe Pesticide Residues Analysis in Food and Feed*; European Commission: Brussels, 2007. [Document no. SANCO/2007/3131].

24. Edelgard, H.; Luc, M. D.; Johanna Smeyers, V. Review inter-laboratory studies in analytical chemistry. *Anal. Chim. Acta.* 2000, 423, 145.

25. Mol, H. G.; Plaza-Bolanos, P.; Zomer, P.; de Rijk, T. C.; Stolker, A. A.; Mulder, P. P. Toward a generic extraction method for simultaneous determination of pesticides, mycotoxins, plant toxins, and veterinary drugs in feed and food matrixes. *Anal. Chem.* 2008, 80, 9450.

26. Charan, P. D.; Ali, S. F.; Kachhawa, Y.; Sharma, K. C. Monitoring of pesticide residues in farmgate vegetables of central Aravalli region of western India. *Am. Eur. J. Agric. Environ. Sci.* 2010, 7, 255–258.

27. Garcia-Reyes, J. F.; Hernando, M. D.; Ferrer, C.; Molina-Díaz, A.; Fernández-Alba, A. R. Large scale pesticide multiresidue methods in food combining liquid chromatography—time-of-flight mass spectrometry and tandem mass spectrometry. *Anal. Chem.* 2007, 79, 7308.

28. Ferrer, I.; Thurman, E. M. Analysis of 100 pharmaceuticals and their degradates in water samples by liquid chromatography/quadrupole time-of-flight mass spectrometry. *J. Chromatogr. A.* 2012, 1259, 148–157.

29. Barr, D. B.; Barr, J. R.; Maggio, V. L.; Whitehead, R. D. Jr.; Sadowski, M. A.; Whyatt, R. M.; Needham, L. L. A multi-analyte method for the quantification of contemporary pesticides in human serum and plasma using high-resolution mass spectrometry. *J. Chromatogr. B.* 2002, 99, 778.

30. Taylor, M. J.; Keenan, G. A.; Reid, K. B.; Fernández, D. U. The utility of ultra-performance liquid chromatography/electrospray ionisation time-of-flight mass spectrometry for multi-residue determination of pesticides in strawberry. *Rapid Commun. Mass Spectrom.* 2008, 22, 2731.

31. Campos, A.; Lino, C. M.; Cardoso, S. M.; Silveira, M. I. Organochlorine pesticide residues in European sardine, horse mackerel and Atlantic mackerel from Portugal. *Food Addit. Contam.* 2005, 22, 642–646.

Index